ENERGY · FOOD · ENVIRONMENT

ENERGY · FOOD
ENVIRONMENT

Realities · Myths · Options

VACLAV SMIL

CLARENDON PRESS · OXFORD
1987

Oxford University Press, Walton Street, Oxford OX2 6DP

Oxford New York Toronto
Delhi Bombay Calcutta Madras Karachi
Petaling Jaya Singapore Hong Kong Tokyo
Nairobi Dar es Salaam Cape Town
Melbourne Auckland

and associated companies in
Beirut Berlin Ibadan Nicosia

Oxford is a trade mark of Oxford University Press

Published in the United States
by Oxford University Press, New York

British Library Cataloguing in Publication Data

Smil, Vaclav
Energy, food, environment: realities.
myths, options.
1. Power resources—Environment aspects
2. Food supply—Environment aspects
I. Title
333.79 ' 16   TD195.E5
ISBN 0-19-828510-8

Library of Congress Cataloging in Publication Data

Smil, Vaclav
Energy, food, environment
Includes bibliographies and index.
1. Power resources.   2. Food supply.   3.Environmental
protection.   I. Title.
TJ163.2.S62 1987   333.79   86-28459
ISBN 0-19-828510-8

Typeset by Joshua Associates Limited, Oxford
Printed and bound in
Great Britain by Biddles Ltd,
Guildford and Kings Lynn

CONTENTS

ABBREVIATIONS

Units and their conversions

barrel	bbl	159 L/ 42 US gallons/ 35 Imperial gallons
calorie	cal	4.186 J
hectare	ha	10^4 m²/ 2.469 acre
joule	J	1 Watt·second/ 0.239 cal
kilogram	kg	10^3 g/ 2.205 lb
litre	L	10^3 cm³/ 0.264 US gallon
metre	m	3.281 ft
metre, cubic	m³	10^3 L/ 264 US gallons/ 220 Imperial gallons
metre, square	m²	10.76 ft²
tonne	t	10^3 kg/ 2205 lb/ 0.984 long ton/ 1.102 short ton
watt	W	1 J/second/ 0.0013 horsepower

Prefixes used in the international system of units

Prefix	Multiplication	Symbol
deka	10	da
hecto	10^2	h
kilo	10^3	k
mega	10^6	M
giga	10^9	G
tera	10^{12}	T
peta	10^{15}	P
exa	10^{18}	E
deci	10^{-1}	d
centi	10^{-2}	c
milli	10^{-3}	m
micro	10^{-6}	μ
nano	10^{-9}	n

Chemical symbols used in the text

Al^{3+}	aluminium cation
C	carbon
CO	carbon monoxide
CO_2	carbon dioxide
Ca^{2+}	calcium cation
$CaSO_4$	calcium sulphate
H^+	hydrogen cation
HCO_3^- -	bicarbonate anion
HNO_3	nitric acid
H_2SO_4	sulphuric acid
N	nitrogen
N_2	dinitrogen
NH_3	ammonia
NH_4^+	ammonium cation
NO	nitric oxide
NO_x	nitrogen oxides (NO and NO_2)
NO_2	nitrogen dioxide
N_2O	nitrous oxide
NO_3^-	nitrate anion
S	sulphur
SO_2	sulphur dioxide
SO_4^{2-}	sulphate anion

Acronyms used in the texts

(definitions or explanations are given in appropriate contexts)

CHD	Coronary heart disease
CFC	Chlorofluorocarbons
EEC	European Economic Community
FAO	Food and Agriculture Organization
FBC	Fluidized bed combustion
FGD	Flue gas desulphurization
GDP	Gross domestic product
GNP	Gross national product
HDL	High-density lipoproteins
IIASA	International Institute for Applied System Analysis
LDL	Low-density lipoprotein
NAS	National Academy of Sciences
NFCS	Nationwide Food Consumption Survey
NHANES	National Health and Nutrition Examination Survey
OAPEC	Organization of Arab Petroleum Exporting Countries
OECD	Organization of Economic Cooperation and Development
OPEC	Organization of Petroleum Exporting Countries
PEM	Protein-energy malnutrition
RDA	Recommended daily allowance
USDA	United States Department of Agriculture

1

How a Civilization Survives

For the improvements of ages had but little influence on the essential laws of man's existence . . .

Henry David Thoreau, *Walden* (1854)

First of all, there must be a habitable planet, a celestial body where life can be sustained in a comfortable environment without any dependence on materials brought in from outer space. The Earth is the only such planet we know of. Its environment—moderate gravity, tolerable means and extremes of temperature, adequate light intensity, steady atmospheric composition, an abundance of water, low rates of meteorite infall, volcanic eruptions and earthquakes, bearable wind speeds and scores of other appropriate conditions—has made possible billions of years of evolution, culminating in the emergence of man. However, while the maintenance of a suitable environment with its myriads of irreplaceable services, remains a critical precondition of human existence, much more is needed for the development of civilizations.

The essence of these additional requirements is the appropriation of energies, first merely by better management of human labour, later by extensively harnessing various renewable transformations of solar radiation (ranging from draught-animals and fuelwood to sailing ships and water mills), finally by extracting fossil fuels and generating electricity.

The dominant social, economic, political, cultural, and military force of this century is the industrial civilization which originated a few centuries ago in Europe and has since diffused rapidly around the world. In the process it has been embraced eagerly by some (Japan), resisted, and only reluctantly and partially malabsorbed by others (China), advanced astonishingly in material terms (the United States) or been heavily circumscribed by ideological constraints (the USSR). Regardless of these fundamental societal differences, the key condition had to be preserved: it is a civilization energized by fossil fuels. Growing flows of these energies have brought greater material benefits and enabled an ever larger diversion of human effort from manual work to intellectual pursuits, thus setting up through a series of deviation-amplifying feedbacks the steps for further material and scientific advances.

But even huge sources of fuels, suitable to energize the extraction and conversion of raw materials and the provision of physical comforts, would mean little without the securing of abundant flows of energies for the direct sustaining of mankind.

The continuous security of food supply has been achieved only since industrial civilization started to augment indispensable solar radiation with considerable energy subsidies, above all in the form of fertilizers, machinery, and management skills.

How does a civilization survive? It survives by harnessing enough energy and providing enough food without imperilling the provision of irreplaceable environmental services. Everything else is secondary—and so it will be energy, food, and environment which will be the multidisciplinary focus of this book's attention.

1.1 Energy flows

> Our present civilization differs from all previous ones in the possession of mechanical power. This has in the last century brought about more rapid and more profound changes in human life and relations than has occurred in all the previous centuries combined.
>
> Morton Mott-Smith (1934)

During the next hour you will spend reading this book—or, tiring of it, listening to some music, strolling down a street, or falling asleep—500 000 t of coal will come out of the bituminous and lignite mines of the world. If this coal were to be carried by large American unit trains, each composed of one hundred cars carrying altogether 10 000 t, the trains would form a line more than 50 km long—so if they moved at 50 km/hr the procession would never end.

During the same hour over two million barrels of crude oil, or about 300 000 t, will be extracted worldwide, with one giant tanker leaving a port every hour; and nearly 200 million m³ of natural gas will be gathered from wells or separated from crude oil—enough fuel to heat 50 000 homes through the whole Canadian winter. And one billion kWh of electricity (1 TWh) will be transmitted from power plants—energy which could light-up 10 billion 100 W bulbs during that hour. Valued in per capita terms, these energy flows come annually to nearly 70 GJ, equivalent to about 2.5 t of hard coal or 1.6 t of crude oil for every inhabitant of this planet.

A century ago the global output of fossil fuels—coal, crude oil and natural gas—was just 1/30 of the present flow (Fig. 1.1), and only the first steps had been taken towards the commercial generation of electricity. Since then far-flung exploration, construction, extraction, conversion, transportation, and transmission efforts have resulted in a huge infrastructure of mines, oil and gas fields, power plants, pipelines, railways, refineries, and high voltage lines—and in a critical dependence on incessant flows of fuels and electricity. Kinds, modes, and rates of energy consumption are, together with the effects of the natural environment, the most pervasive factors determining the physical amenities of a civilization and as such strongly influence its social organization and intellectual search.

Indeed, the rich world's highly urbanized society, mass-producing a galaxy of consumer goods which are eagerly bought by people who live comfortably, eat

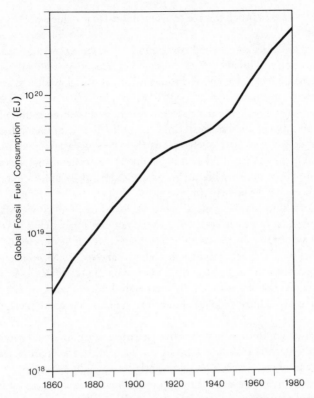

Fig. 1.1 Semi-logarithmic chart of global fossil fuel consumption between 1860 and 1980 (in 10^{18} J), perhaps the single most revealing trace of the rising industrial civilization.

well, expect to live almost to four-score years, like to move about, and to indulge in many exciting pastimes, can be seen as nothing other than a product of fossil fuels employed to liberate human muscles and minds for finer pursuits. Gains in convenience, speed, and reliability, resulting from this transformation, have been enormous but we have become so accustomed to this new situation that in the rich countries people are no longer aware of the gap separating them from the former modes of work. I will illustrate this point with an example bearing on both of the two other subjects of this book's inquiry: the pumping of water to irrigate crops.

The traditional way of lifting water is largely by a variety of muscle-powered devices ranging from the Egyptian *tablia* (a rim discharge wheel) to the Chinese *long gu che* (dragon-wheels): men, women, children, donkeys, bullocks, camels, and horses exerting themselves in slow repetitive movements, tiring under the scorching sun. The modern way is a small pump, usually powered by liquid fuel or electricity. To compare the performance I will use the same 9 m deep well and the same task: to deliver just one cm of water to one ha of cropland (as irrigation requirements vary so much this is a convenient way to provide a base figure for easy multiplication).

A pair of bullocks and their driver pulling up a self-emptying bucket into an irrigation trough will lift about 7 m³ of water per hour so it would take them nearly 14.5 hours to deliver 1 ha·cm. A little centrifugal pump with a shaft power of 3.75 kW (that is just five horse-power) and a pumping efficiency of 60 per cent will lift 1 ha·cm from the depth of 9 m in just one hour and seven minutes. The huge difference in time—it takes thirteen times longer to do it with the bullocks—means that a farmer gains at least twelve hours (I am taking a generous hour off for fuelling and maintenance of the pump) in which to do other things: he can tend his crops or animals, or improve his ditches or sheds, or read about how to raise rabbits, or make a bee-hive, or he may simply rest.

The relationship may be generalized in terms of daily units of useful work. A healthy adult man can work continuously at a rate not surpassing 8 kg·m/sec, that is about 78 W or roughly one-tenth of one horsepower. If he puts in an eight-hour day, the total useful work output will come to about 2.25 MJ. Draught-animals deliver, depending on their size and speed of work, 250–750 W (one-third to one horsepower). Taking 500 as the mean, their eight hours of work will be equivalent to 14.4 MJ, or a rough equivalence of six strong men for one good, average draught-beast.

The smallest common machine to displace men and animals is a garden tractor, now the cornerstone of Chinese farming mechanization. Its typical 4 draw-bar hp translates into about 85 MJ of useful work in an eight-hour day, or six good bullocks, or 36 men (for a small machine). A medium tractor would have, once again, a useful power rating six times as large as its small garden counterpart so that a 24 draw-bar hp machine would deliver about 510 MJ in a day. The big tractors, now increasingly working on the American Plains and Canadian Prairies, rate over 200 hp so that it is not uncommon to see a machine of 144 draw-bar hp, again six times more powerful than the medium tractor.

By this time we are up to 6⁴, or 1296 hard-working men, or 216 vigorous draught-animals. What 220 horses and their masters would do in a day, a single driver, sitting high in the air-conditioned cab, bathed in stereo sound, does alone with his Versatile or International Harvester tractor. Nothing else so reduced the need for farmhands as this transformation, charted for the United States in Fig. 1.2. Tractors displaced horses, bigger tractors displaced smaller ones (the average size of a new tractor in the United States is now around 100 hp!), the man-hours needed to produce crops plummeted, and the rural surplus labour moved to towns and cities to man the industrial production which was being equally revolutionized by fossil fuels and electricity.

Instead of being the energizers of production, workers became the controllers of fuel and electricity flows. Again, a historical comparison best illustrates the gains. A supervisor during the building of the Pharaonic pyramids would direct work gangs of up to 200 men whose short-term straining to pull the wooden sleds loaded with huge stones could combine to produce, for spells of a few minutes, power of over

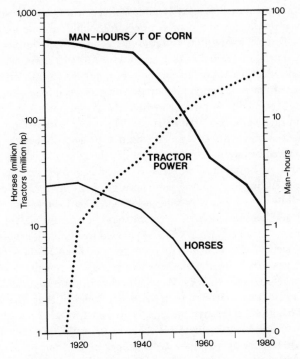

Fig. 1.2 Data from the United States Department of Agriculture document how the availability of cheap fossil fuels revolutionized field farming. Semi-logarithmic graph charts the rise of tractor power and the concurrent demise of working farm horses and the large reductions in labour needed to grow corn, North America's foremost crop.

30 kW. A century ago a Californian farmer, owning one of the large Houser combine harvesters drawn by 24 big horses and crewed by four men, controlled about 18 kW of continuous power.

In contrast, a farmer, ploughing with a Versatile 950, controls over 200 kW— and even higher power flows are managed by single operators in countless tasks in factories. But energy industries themselves offer the best examples of the amazing levels reached in power control. For example, in the world's largest walking dragline (a machine weighing over 12 000 t with a reach of nearly 125 m and a bucket of 170 m³), a single operator governs 33.5 MW of power flowing through the machine's electrical equipment. And in a large power plant, a chief engineer in the main control room can flip the switches and press the keys to govern power flows of several GW, the equivalent of total energy flow in a small European nation until a few generations ago!

What is no less notable than the scale on which we have surpassed the delivery of animate energies and the magnitude of power flows controllable by a single brain, is the cheapness of fossil fuel energies without which it would be impossible to have mass industrial production at accessible prices. Although the common perception, ever since the first round of OPEC's oil price increases in 1973-4, is that this energy became very costly, the reality is that its cheapness is still stunning. The proper yardstick with which to assess the expenditure is not yesterday's price but, once again, the cost of the basic option predating the rise of fossil fuels—human labour.

The prices of refined fuels vary throughout the Western world but 50 cents per litre of liquid fuel was a good 1985 average. This fuel contains about 40 MJ and when it is used to power directly one of countless gadgets (i.e. largely gadgets and machines powered by internal combustion or diesel engines) which we use to move loads, to scrape, cut, or dig the ground, or to make any of the numerous changes to materials (from cutting to moulding, from boring to polishing), the efficiency of the operation will be most commonly between 10 and 30 per cent. Converting, more conveniently and hence in modern factories much more commonly, to electricity results in similar overall efficiencies (30-40 per cent for generation, 60-90 per cent for motor efficiency, with a final useful conversion of about 15-35 per cent).

Taking 15 per cent as a rather conservative average, means that 1 MJ of useful energy derived from oil would cost about 8.5 cents. In contrast, unskilled manual labour came in 1985 at about $5 an hour (with surprisingly little difference between construction labourers in Japan or in Canada). As noted, an adult man cannot work continuously at a rate surpassing 80 W, and so it would take him 20 hours to deliver the useful energy provided by a 15 per cent efficient diesel-powered machine and it would cost the employer $100, or about $17 per MJ—200 times more than liquid fuel! Naturally, the rapid drop in crude oil prices in early 1986 only accentuated this large difference.

But, as many a reader will object immediately, this comparison ignores the capital cost of the gadget, machine, or engine as well as maintenance expenditures incurred during its lifetime. Indeed, these expenses are not included but not because they differ so considerably from item to item. The surprising reason is that their exclusion makes little difference to the order of magnitude established in the comparison. Even for such a relatively little used machine as a small English passenger car, *The Economist*'s account showed that gasoline cost about one-half of the total motoring expense (and this included not only depreciation and repairs but also licence and insurance). In this case the full cost of employing a fossil-fuel converting gadget would double—but that would still leave a 100-fold gap in comparison with human labour.

Obviously, in practice it matters little if the expense gap between human labour and a fossil fuel or electricity is 200 or 150 or even 'just' 100-fold: the cost advantage remains overwhelming. Together with greater convenience, versatility, and reliability, it is at the root of our dependence on fossil fuels whose continuing flows underlie modern industrial civilization—not post-industrial as some seers and sociologists would have it. Their glib characterizations do not recognize that all higher levels of the service sectors (lawyers advising lawyers on how to sue other lawyers, fast food eateries, video empires, and computerized nirvana) would flicker into instant demise if the rich world did not digest a ceaseless diet of fossil fuels and electricity.

The reliable and relatively cheap provision of these fuels remains the foundation of our civilization—now even more than a century ago when the transition from animate energies started to gather speed. But we are profoundly indebted to that period during the latter half of the last century when an unmatched intellectual explosion introduced virtually all the technical innovations which to this day constitute the mainstay of our energy base.

The quantitative growth of individual technologies and their qualitative improvement have been amazing—Fig. 1.3 charts the exponential rise of three common energy extraction, transportation and general processes—but Edwin Drake, George Westinghouse, William Stanley, Rudolf Diesel, Ferdinand Carré and Thomas Edison would not need much time to recognize the core of their inventions or the improvements in every oilfield, power plant, transformer, car, refrigerator and lightbulb.

With a few explanations, these men would not be lost in our society: their

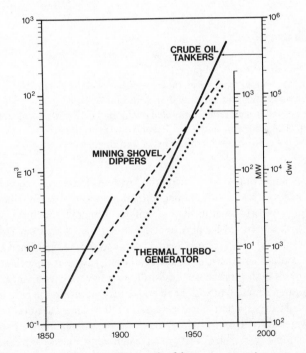

Fig. 1.3 To chart the steep exponential growth of three representative energy technologies during the past hundred years would require semi-logarithmic paper with four cycles! Fundamentally, these are the 19th century technologies, but current performances are orders of magnitude higher.

achievements initiated the great liberation of solar energy stored as solid, liquid, and gaseous fuels in the ground 10^1–10^2 million years ago—and hence they stand so much closer to us than to their great-grandfathers of the eighteenth century. Conversely, our world is fundamentally still made so much in their image—precisely because it is powered by the same sources of energy whose large-scale harnessing and consumption took off a century ago. Adornments, eye-catchers and mind-benders do differ but this is still the era of fossil-fuelled civilization (a much better term than industrial society), in the early decades of its second century.

The only fundamentally new energy conversion introduced on a large commercial scale during the present century has been nuclear fission but it has had so far only a very limited effect and its outlook is very uncertain. The eventual transition from fossil fuels is unavoidable but I will not speculate here on the pace and timing of the switch to a new dominant source, or a mixture of sources. The rational human planning horizon is no longer than 50 years—and during that time the fossil-fuelled era will continue. To be sure, there will be many changes, modifications and new challenges. The greatest of these challenges will be how to spread much wider the benefits of a high-energy society to the world's poor nations, a process which must start with their having always enough to eat.

1.2 Food harvests

> Rice, broom corn, early wheat, mixed all with yellow millet;
> Bitter, salt, sour, hot and sweet: these are dishes of all flavours . . .
> Stewed turtle and roast kid, served up with yam sauce;
> Geese cooked in sour sauce, casseroled duck, fried flesh of the great crane . . .
> Plump orioles, pigeons, and geese, flavoured with broth of jackal's meat . . .
> Pickled pork, dog cooked in bitter herbs, and zingiber-flavoured mince . . .
> Chu Ci, *The Songs of the South* (D. Hawkes translation)

When the ancient Chinese were luring the lost souls to return home, they did not offer precious possessions—but rather enticing choices of meals, the stanzas of poetry adding up to fascinating lists of edibles. But such fanciful combinations come only with diversified farming and the emergence of complex cultures—and they have been, until very recently, available only to those rather small segments of society who could afford them. In isolated gathering and hunting societies as well as in grand empires built on crop surpluses, all but a few men, women, and children have always eaten simple fare with a few foodstuffs supplying the bulk of all needs.

And in civilizations where élites savoured multi-course banquets for millenia, peasants always subsisted on combinations of one or two staple cereals, some tubers, and a few simple vegetables, with meat eaten only during infrequent festive occasions. Only the ingredients differed in the France of Louis XVI (1774–92), on the plains of central Russia under Alexander I (1801–25), in the Japan of the last Tokugawas (before 1868), or in the dusty villages of northern China during the last years of the Qing dynasty (before 1911): the pattern was the same, dominated by staple cereals and tubers, almost meatless, almost sugarless, morning meals differing little from the evening ones.

In the early decades of the nineteenth century European peasants consumed each year no more than 1–2 kg of sugar per capita, less than 5 kg of meat, but around 200 kg of staple grain and tubers—quantities almost identical with those still eaten by Chinese peasants a century leater. But while the rest of the world continued to eat as they had for millenia before, typical European and North American diets started to change in a transformation no less far-reaching, but incomparably more rapid, than the one brought about thousands of years ago by the transition from gathering and hunting to regular crop farming.

Certainly the most telling characterization of this great dietary transformation was a steady shift toward animal foods, above all more eating of meat; a shift which may be only now, 100–150 years later, reaching a plateau and relative saturation. During this period of great dietary change the sciences of organic chemistry, experimental human physiology, and nutrition were also born and rapidly expanded. Food came to be seen in terms of essential nutrients, and efforts at defining their best daily intakes became a preoccupation of theorizing nutritionists stirring up many controversies whose end is not yet in sight (see section 3.1).

When seen in nutrient terms, the dietary transformation brought about by industrialization, urbanization, and rising personal affluence in Europe and North America can be best described as a rather rapid decline of complex carbohydrates and a fast rise of animal proteins and refined sugar (Fig. 1.4). While traditional peasant diets everywhere derive just around one-tenth of all protein from meat, dairy products, and eggs, in industrializing nations this ratio almost doubled before the end of the nineteenth century, then it doubled again before the beginning of World War II and, after a temporary setback, has continued to rise since the early 1950s. By the early 1980s the French were getting a little over 70 per cent of their protein from animal foods, the Americans about 68 per cent—while Indian diets changed little, still averaging no more than about 10 per cent of animal protein.

But not all carbohydrates went down: as diets moved into more meat and fat, sugar consumption rose everywhere, increasing ten, twenty and even thirty-fold from its pre-industrialization levels. So it can be used today, together with animal proteins, as another quick indicator of a country's nutritional make-up: rich countries eating 30–60 kg a year per capita, and poor ones mostly 5–20 kg. And there have been also qualitative changes where the consumption totals have shifted little but variety has grown greatly: the eating of vegetables and fruits is certainly the most notable example of such a change.

These dietary changes have resulted in the conspicuously increasing heights and weights of maturing humans (Fig. 1.5 is a spectacular Japanese example), and in the steady declines of age at menarche (Fig. 1.6). Nevertheless, the splendid advances in public health and the unprecedented improvements in the prevention and treatment of formerly fatal infectious diseases—the two great transformations whose large-scale introduction in the industrializing countries coincided with better nutrition—must be given a large share of the overall credit. Although better

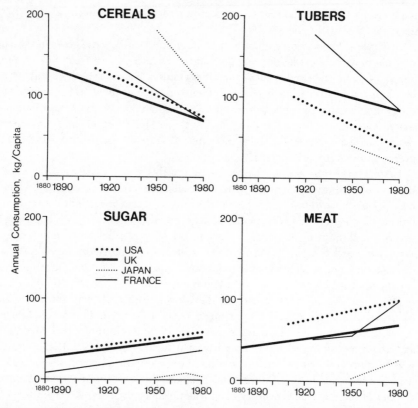

Fig. 1.4 The rates differ but the trends are uniform: historical statistics show how increasing affluence is accompanied by declines in cereal and tuber consumption while average intakes of sugar and meat are rising.

nutrition can be definitely associated with faster maturation, and with the taller and heavier bodies of children and young adults, the link with improved life expectancy is much less obvious.

A comparison of the life expectancies at birth for those European nations which have long and fairly reliable series of demographic data, shows very impressive improvement during the past 150 years, from around 40 to more than 70 years. However, disaggregation for different ages, here illustrated with English and Welsh males for the years 1841–1981 (Fig. 1.7), shows plainly that the bulk of the gain is due to a drastic reduction in child mortality and that the gains in life expectation for adults over 20 years of age have been surprisingly modest. Looking at the causes of death, one quickly discovers that it was the virtual elimination of typhoid, tuberculosis, diphtheria, and later a sharp reduction in pneumonia mortality which allowed the children to survive—not primarily better nutrition.

China provides an outstanding example of the combined power of better

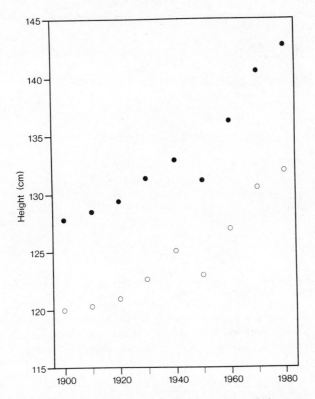

Fig. 1.5 Secular trends in the height of nine (empty circles) and eleven-year old (black circles) Japanese boys reflect the slowly improving nutrition during the first decades of this century, then the surprisingly large setback during World War II, and finally the renewed steeper growth.

sanitation and wide-spread inoculations· in spite of the country's decidedly modest food intakes and the predominance of a pre-industrial nutrient pattern (for details see sections 3.1.1 and 3.2.2), life expectation at birth is now 67 years for men and 70 years for women which is much closer to the means of rich countries than an educated guess would have supposed it.

Clearly, the case of food energy is unlike that of fuels and electricity which provide heat, light and motive power: a vastly higher consumption of the latter in rich countries is the very essence of industrial civilization and, although there is no simple linear relationship between economic development and the use of energy (see section 2.3.3 on this), there is also little doubt that the poorest nations must multiply their combustion of fossil fuels and generation of electricity anywhere between five to ten times before their economies will provide a comfortable life for most of their citizens.

But securing enough food for healthy, active life does not call for even a

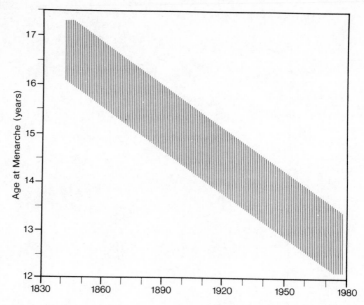

Fig. 1.6 Better nutrition is also reflected in the faster maturation of girls. Virtually all national trends in Europe and North America would fit inside the broad diagonal band showing a decline of menarche from an average age of about 17 years in 1830 to about 13 years by 1980.

doubling of the poor world's food consumption. Considerable attention will be given later to our imperfect knowledge of human nutritional requirements (section 3.1) but one can get a correct global appraisal of basic food energy needs by taking the standard daily requirements of a thirty-year old 60 kg man and a 50 kg woman (these body weights are much closer to whatever the real mean is rather than the commonly used weight means of rich nations), calculating their average intake and reducing it by 10 per cent to account for lower food consumption by smaller children and people over 40 years.

The result: no more than 2150 kcal a day per capita. Rounding upwards, these 2200 kcal should suffice to keep a population healthy and active and, scanning the Food and Agriculture Organization food balance sheets, show not only that the global mean availability of food is well above that figure (hardly surprising with the rich nations included) but that of 150 nations 125 stand above the mark and 15 miss it by less than 10 per cent. Only ten countries have shortfalls greater than 10 per cent but virtually all of them—Afghanistan, Bangladesh, Chad, Ethiopia, Kampuchea, Laos, Mozambique, Uganda, Zimbabwe—have been the victims of recent wars, civil wars, military take-overs, and recurrent natural disasters.

This simple calculation and comparison were not made to justify the assertion

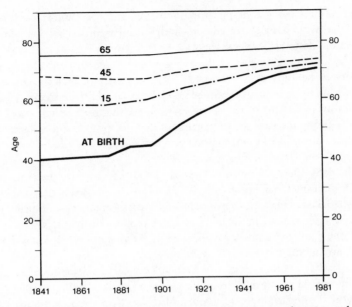

Fig. 1.7 Life expectancy at birth for English and Welsh males rose by 30 years between 1841 and 1981 but the rise had clearly more to do with infant and childhood survival than with a lifetime of better nutrition as the expectancy at age 15 increased by only about 10 years.

that nearly all is well (discussion in the food chapter will show how uncertain are *all* these values) but rather to show how relatively little needs to be done to obviate the proposition that sharing by rich countries (very unlikely) is the only way to achieve sufficient food intake everywhere. A mere 10–15 per cent increase in average per capita food intakes would, with equitable distribution, put most of the poor countries above the level of malnutrition. The diet would still be simple but if dairy production, poultry, and fish breeding were promoted, animal foods could easily supply one-tenth of all energy and one-fifth of all protein—a fare far from the aspirations of average Western man but far closer to what current scientific consensus sees as healthy eating.

If the rich nations would adopt similar diets while continuing to produce major cereal crops at about the current level, the world would be awash in grain and, as will be shown in some detail later (section 3.2.1), these surpluses would be still very large if the average meat consumption in the rich countries were cut by just one-third, hardly a drastic change in existing diets. This realization is fundamental in appraising all those countless writings on the global food prospect: at what level do we pin our assumptions? The great difference in requirements is not between the current food availability in the poor countries (ranging from having just about enough to there being widespread malnutrition) and the comfortable level fully compatible with healthy growth of children and rigorous life for adults, but between this desirable level and the way the rich nations eat.

As the interesting reviews of the eating habits of centenarians prove, extreme longevity is typically associated with very simple diets; in terms of essential nutrients hardly more than would be good Asian fare. Little indeed is needed for healthy survival and the world has moved closer to this most desirable goal than at any time during the past: the proportion of well-fed people is now definitely the highest in human history and the past generation, contrary to some sensationalist claims, has seen notable improvements.

In the early 1980s, 90 out of 120 poor nations had on the average more food available per capita than in the early 1960s. As will be stressed in the food chapter, those who failed to make this advance failed primarily not because of their reaching insurmountable natural limits (although this too, obviously, plays an important part in marginal environments) or because of the lack of specific know-how and needed farming inputs (although, again, these mattered greatly).

Global adoption of American or French diets could not be sustained for the nearly five billion people inhabiting this planet: it could not be supported even for a single year. These rich diets, very much like North American and European per capita levels of fossil fuel consumption, are not desirable goals whose eventual achievement should be seen as the consummation of the developmental process. Rather, they are aberrations, made highly appealing as a part of the vigorously promoted carefree lifestyle symbolized so perfectly by the worldwide diffusion of hamburgers and soft drinks.

Yet arguments about how relatively little is needed to keep the Earth's population satisfactorily fed, and irrefutable calculations showing that, with these modest diets, the global food prospect is not at all discouraging, have only a limited value in a world where eating is much more than supplying needed amounts of essential nutrients and where exports play a very important part even though most of the food must be produced where it will be eaten.

Here the complex interplay of culture (traditions, religions, attitudes, habits, experiences, preferences), environmental conditions (soil quality, erosion, precipitation, droughts), and economic management (treatment of farmers, pricing, long-term modernization policies, research, investment) determines the outcome. Any individual component, taken in isolation, is a poor predictor of the overall possibilities and only very rarely are a few factors so detrimental that a better performance appears foreclosed forever.

The problem needs restating. Feeding the world is much more of a qualitative than a quantitative challenge, both in terms of how to go about it and what to eat, and the latter concern should be the starting point. The conventional worry is about pushing up the yields of the crops we plant, increasing the numbers of livestock we like to eat, raising the catch of fish we have always been netting. But the essence of the quest is elsewhere and it does not fit into the single adjective: more.

How much of what to eat for healthy growth, active adulthood, and contented retirement should be the first question we ask—and some detailed discussions in chapter three will show how surprisingly uncertain we are about these basic recommendations. Only then can harvests be planned with the one overriding priority always in mind: how to make the process as sustainable as possible. Practically, this means how to optimize energy costs and preserve the natural productive capacity of agro-ecosystems and the provision of irreplaceable environmental services.

1.3 Environmental services

Then followeth thus, that nature doth dissolve
Each thing again to its own elements . . .

Lucretius, *De rerum natura* (C. E. Bennett translation)

The goods might seem all-important at first glance—what would we do if we were not able to tap water, extract minerals, cut wood, catch fish, domesticate grain plants—and, indeed, no civilization on this planet could do without them. So to speak, as ecologists do, about the 'benefits' derived from these 'free' goods is using a term too weak to express the irreplaceable dependence.

Yet within these confines there is a surprisingly large amount of room for manœuvre: drinking water can be brought from long distances, even distilled from the ocean; many minerals are substitutable and technological advances make it possible to recover deposits which a short time ago would not even be seen as a useful resource; managed forest plantations can yield much more than natural forests; agriculture, with selection, and even more so with the coming manipulation of genes, can result in crop species superior in many ways to natural ones.

But there is no practical range of options where nature's numerous services are concerned: here our dependence is total since the functioning of any ecosystem, with or without human interference, is utterly dependent on complex chains of chemical and biochemical reactions and interactions. These services, which human civilization cannot replicate, assure above all the smooth functioning of grand biogeochemical cycles, the degradation of wastes and the perpetuation of the atmosphere's composition.

For a detailed illustration of this dependence I will turn to the services provided by the organisms which are largely invisible and whose prodigious efforts (Delwiche's characterization is 'a work of cosmic importance') make possible the continued growth of forests and grasslands, the harvests of crops and the degradation of organic wastes. Life in the soil will take over the next few pages. As soil studies moved from their early physical beginning towards deepening recognition of soil as a living assemblage, microbiology accumulated tomes of fascinating details about seemingly countless numbers of soil organisms whose activities make it possible to close the critical connections between the three great natural cycles I will concentrate on later in this book.

Walking across a pleasant summer meadow it is the greenness of the vegetation carpet, the variety of flowering species (sometimes their scents) and the flight of the insects which attract most of our attention, and there will be also an occasional bird or a small mammal running away. Yet these attractive sights will encompass only a minority of the living organisms which make up the grassland. Most of the life is hidden, and much of it is never seen by human eyes even where a burrowing animal or flowing water expose the soil and plant roots.

Thanks to the fascinating studies done during the International Biological Programme I can cite some admirably detailed and reliable values concerning the partition of these captivating and invisible realms. Every grassland will have more dead phytomass—in withered shoots in its canopy, in litter on the ground, and among its roots—but I will make comparisons only among the living biomass. In natural grasslands roots weigh usually almost 4.5 times as much as the above-ground shoots so the pleasing greenness of blades and stalks swaying in the wind may not represent more than one-tenth to one-seventh of all the living mass while about one-half is in thick, intertwined root mats.

This still leaves one-third to two-fifths of all the living mass unaccounted for. Buzzing and fluttering insects will account for a negligible total of less than 0.05 per cent and even the invertebrates underground will add no more than 0.5 to 1.0 per cent. The remainder, 35–40 per cent, are the micro-creatures; single-cell bacteria, fungi, algae, and protozoa. In a breakdown shown in Fig. 1.8 depicting the distribu-

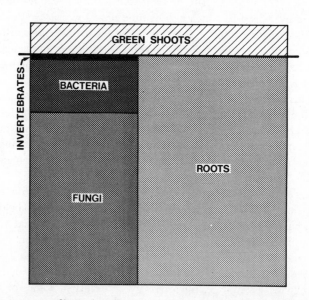

Fig. 1.8 Partitioning of living biomass in Canadian Prairie grassland (Matador, Saskatchewan). Underground biomass predominates and bacteria and fungi account for no less than 30 per cent of the total.

tion of living mass in natural grassland at Matador (Saskatchewan), fungi are dominant and the total mass of microbial life is in excess of 220 g/m^2, or more than 2 t/ha, three times as much as the biomass of green shoots.

But in all terrestrial ecosystems bacteria are by far the largest group of micro-organisms; fungi have fewer cells present but their larger size gives them often greater total biomass. Typical bacterial counts in one gram of soil are between 10^9–10^{10} and each of these unicellular microbes occupies a mere 10^{-11}–$10^{-12} cm^3$ and has a live weight of only 10^{-10}–10^{-13} g—but their huge numbers mean that one hectare of soil just 20 cm deep (i.e. a total of 2×10^9 cm^3 weighing, with a mean density of 1.25, about 2500 t) will contain, taking 2.5×10^{-13} g as the mean mass, roughly between 500 and 5000 kg of bacteria. Investigations in many different locations show that in poor soils bacterial mass per hectare may not be even 100 kg (almost all of it in the rhizosphere, the zone of much enhanced microbial activity around active roots), while under alfalfa in good soil it may be 6–7 t per hectare in the root zone and another 1.5–2.0 t outside it for a total of 7.5–9.0 t of cells.

Fungi, with usual counts between 10^4–10^6/g of soil, are much less abundant than bacteria but a relatively large diameter and often very extensive networks of their filamentous mycelia make them frequently a dominant part of microbial proto-plasm, especially in forest soils and in wetter and acid environments. Those fungi which sporulate profusely, such as *Penicillium* and *Aspergillus*, are encountered most frequently. In contrast to heterotrophic fungi, soil algae can obviously grow only in the surface layer where the light is sufficient for photosynthesis but even so they may number anywhere between 10^2–10^4/g of soil, being mostly species of *Chlorophyceae* (green algae), *Xanthophyceae* (yellow-green algae) and *Bacillario-phyceae* (silica-encrusted diatoms).

Unicellular protozoa are almost as abundant as fungi (10^4–10^5/g of soil) and they, too, are heterotrophic but mostly motile, moving with the help of finger-like extensions, flagella or cilia, as they prey on bacteria and other unicells. In turn, these soil amoeabae (*Rhizopoda*), *Ciliata* (including the much-studied *Paramecium*), and *Mastigophora* are eaten by numerous micro-arthropods. Particularly abundant micro-arthropods in soil and litter are eight-legged mites (*Acari*) and six-legged springtails (*Collembola*). Barely noticeable are common eelworms (*Nematoda*), usually less than 0.5 mm long, either predacious or saprophytic and many of them crippling parasites of crops.

Soil macro-fauna is conspicuous, largely owing to the swift *Myriopoda* (which include both the saprophytic millipedes and the predacious centipedes) and to the longest soil invertebrates, earthworms (*Lumbricidae* in temperate latitude). There are also many insects and their larvae, as well as burrowing mammals (of which only moles live solely wihin soil) but their biomasses are very small compared to those of micro-organisms. Table 1.1 summarizes typical orders or magnitudes of abundance, average weight, and total living biomass of all important soil creatures and contrasts them with one of the world's highest human population densities, one person per one-tenth of a hectare of arable land in China.

Even at that high density human biomass adds up to no more than the mass of

Table 1.1 Soil biomass contrasted with a densely distributed anthropomass

Biomass	Abundance (number/m²)	Mean live weight (g)	Live biomass (kg/ha)
Microflora			
Bacteria	10^{14}	10^{-12}	10^3
Fungi	10^9	10^{-7}	10^3
Algae	10^6	10^{-6}	10^1
Microfauna			
Protozoa	10^8	10^{-7}	10^2
Mesofauna			
Nematodes	10^7	10^{-6}	10^2
Acari	10^5	10^{-5}	10^1
Collembola	10^4	10^{-4}	10^1
Macrofauna			
Enchytraeidae	10^4	10^{-4}	10^1
Myriopoda	10^2	10^{-2}	10^1
Megafauna			
Earthworms	10^2	10^{-1}	10^2
Chinese peasants	10^{-3}	10^4	10^2

earthworms and it will be with earthworms, Darwin's great favourites, that I shall start the short descriptions of invaluable services provided by soil organisms. The great scientist had a life-long interest in earthworms: he first lectured on them in 1837 before the Geological Society and in 1881 he published a book of more than 300 pages on *The Formation of Vegetable Mould through the Action of Worms* which ends with an exalted praise: 'It may be doubted whether there are many other animals which have played so important a part in the history of the world, as have these lowly organized creatures.' The main contributions of these lowly 2.5–25 cm long creatures to soil fertility are, of course, incorporation of dead surface phytomass into the deeper soil layers and recurrent 'ploughing' of the land. In Darwin's words, it is 'a marvellous reflection that the whole of the superficial mould . . . has passed, and will again pass, every ten years through the bodies of worms'.

Arthropods, usually the most abundant detritivores in terrestrial ecosystems, are major processors of dead phytomas and animal faeces—in turn, faeces of mites and springtails make up a large portion of soil humus—and their actions are critical in determining mineralization rates of such key elements as nitrogen, phosphorus, and calcium. By speeding up decomposition rates, they enhance the nutrient concentrations of the whole litter–microbe system and their contribution is especially important in reducing the standing mass of woody litter whose high C/N ratio makes it relatively resistant to microbial decay.

Fungi are rapid, pioneering colonizers of plant residues and their growth accelerates the decomposition and eventual mineralization of all major nutrients needed by photosynthesizers. Colonization of virgin plant tissues usually proceeds in sucessive overlapping stages with weak parasitic species present on senescent tissues even before they reach the ground, followed by fast-growing saprophytic fungi living on sugars and carbon compounds simpler than cellulose, then by species feeding on cellulose and its associates, and finally by cellulose and lignin decomposers. In contrast, more complex substrates, such as herbivore faeces, are colonized completely by all kinds of fungi.

Adept as the fungi are at decomposing organic matter with the ultimate release of ammonia, they cannot do the next key step in recycling nitrogen for plants—the oxidation of ammonia to nitrate which, unlike readily immobilized ammonia, is free to migrate to roots and to sustain plant growth. This dehydrogenation process can be handled only by nitrifying bacteria, chemoautotrophs which receive the needed energy by the exothermic oxidation of ammonia to nitrite (*Nitrosomonas, Nitrosospira, Nitrosolobus*), and then nitrite to nitrate (*Nitrobacter*). Dehydrogenations of sulphur compounds producing elementary sulphur, sulphites but most often sulphates necessary for plant growth, are carried out by various species of *Thiobacillus*, one of which also oxidizes iron (*Thiobacillus ferrooxidans*).

A host of reductions, critical in closing major nutrient cycles, is also carried out either solely or predominantly by specialized bacteria. Staying with the nitrogen cycle, the opposite process to nitrification (reduction of nitrate to nitrite and then nitrous oxide and nitrogen) is the key method of returning nitrogen to the atmosphere and is carried out by denitrifying anaerobic bacteria (*Thiobacillus denitrificans*, and some species of *Bacillus, Micrococcus, Achromobacter*). And, in the best-known and most studied conversion of the whole nitrogen cycle, atmospheric dinitrogen is reduced to ammonia by a variety of free-living and symbiotic bacteria, although blue-green algae (*Cyanophyta*) can also do the reduction and their contribution is very important in shallow waters and waterlogged soils (above all in rice paddies).

But by far the most important pathway of nitrogen fixation is through a nodule-forming symbiosis of heterotrophic *Rhizobium* bacteria with leguminous plants ranging from common temperate crops (alfalfa, soybeans) to numerous tropical trees. Non-nodulating symbioses involve above all *Azotobacter* and *Beijerinckia*, and these two are also among many free-living bacterial fixers. Other frequently encountered free-living reducers of atmospheric dinitrogen include anaerobic *Bacillus, Klebsiella* and *Clostridium*. Analogically, various bacteria perform the reduction of sulphur (*Desulfovibrio, Desulfotomaculum*) and iron.

And the largest number of soil bacteria are keeping the nutrients in circulation by breaking down organic matter. The most commonly isolated motile bacterium in soil, *Pseudomonas*, uses just about any carbon substrate, as does the very common *Arthrobacter* and *Nocardia*. On the other hand, species belonging to such genuses as

Cellulomonas, Micromonospora or *Cytophaga* specialize in decomposing cellulose or other polysaccharids except for lignin whose breakdown is left to fungi (commonly *Coprinus, Agaricus, Poria*).

Moreover, many bacteria, as well as fungi, are able to attack and decompose different kinds of hazardous man-made compounds and hence could be used for biological waste treatment. The use of selective types of micro-organisms in degrading wastes has of course long been common in the removal of nitrogen through sequential nitrification and denitrification in water treatment plants but large-scale commercial applications to disppose of many hazardous organic synthetics are not imminent.

Still, the list of proven possibilities has grown very long and it includes biodegradation of nonhalogenated aromatics such as benzene, toluene, creosol and phenol (by *Pseudomonas, Bacillus, Vibrio* etc.), halogenated compounds, polycyclic aromatics (benzo(a)pyrene, naphthalene) and also the polycyclic halogenated aromatics, including polychlorinated biphenyls (by *Pseudomonas, Vibrio, Spirillum*) whose presence in the environment has caused so much apprehension during recent years.

Finally, it should be at least mentioned that many soil bacteria produce a wide range of antibiotic compounds and a huge industry sprang from the *Streptomyces* species isolated from soil as the antibiotics found widespread applications not only in human and veterinary medicine but also in animal nutrition, plant disease control, and food preservation.

Expressing the value of these microbial services in monetary terms is rather easy in some instances but it is not only impossible but simply inappropriate to try to do so in most other cases. For example, estimates of nitrogen fixation by symbiotic and free-living bacteria show that they add each year about ten million t of nitrogen to the croplands and pastures of the United States, a mass identical to the application of synthetic nitrogenous fertilizers.

This is one of the few instances where we can do what bacteria can—although in a way which looks embarrassingly awkward compared to their capabilities. While they reduce dinitrogen to ammonia at ambient temperature and pressure, unseen, underground, filling leguminous nodules or dispersed in the soil and catalyzing the reaction with such minute quantities of an enzyme (nitrogenase) that its total planetary mass most likely does not surpass a few kilogrammes—we have to build large plants where we force the synthesis of dinitrogen and ammonia at pressures of 10^2 atmospheres.

The only surprising advantage on our side: we actually use less energy to do it. Modern, efficient Haber–Bosch synthesis needs no more than about 30 MJ/kg of nitrogen—while the symbiotic *Rhizobium* fixation consumes at least 12 g of glucose/g of nitrogen, a value equivalent to 180 MJ/kg N. But this advantage appears in an entirely different light once we remember that the feedstocks we use for the synthesis (mainly natural gas, some crude oil) are non-renewable, as is most of the electricity to drive the reduction (hydroelectricity provides only a small part).

Consequently, even this comparison rests on an unequal basis (renewable vs non-renewable energy source) but if American farmers had to double their use of synthetic nitrogen to maintain current productivity it would cost them almost $(1985) 5 billion a year, and it would require additional energy consumption (mostly as natural gas) equivalent to nearly 15 million tonnes of crude oil. Adding 3.5 million tonnes of nitrogen which are recycled by decomposition of crop residues left on the fields raises these sums, respectively, to nearly $7 billion and 20 million tonnes.

But, as noted, such calculations are exceptions. How can one value the fact that, without recycling, current rates of primary production would deplete all of the biosphere's mobile nitrogen in just four million years and that, without denitrification, even the atmosphere's huge reservoir of nitrogen would be exhausted in some 70 million years? For those who may object that such time spans are irrelevantly long for any discussion involving human civilization (of course they are not, as increasing evolutionary complexity leading to our civilization took much longer) let me take just one more example: this one about decomposition of waste in the United States.

Every year, no more than 30 per cent of crop residues are removed from the country's fields, leaving about 400 million dry tonnes of straw, stalks, and vines on the ground. Unlike human wastes which end up mostly in water, animal wastes are largely deposited on land, adding 200 million dry tonnes a year; and about 100 million dry tonnes are dumped on the land each year as various urban and industrial organic wastes. This total of about 700 million dry tonnes a year corresponds, even if uniformly spread over all of America's fields and rangeland, to nearly 1.5 t/ha and there is no need for calculations to realize that if this huge mass was not decomposed promptly by micro-organisms, more than farming would come to a stop in just a few years.

Clearly, environmental services are truly invaluable and the most worrying forms of human interference are not the acutely harmful local or regional pollution episodes so avidly reported by the mass media but rather the gradual, cumulative changes in the grand biogeochemical cycles without whose balancing work this planet would lose its habitable environment. That is why three key cycles will get much attention later in the book—but first will come three chapters detailing the great rise in interest devoted to energy, food, and environment; discussions of the changing perceptions of these complex affairs; and inquiries into their myths and realities.

2

Fuelling the World

Nature, in providing us with combustibles on all sides, has given us the power
to produce, at all times and in all places, heat and the impelling power which
is the result of it.

Sadi Carnot (1824)

The two decades between 1950 and 1970 were the time of unprecedented
worldwide economic growth powered by record expansion of fossil fuel consumption. In those twenty years global output of fossil fuels rose nearly three-fold and
the rich nations multiplied their already high per capita energy use twice (from
about 85 GJ in 1950 to some 170 GJ in 1970) while the poor ones expanded it from
less than 4 GJ to roughly 10 GJ.

Yet, in what must rank among the most notable curiosities of the period,
precious little public attention was given to the energy problem as a whole during
all those years. For the mass media it was a continuous non-event. Scientific books
dealing with total national or international energy systems were rare. And not even
governmental bureaucracies, those organizations ever ready to proliferate on a new
substrate, caught on: no ministries of energy were established, relatively few low-
key forecast studies were done.

Inexplicably, until the late 1960s, energy affairs were getting systematic attention
only by engineers and technical experts managing the production, distribution, and
conversion industries. Only the early 1970s brought the first stirrings of popular, if
not mass, energy consciousness and OPEC actions in 1973 and 1974 elevated
energy to a place of unprecedented prominence in public affairs. A literal explosion
of media, governmental, and scientific interest in energy resources, supplies,
technologies, and futures made the 1970s into a veritable decade of energy
concerns, a process accompanied by much laudable, and greatly overdue, research
and analysis as well as by the emergence of armies of instant experts and by the
offerings of hasty, dubious curative precepts.

That a stream of misinterpretations, fallacies, erroneous forecasts—leaving a
large mass of plain nonsense and gross factual misinformation aside—accompanied
such sudden, intense outbursts of interest and worry is hardly surprising. Unfortunately, once such myths, or worse, are diffused via newspaper editorials, cover
stories of mass-circulation weeklies or, more frequently than one might think,

through referreed writings in scientific periodicals, they are exceedingly difficult to correct.

The most common and the most persistent items in this long gallery of errors had to do with the time left before we run out of oil; with OPEC's might, invulnerability, and future actions; with the dependence of economic growth on energy consumption; with the potential of nuclear generation; with the possibilities of a rapid transition to renewable energies; and with long-term energy consumption forecasts. All of these high-profile items will get a separate section in this chapter which will open with a recounting of the first century of great energy transitions and with an appraisal of the uncomfortable duality of the current global energy use.

2.1 A Century of energy transitions

> If, on the one hand, our great mechanical inventions owe so much to the abundance and consequent cheapness of our fuel, it is no less true that some of those inventions have, on the other hand, materially assisted . . . in bringing about that abundance.
>
> G. R. Porter, *The Progress of the Nation* (1847)

For millenia all civilizations, no matter how highly structured, energized themselves by a combination of animate energies (human exertion and domestic animals for motive power) and biomass fuels (wood, crop residues and dried dung for heat), while the more inventive societies extended the capture of solar energy by adding some progressively more efficient machines to convert the powers of wind, flowing water, and even ocean tides to rotary motion usable in many tedious tasks (grain milling, oil pressing, lumber sawing). As a result, there was little difference betwen energy flows in Alexandrine Greece (336–323 BC) and in the Spanish Netherlands of the late seventeenth century, or between China under the Han dynasty (202 BC–AD 9) and during the last decades of the last (Qing) empire in the late 1800s.

In each of these comparisons 2000 years meant little in changing the energy bases of these otherwise sophisticated civilizations: planting by hand, weeding by hoeing, harvesting with sickles, threshing by trampling, travelling on foot, on animal backs or in animal-drawn carriages, moving goods in the same ways or by sail ships, cooking and heating by burning wood and straw, smelting ores by charcoal, here and there a windmill or a waterwheel. While we have obviously no statistics of energy consumption by sources in the pre-industrial world it is not difficult to calculate the representative shares.

Minimum per capita fuelwood requirements for heating and cooking, range—depending on climate, level of comfort, and modes of cooking—between 500 and 2000 kg a year. Taking 1 tonne as a conservative average, a five-member family of subsistence farmers would burn annually at least 75 GJ which would provide, with no more than 10 per cent efficiency for simple fireplaces, about 7.5 GJ of useful energy. The family would need annually about 15 GJ of food energy and if its labour equalled that of three fulltime workers their annual useful energy output would add a bit more than 1 GJ. Their only draught animal would eat some 20 GJ of roughages and concentrates and, if working 200 eight-hour days, deliver nearly 2 GJ of useful energy.

Consequently, the family's survival required inputs of about 110 GJ of biomass energies in fuelwood (70 per cent of total energy), feed (20 per cent) and food (10 per cent) and the flows of useful energy would be converted in much the same proportions among wood, animal, and human labour. For a pre-industrial society as a whole, the dominance of fuelwood was even greater as the cities needed large quantities of wood and charcoal for households, commerce and industry. Careful reconstruction of American energy consumption in 1850 shows that some 220 PJ of useful energy were gained from fuelwood burning by households and industries while animal and human labour contributed, respectively, just 12 and 5 PJ, or less than 8 per cent of wood's contribution. Still, the combined total of these animate energies was much larger than the use of wind and water for motive power which was only about 5 PJ.

Breakdown of useful energy flows in the frontier America of the 1850s was thus hardly different from the situation prevailing for millenia all around the world and, no less notably, the main traditional sources of heat and motive power kept on expanding even when the country embarked on its first fundamental energy transition to coal. American fuelwood consumption peaked during the 1870s at about 3 EJ and the useful work output derived from animals exceeded that from all inanimate sources of motive power also until the 1870s! But then came the most rapid phase of change-over to coal, the first phase of continuing energy transitions for whose study the American experience is so convenient owing to the combination of the country's reliable historical statistics, rich resource endowment, vigorous energy demand, and high technological prowess.

By the end of the first decade of the twentieth century, reliance on wood fell to 10 per cent of the total fuel consumption from 90 per cent in 1850 as coal supplied more than three-quarters of the rapidly rising energy needs. But coal stood so high for just a decade as crude oil started to move in, surpassing 10 per cent of the fuel total by the end of the First World War and one-third by the late 1940s. Then natural gas started to gain rapidly, expanding from 10 per cent in 1950 to 33 per cent by 1970.

A graph of these changes (Fig. 2.1) offers surprisingly smooth transition curves conforming, more or less closely, to normal (Gauss–Laplace) distribution and providing an obvious forecasting tool. The global pattern of primary commercial energy substitutions is very similar and hence it is easy to think of wood, coal, crude oil, and natural gas in terms of market competition and to analyse their shifts and to forecast their production within the framework of market penetration analysis (Fig. 2.2). Marchetti's work is perhaps the best example of this approach.

Expressing historical primary commercial energy consumption data as log $[F (1-F)]$ the logistic curves showing the market fractions in Fig. 2.2 then appear as

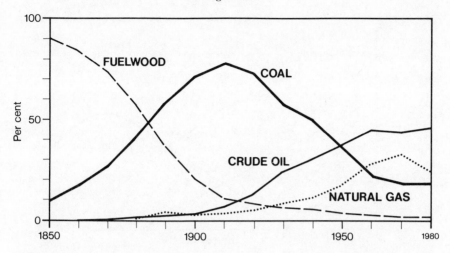

Fig. 2.1 Primary commercial energy transition in the United States since 1850, a sequence of strikingly normal (bell-shaped) curves marking the rise and decline of new sources.

straight lines and give a strong impression of orderly, seemingly inevitable, transition (Fig. 2.3). Rates of substitutions are slow, with about a century needed to penetrate 50 per cent of the market, no new source saturates the demand and penetrations remain amazingly constant regardless of numerous external disturbances. This led Marchetti to comment that 'it is as though *the system had a schedule, a will, and a clock*'.

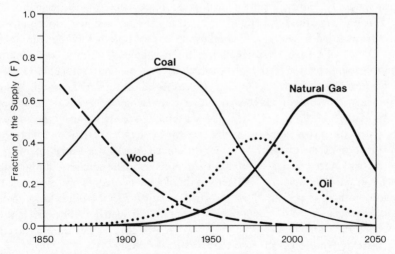

Fig. 2.2 Global primary commercial energy transitions have also followed a normal course. Their shape after 1985 is obviously speculative.

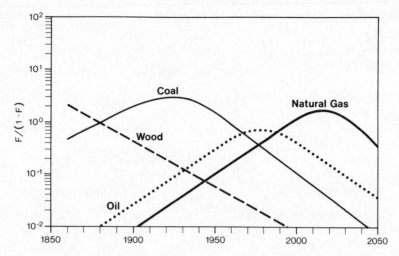

Fig. 2.3 Even more orderly secular substitution emerges when the share of resources in global primary energy consumption is expressed as log F (1-F).

Clearly, there must be strong reasons for these highly orderly transitions, but a closer look reveals that the substitutions have been driven by many time- and place-specific causes rather than by one or two overpowering impulses. On local, regional, and often even on a nationwide level, convenient availability has dictated the transition to a new source, as well as its future persistence. In England relatively high population density, incipient small-scale industrialization, shortage of fuel-wood in lowland areas, and abundant opportunities for opening surface pits or shallow mines led to widespread combustion of coal long before the commonly cited take-off of the industrial revolution in the eighteenth century—and later the country's abundant reserves of high-quality coals kept the fuel dominant for almost two centuries (by 1950 it still supplied 90 per cent of the total fuel consumption).

Similarly, other important coal-mining nations lacking other fossil fuels—most notably Germany, Poland, Czechoslovakia, and South Africa—have had much longer periods of coal domination than shown by the global trend in Figure 2.2. China is in the same category but for a different reason: its potential oil and gas resources appear to be fairly rich but until very recently the country did not adopt the advanced and expensive Western technologies needed to discover and to develop the fuels (especially those in offshore basins). On the other hand, many countries which started to modernize only after World War II did not even bother to develop their known coal deposits, or to look for coal, as they moved directly from wood to crude oil as their principal commercial fuel.

Costs played an obvious role. First, wood transported from the mountains could not compete with coal extracted by tens or hundreds of thousands of tonnes a year from a single mine; then coal from old deep pits where a miner produced, with great risks and by hard manual labour, a tonne of fuel a day became exorbitantly expensive in comparison with refined fuels derived from automated oil wells, some yielding 10^3 t of crude oil each day. But the orderly American experience demonstrates that neither the resource availability nor its cost is the sole determinant of the shift. Crude oil started to move in rapidly after 1900, not because coal was near exhaustion (even today the known reserves are several hundred times greater than annual extraction), or because it was unbearably expensive (coal from old deep mines could not compete with crude oil from rich new fields, but the fuel from large surface mines was always competitive). What made crude oil such a desirable source of energy was its quality and versatility.

All fossil fuels are organic mineraloids, assemblages of organic molecules with minor quantities of inorganic contaminants, but coals vary greatly in quality (poor European brown coals may have only 8.5 MJ/kg; excellent American bituminous coals may surpass 37 MJ/kg; typical, so-called standard coal has 29 MJ/kg). They may have high amounts of moisture (up to 60 per cent), and uncombustible ash (up to 45 per cent), as well as a rather large share of difficult-to-remove sulphur (up to 7 per cent, with 2–3 per cent being most common). And, of course, coals are not easily extracted (especially from thin, inclined, broken, deep, underground seams), are cumbersome to store and to transport, dirty and inconvenient to handle and to use.

By contrast, ultimate analysis of crude oils shows remarkable consistency (owing to the dominance of a few homologous hydrocarbon series). They contain typically 1.5 times more energy per unit mass than even very good coals (around 43 MJ/kg). They have hardly any moisture, mere traces of ash, and their sulphur can be relatively easily removed during refining. Their homogeneity makes for easy interchangeability, their extraction requires very little labour, their water-borne and pipeline transportation is much cheaper than shipments of coal. They can be conveniently stored in above-ground tanks or in underground reservoirs for long periods of time. Their handling is smooth and modern refining turns them into several hundred end-products having an amazing range of final uses.

Several of these refined products make the world's most universal and most convenient transportation fuels, others dominate the urban space-heating market, contribute heavily to electricity generation, and provide essential feedstocks for a multitude of chemical syntheses. Liquidity of these fuels is of such importance that much research has gone into extracting liquid oil from solids (tar sands, oil shales) and into liquefying coals (see section 2.4.3 for details). Natural gases share all the advantages of crude oils except that their lower energy density (1 kg of typical natural gas contains over 45 MJ, but under normal atmospheric pressure it occupies 1.3 m^3) makes for higher transportation costs (overseas shipments require costly liquefaction at $-162°$ C, and pipelines need high-performance compressors); however, they make even better (more convenient, cleaner) heating fuels and unsurpassed feedstocks for such critical syntheses as production of nitrogen fertilizers.

There is also a fundamental thermodynamic advantage favouring hydro-carbons: their inherently better performance in all conversions where they are substituted for coal. A diesel engine converts about 35 per cent of the fuel's energy into motion compared to the usual 5–8 per cent for a coal-fired steam locomotive; a well-tuned gas furnace will use up to 80 per cent of the fuel in heating a house, while a typical coal stove converts no more than 25 per cent into useful energy. And, at the same time, the hydro-carbons are much cleaner and immeasurably more convenient.

And, finally, there is the important consideration of safety. National statistics differ in particulars but not in orders of magnitude: in industrialized nations incidence of accidental death for coal miners working underground is at least 100 times, but more likely 1000 times greater than for oil and gas drillers and fieldworkers, and the same difference of two to three orders of magnitude applies to injuries. About 10^3 higher death and injury risks also accompany coal trans-portation by trains compared to oil and gas shipments by pipelines, the safest and most reliable bulk carriers available. And the huge safety gain is also apparent when comparing fire risks associated with coal burning in household stoves to those arising from oil or gas combustion in modern furnaces.

And it has been the convenience and versatility for final uses which have led increasingly to indirect utilization of fossil fuels as electricity. This important transition best proves that neither much lower thermodynamic efficiency (a typical thermal power plant will waste two-thirds of the fuel during generation), nor much higher costs (an inevitable consequence of the large conversion losses) were a deterrent to this energy transition the extent of which serves perhaps as the best indicator of the energy and economic maturity of a nation.

I know of no better description of the benefits of electriciy than Robert Caro's story of Lyndon Johnson's fight to electrify the poor farms of his Texas Hill Country district in the late 1930s (a small part of a grand, and generally hardly flattering biography). As late as 1935 nearly 90 per cent of America's farms had no electricity, decades after the cities started to enjoy its versatile services. Caro's moving, truly unforgettable account of the everyday drudgery—from exhausting pumping and hauling of water to tedious washing and dangerous ironing with scorching six- or seven-pound wedges of iron—should be required reading for any energy analyst obsessed with maximization of thermodynamic efficiency: only electricity could have been so almost unbelievably liberating!

Rapid gains in electrification of economies came only during the post-World War II period. Before that less than 5 per cent of the global production of coals, oils, and gases was converted to electricity—while by the early 1980s the fraction rose to about 25 per cent and national shares among the richest countries are considerably higher. Not surprisingly, the United States leads with some 40 per cent and, although the trend is still in its initial stages in most poor and industrializing countries, even there it is unmistakably discernible. China, representing the poor populous nations, increased its share from 5 to 20 per cent between 1950 and 1983. This growing electrification of rich as well as of poor economies appeared to offer a perfect niche to nuclear power generation but, as will be seen shortly (in section 2.4.1), actual achievements and current outlook look pitiful in comparison with earlier expectations.

Inserting another take-off curve into the transition chart of global energy sources—as many analysts have done repeatedly—would be thus nothing but wishful thinking. And if nuclear electricity is not taking over, other alternatives are not doing so either (as will be detailed in section 2.4.2 and 2.4.3). Orderly as they may be, past market penetration curves tell us nothing about what the next major source will be and when it will start coming on strong: they just provide solid evidence that it will take many decades before such a source will diffuse widely enough to make a global difference.

Of course, this critical conclusion about slow penetration is based on just a century of shifts involving the emergence and diffusion of just three major fossil energy sources. The future may not replicate the past: the possibility of a much more rapid diffusion of a new source should not be discounted but it is much more unlikely than a rather long period of a 'holding pattern', continuation of the mixed supply characterizing the current situation, prolongation of the trend evident in Fig. 2.4 for the 1970s when coals, crude oils, natural gases, as well as biomass fuels (mostly wood) held concurrently important shares of the global energy consumption total.

This chart is, obviously, different from Marchetti's rendition of energy transitions in Figs. 2.2 and 2.3 which was limited to comercially extracted and delivered primary energies. Looked at within such a framework, fuelwood nearly

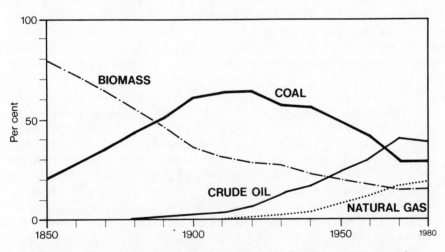

Fig. 2.4 The three previous graphs disregarded the still important contribution of non-commercial biomass fuel throughout the poor world. My best estimates put the share of forest fuels, crop residues, and dried dung at no less than 15 per cent of global energy consumption.

disappeared to be displaced by coals or liquid fuel as the extraction of fossil fuels went much beyond the needs of successive stages of displacement to energize the expansion and advances of industrial, machine, civilization.

Hence the last 100 years were both a period of fascinating energy transitions (driven by complex and often contradictory considerations of resource abundance, production cost, thermodynamic efficiency, versatility, and utilization convenience) as well as an era of unprecedented expansion of energy use—but these transitions and expansions have not been a truly worldwide phenomena such as the eradication of smallpox: all countries have been affected but the lives of only a minority of the world's population have been profoundly changed by these changes, and an uncomfortably deep duality of energy consumption remains the hallmark of today's global energy use.

2.2 Duality of global energy consumption

> Thus I do fully persuade myself that no equal and just distribution of things can be made . . . unless their proprietorship be exiled and banished. But so long as it shall continue, so long shall remain among the most and best part of men the heavy and inevitable burden of poverty and wretchedness.
>
> Thomas Moore, *Utopia* (1516)

The spectacular rise of fossil fuel consumption and the even faster growth of electricity generation have been the critical moulding factors of modern industrial civilization whose machines, methods, and consequences now extend, a few centuries after its gradual European emergence, to every country on this planet. Americans—who have done more than anybody else to advance this civilization both in terms of fundamental scientific and technological inventions and processes as well as along the line of mass consumption—are now consuming every year about 300 GJ of primary commercial energy per capita, an equivalent of just over 10 t of bituminous coal, or about 7 t (50 barrels) of crude oil.

The Japanese, relative latecomers thrust into industrialization almost literally overnight by the Meiji restoration in 1868, have come to be perceived (nearly universally and far from correctly) as a new cutting edge of machine civilization and, industrially efficient and domestically frugal as they may be, they still need annually about 100 GJ of fossil fuel energy per capita to keep the rice-cookers, kerosene-heaters and footwarmers energized, and to keep the world stocked with Toyotas, Yamahas, Minoltas, Sonys, Fujitsus. . . . The old large European industrial powers, England, France, and Germany, which were a generation ago two to three times ahead of the Japanese level, still use more fuel per capita—France about 120, UK 140, and West Germany 160 GJ—but all of them were surpassed by the Soviet consumption (about 180 GJ) which reflects less a personal well-being than the general economic might of the country (inefficient in so many ways but the world's top producer of crude oil and steel, as well as nitrogenous fertilizers, and tractors), as well as its sustained military build-up initiated after Khrushchev's fall.

Besides these six industrial giants all other affluent countries—Canada, Australia, New Zealand, and the European nations (except for Albania, Greece, Portugal, Spain, and Yugoslavia)—consume each year at least 100 GJ of commercial energy per capita. The weighted mean for the world's 30 top energy-consuming nations—a group including also a handful of major oil producers with small populations (United Arab Emirates, Kuwait, Saudi Arabia) whose prosperity has been rising rapidly but whose modernization is still far from complete—then comes to almost exactly 190 GJ a yer per capita which means that, although these affluent nations contained in the early 1980s only about one-quarter of the global population, they consumed nearly four-fifths of the world's output of coals, crude oils, natural gases, and primary (hydro and nuclear) electricity.

Conversely, all of the remaining countries (about 100 of them if one leaves new tiny island nations aside) with three-quarters of the world population used just one-fifth of the global primary commercial energy flow, and their average per capita consumption rate is, at less than 20 GJ a year, an order of magnitude below the rich countries' mean. And consumption differences within this large group of poor nations (a straightforward, although admittedly 'loaded' term I prefer to euphemisms such as 'developing', 'emerging', 'Third World', or, in the World Bank parlance, 'low income' and 'middle-income economies'; needless to say, my usage of 'poor' refers only to the countries' current economic status, not to their resource endowment or developmental prospects) are much larger than within the rich class.

There the extremes differ about four-fold while among the poor countries there is a 100–150-fold span between Nepal, Burundi, or Chad consuming just around 0.5 GJ at the lowest end and North Korea, with some 80 GJ a year per capita at the top. Figure 2.5 summarizes this grand duality of global commercial energy consumption by charting the hyperbolic frequency distribution of per capita commercial energy. Of the 160 nations, roughly one-third averages less than 10 GJ a year per capita, one-half less than 20 GJ, two-thirds below 40 GJ—while only one-tenth surpasses 150 GJ.

Other ways illustrate the duality even more persuasively. Perhaps the simplest one is to compare the average consumption levels of the rich and the poor group with the global mean. This value of about 60 GJ a year per capita is more than three times larger than the poor world's mean but only a third of the average for the affluent nations—one of the perfect cases of frequently meaningless averages! But certainly the most stunning representation of the inequality is through a Lorenz curve (Fig. 2.6) which shows that the top ten industrialized consumers (the two superpowers, Japan, West Germany, Canada, France, UK, Italy, Poland, and East Germany), or less than 7 per cent of the world's nations, account for some 70 per cent of the global energy consumption total.

Energy consumption is far from being a perfect indicator of economic achievements and quality of life (more about this in the next section), but the division between affluence and subsistence based on per capita use of commercial energy is unmistakeable. But the duality does not concern merely the level of consumption, it surfaces also in its composition and in sectoral breakdown. The compositional

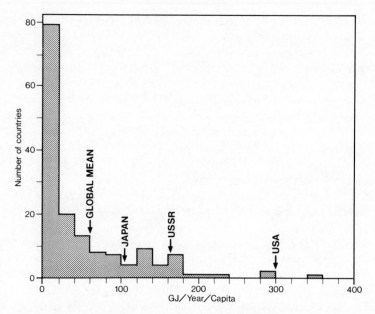

Fig. 2.5 Out of the world's 160 nations for which reliable energy consumption statistics are available (excluding the non-commercial biomass fuels) nearly 80 consume less than 20 GJ/capita a year, an amount equal to only one third of the global average and an order of magnitude smaller than the rich world's mean.

duality is reflected in the dependence on traditional biomass energies: in rich countries fuelwood is a marginal, often fashionable (at least one fireplace in two-thirds of all new American houses!) source, accounting for less than one-tenth and most frequently for less than one-twentieth of total energy supply.

Throughout Asia, Africa, and Latin America biomass fuels—woody matter in stems, branches, bark, roots, and shrubs, crop residues including above all cereal straws, corn stover, cane, millet, and cotton stalks, jute sticks and tuber and legume vines, dry animal dung, as well as leaves and grasses—continue to provide critical portions of the meagre energy supply. No precise nationwide figures about their consumption can be ever available as most of these fuels are gathered by the users (the work is done often by women and children) and burned in simple stoves or in open fireplaces.

My estimates, based on a wide variety of the best fragmentary evidence from various countries, suggest that world-wide these forest fuels, crop residues, and dung contain energy equivalent to nearly 900 million tonnes of crude oil (over 40 EJ), compared to the poor world's annual use of all commercial energies which now surpasses 1.4 billion tonnes of oil equivalent (almost 60 EJ). Consequently, the poor world's consumption of fuels is about 70 per cent higher than is indicated by the available commercial energy statistics (confined to fossil fuels and primary

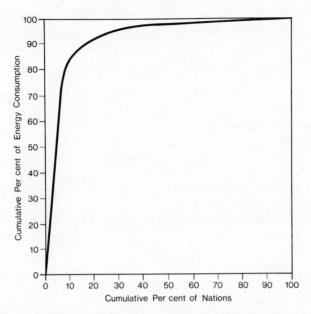

Fig. 2.6 Extremely uneven distribution of global primary commercial energy consumption is perhaps best illustrated by a Lorenz curve: a mere one-tenth of the world's nations uses more than four-fifths of the planet's energy.

electricity); in other words, about two-fifths of the poor world's total fuel consumption comes from biomass.

In most of the African countries and in many Asian nations this dependence is much higher than the global average. Among the most populous poor nations the share of biomass in the total fuel consumption is nearly 90 per cent in Bangladesh, 80 per cent in Nigeria, 66 per cent in Indonesia; in India the figure is close to 50 per cent, in Brazil it is about 33 per cent, as it is in China in spite of the country's huge fossil fuel output.

An approximate classification of developing nations according to their primary energy base shows that in the early 1980s some 750 million people, nearly a quarter of the poor world, lived in countries where biomass provided more than two-thirds of fuel energy; one billion people lived in countries where biomass supplied 40–60 per cent of the total; and only some 300 million people, less than a tenth of the poor world's population, lived in the nations that have shifted to almost exclusively fossil-fuelled economies (South Korea and Argentina are good examples). Clearly, the biomass fuels, now so marginal in the industrialized nations are still a critical, and often dominant, source of energy in the poor world—for everyday household use, for local small workshops, and even for large industries (charcoal for iron-making, and fuel for brick-firing are the two leading uses).

The poor world's continuing dependence on biomass energies has profound environmental and socio-economic implications. Growing demand for fuelwood is one of the principal causes (together with commercial lumbering, shifting agriculture, and conversions to farmland and pastures) of rapidly advancing deforestation in virtually every tropical nation (section 4.2.2). Local effects are already felt in countless places throughout the three poor continents. Deforestation changes local climate by changing the surface albedo, reducing water retention, and raising ground temperatures and wind speeds, and it causes increased soil erosion and flooding. Ever-widening tree-less circles around settlements force people to take longer, more time-consuming and energy-intensive trips to collect the fuelwood, leaving less time for farm, household, or shop work. The preference of relatively richer urban dwellers for charcoal, which contains more energy in a unit mass, is cleaner to burn and easier to transport and store than wood, is even more destructive: each volume unit of charcoal requires at least two units of wood to prepare.

A very large portion of crop residues burned as fuel should not be so consumed at all. Rather, it could be fed to livestock, turned into many household articles, used for construction, papermaking, or crop mulching. Best of all, they should be recycled, by decomposing in and on the ground, or by being returned to the soil as a part of fermented manure. All animal dung should also be fermented and recycled. Residues left on the fields, or recycled with manures, are invaluable sources of nutrients that the poor nations cannot afford to supply in chemical fertilizers. Moreover, they lower soil erosion rates and increase the soil's organic content.

Deforestation and burning of crop residues or dung thus inevitably lower the food producing capacity through climatic change, erosion, and soil deteriorioration. Poor peasants, who cannot afford any other source of energy (kerosene, fuel oil, electricity), and who spend increasing amounts of time or money to collect or to buy local biomass fuels, are thus leaving behind a degraded environment while drawing not one but two circles of deepening poverty—those of fuel and food.

In many places since the onset of the high oil prices in 1973 this situation has only worsened. Imported fuel is now beyond the reach of all but the relatively richest peasants. Many small local industries which converted previously to liquid fuels are reconverting to fuelwood (which, though also more expensive may be still more readily available), and the ever-growing numbers of urban poor are now paying substantial portions of their low wages just to purchase minimum quantities of fuelwood or charcoal for cooking. Given the already high indebtedness of most poor nations, as well as their frequent internal instability, further general deterioration of biomass fuel supplies must be expected. This outlook is most alarming because biomass fuels are already the only energy sources for the poorest villagers and urban residents of Asia, Africa, and Latin America.

Contrasting poverty and richness of the supply are also mirrored in the utilization breakdown of the available energies, a duality which can be perhaps best illustrated by comparing the breakdowns in the world's strongest economy and in the most populous poor nation. China's mid-1980s per capita consumption of commercial energy averaged almost exactly 20 GJ annually (very close to the poor world's mean) but as industrial production used nearly two-thirds of the total, transportation and pre-conversion losses each accounted for about 10 per cent, and farming took some 5 per cent, direct energy use by households amounted to no more than about one-tenth of the total flow.

This minuscule sum of 2 GJ is, however, yet another unrepresentative statistical mean rather than a reflection of typical experience. Four-fifths of the Chinese live in villages where direct household consumption of fossil fuels and electricity is a small fraction of urban averages. Rural energy surveys have shown that no more than 30 million tonnes of coal and a mere one million tonnes of liquid fuels are used annually for cooking, heating, and lighting and that biomass energies still supply the bulk of everyday fuel use in rural households—and that at least half of these households (or some 400 million people) are short of wood and straw for four to six months of each year!

China's annual biomass energy consumption now includes about 250 million tonnes of crop residues, about 350 million tonnes of woody matter, leaves, and grasses, and at least 10 million t of dry dung. Converting at 15 MJ/kg this adds up to roughly 9 EJ (equivalent to slightly over 300 million tonnes of coal) or to no more than 11 GJ a year per capita. Addition of the small amounts of fossil fuels and electricity (just 10 TWh a year used for lighting) increase this mean by barely 1 GJ.

In contrast, in the United States industrial production will claim about 30 per cent of the final consumption, transportation a third, and residential consumption nearly a quarter; however, as almost half of the transportation share is in private cars, direct consumption of commercial energy in American households comes to about 35 per cent of the total energy use, or just over 100 GJ a year per capita. An average American, that is a city-dweller, would thus seem to consume directly about eight times more energy than an average Chinese subsistence peasant. However, detailed comparisons in Table 2.1 show that the real difference is at least three times larger (that is about 25 times).

This is explained by drastically different conversion efficiencies, yet another instance of the rich–poor energy consumption duality. Obviously, we use energy to perform specific tasks and traditional ways may use several times more energy than efficient, modern devices: what matters, of course, is the quantum of usefully converted energy rather than the total amount of energy entering the process. Stoking a simple mud or brick stove with straw or shrub branches may convert just one-tenth of the fuel to the desired final use of cooking the rice or heating the main family room; in contrast, new flue-less natural gas furnaces have efficiencies in excess of 90 per cent and cooking with electricity, even when considering the conversion losses during its generation from fossil fuels, is at least three times as efficient as cooking on simple stoves.

So, even when using fairly conservative average efficiencies for the United States conversions, the difference in the amount of direct effective household energy between an American and a Chinese is more than twenty-fold! A few comparisons deserve a closer look in order to provide further illustrations of the rich–poor

Table 2.1 Comparison of total and effective energy consumption in average households of rural China and urban United States

	Average annual household energy consumption (GJ/capita)		Typical conversion efficiencies (%)		Effective energy (GJ/capita)	
	China	USA	China	USA	China	USA
Heating	8	42	12	50	0.96	21.00
Cooking	4	8	12	40	0.48	3.20
Lighting	0.1	5	5	10	0.00	0.50
Transportation	0.0	50	–	25	–	12.50
Total	12	105			1.45	36.70

energy gap (keeping in mind that about half of the world's poor nations use even less energy than China does).

Although early in 1984 a newly-rich peasant family in a Beijing suburb bought the country's first privately-owned car, such a feat gets lost among one billion people. Overcrowded buses and long-distance bicycle commuting of many urbanites in muggy August days or (in the North) against cutting January winds aside, the absence of private cars in China does not preoccupy peasants who often do not have even enough straw or wood to cook their three daily meals. Contrasts between the levels of essential energy use in Chinese rural households and lavish American flows of natural gas and electricity are hard to comprehend for those of us used to plentiful energy supplies.

Many South Chinese families without access to fuelwood decided to install biogas digesters where crop residues, grasses, animal, human, and household wastes are anaerobically fermented to yield a mixture of methane and carbon dioxide which can be used for clean and relatively efficient cooking. A typical family digester will produce aout 1 m³ of biogas a day (energy equivalent of 24 MJ), an amount considered by the Chinese to be just sufficient to fuel the cooking of three meals and heating of water for a family of five or six people. Consequently, even such a 'modernized' family will not have more than 4–5 MJ per capita a day to cover all cooking and hot water needs.

In contrast, an urban American will consume the same amount of energy in natural gas by just taking a light four-minute shower (if she indulges in a bathtub filled to the brim with hot water the consumption doubles) and the daily per capita use of energy for cooking, refrigeration, freezing, water heating, dish and laundry washing adds up to at least 25 MJ in an unlikely case where all the devices would use natural gas and to some 75 MJ in an all-electric home.

Differences in lighting consumption are even more telling. In China 500 million peasants (60 per cent of all the rural population) do not have any electricity in their homes, and those who do have just two or three bare 25 W bulbs and consume daily a per capita equivalent of a single 100 W bulb lit for forty minutes—while in an American house installed lights will easily add up to 500–700 W per capita and a daily use will be equivalent to ten 100 W bulbs burning for over one hour.

Implications arising from these differences for the quality of life are enormous: the substitution of time-consuming, tedious, tiring tasks such as fuel gathering, kindling, and stoking of fires, heating water, hand-washing and pressing clothes with heavy, scorching irons by automatic water heaters, washing machines, and lightweight electric irons did more for the true liberation of women than legions of Gloria Steinems and Germaine Greers ever could. Extension of the day by electric light, a revolutionary feat we readily acknowledge but whose significance we cannot anymore really comprehend after so many decades of reliable supply, is still to bring its magic, leisure-time and educational benefits to most of the world's poor subsistence peasants.

Consequently, the post-1973 developments have been all the more unfortunate: doing away with the pervasive duality of global energy consumption would be a frustratingly protracted process even under fairly favourable circumstances, but sharp rises in the price of crude oil, the leading source of fossil fuel energy in all but a couple of countries in the poor world, the subsequent global economic slow-down and deepening debt crisis have made any early, decisive narrowing of the gap much more unlikely. With lower oil prices in 1986, the outlook for many poor nations is better but the underlying economic malaise, especially in Africa and Latin America, remains.

Rich countries suffered too, though in very different ways. Several critical energy-economy relationships and trends which came to be seen as immutable during the rapid growth decades of the 1950s and 1960s have been greatly altered since then and many perceptions concerning resource availability, technological capacities, future consumption rates, and long-term energy strategies have undergone often bewildering and always fascinating developments and reversals. To understand where we have come from and where we are headed a review and a critique of the post-1973 energy situation is a must.

2.3 1973 and after

> Vain are the prayers of the unbelievers . . . Allah gives abundantly to whom he will and sparing to whom he pleases.
>
> *The Koran*, 13th Chapter (Thunder) (N. G. Dawood, translation)

With every receding year the impact, the panic, the fear of a permanent crisis brought by the first round of rapid crude oil price increases in 1973–4 and, even more incredibly, by the more recent second round in 1979–80, seem not only duller but somehow increasingly, almost embarrassingly, incomprehensible. Queues of cars winding around several city blocks, the obsession with topping up tanks, rationing schemes matching even- and odd-numbered cars and days, occasional violence in the slow-moving queues, tales of worsening heating oil shortages, and predictions of America once overflowing with cheap Texas oil at the complete mercy of greedy Arabs—tiresome cartoons of sinister looking men in flowing robes and headgears having us over the barrel.

In 1985 the headlines did not promise Western civilization freezing in the dark, and did not lament any more extortionary transfers of the industrialized world's riches to the princely mansions and royal accounts of the Middle Eastern oil producers. In 1986 oil prices tumbled to between $(1986) 10–15 per barrel as global oil production was bumping along near the bottom of a ten-year low and the informed consensus was that at least for the rest of the 1980s there was no need to worry either about the availability or the price of oil. In a way, it as if the oil saga of 1973–81 did not happen: but it did and the changes it brought have been profound, and many of the concerns real.

The first price jump in 1973 helped to bring an end to a generation of unprecedented global economic growth. Worries about the future crude oil price and supply were perhaps the leading, unabating international concern for the rest of the decade. Western governments changed, African *coups d'état* were staged with their accustomed frequency, Communist countries kept on laying out their five-year plans—but nobody's dealings and plans were the same as before 1973. Losses of the losers looked dismally painful and endless, the winners were gaining grandly; schemes put forward to deal with the changed situation ranged from the ingenious technical fixes (more on them in section 2.4) to desperate military moves. And then came suddenly an unexpected, imperceptible easing of tensions in the early 1980s and feelings of a more comfortable outlook and then the crude oil price collapse of 1986.

But the account must start with the chronological review of price changes. The 1973 price rise was so shocking not only because the increase was so unusually great for a commodity so basic, but perhaps even more so because it came after a century of steadily declining prices (Fig. 2.7). Measured in constant US (1970)$, an average barrel of traded light crude oil was selling in 1970 for less than a quarter of its price a century before, and at a mere $1.8/bbl it had to be one of the greatest all-time bargains (regardless of the fact that its average production costs were only about a tenth of the total cost).

The Organization of Petroleum Exporting Countries (OPEC) was set up in Baghdad in 1960 by three Arab nations, Saudi Arabia, Kuwait, and Iraq, with Iran and Venezuela also as its founding members. Later admissions—Qatar (1961), Libya and Indonesia (1962), Abu Dhabi (1967), Algeria (1969), Nigeria (1971), Ecuador (1973), and Gabon (1975)—raised the total to thirteen countries and the steadily increasing worldwide demand and growing assertiveness of major exporters (above all Iran, Libya and Saudi Arabia) began to be translated into higher posted prices in 1973.

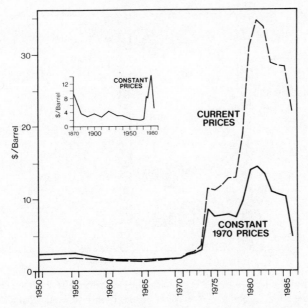

Fig. 2.7 More than a century of crude oil prices is charted here in constant 1970 dollars, and since 1950 also in current prices. Rapid inflation during the 1970s and early 1980s made the price rise vastly more impressive than it appears in constant terms.

OPEC's posted price, the fictitious level used for calculating the taxes and royalties owed to the governments, and expressed in terms of the Saudi Arabian light oil, rose to US (1973) $2.591 per barrel at the beginning of 1973, and to $3.011 a barrel by 1 October 1973. Five days later, on the Yom Kippur holiday, Sadat's army crossed the Suez Canal and broke the Bar Lev line in its central sector. With much difficulty Israel reversed the tide and started to move deep inside Egyptian territory west of the canal before the United States stepped in to save the encircled Egyptian Third Army and Henry Kissinger negotiated the cease-fire and dis-engagement of forces. The war on the battlefield ended, with Sadat's prestige much enhanced and with his troops back in a sliver of the Sinai, but the oil fight had just begun.

On 16 October 1973 the OPEC nations raised the posted price from $3.011 to $5.12 per barrel and on 17 October the Organization of Arab Petroleum Exporting Countries (OAPEC) agreed to embargo oil exports to the United States and the Netherlands and to initiate gradual reductions in crude oil extraction until Israel's withdrawal from occupied Arab territories. On 23 December 1973 OPEC announced that the posted price would go to $11.651 per barrel on 1 January 1974, a 4.5-fold increase since January 1973 (Fig. 2.8). The oil export embargo proved ineffective (as the multinational companies just redirected their tankers) but any hopes that the new high prices were only temporary and that, in the long run,

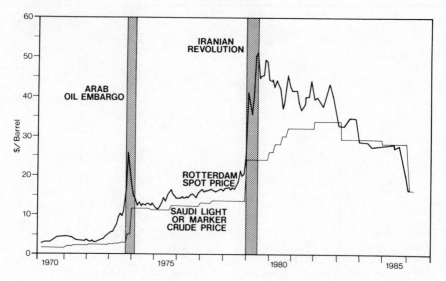

Fig. 2.8 OPEC the price-setter or just a follower? Comparing the course of crude oil spot prices and the Saudi Arabian marker price makes it easy to argue that OPEC merely followed the levels established in free trading and driven high by uncertainties during the 1973–74 OPEC embargo and by the Iranian revolution of 1979–80.

OPEC would not be able to sell oil costing about ten cents to produce for a hundred times as much proved illusory.

OPEC's official sale price, which averaged $11.28 per barrel in 1974 (as oils lighter than the Saudi light—such as Nigerian and Algerian crudes—always sold for up to 15 per cent above the benchmark, while the prices of heavy, high sulphur crudes have been at least 5 per cent lower, the mean price reflects the quality differentials of the exports), changed very little during 1975 and 1976 (averaging, respectively, $11.02 and $11.77 per barrel), rose slightly in 1977 ($12.88), held in 1978 ($12.93), and then, in the second, Iranian, round of rapid price escalation, took off again during 1979 (Fig. 2.8).

I will have much more to say about OPEC's behaviour in the following section but while you are looking at Fig. 2.8 note how easily one can argue that the cartel merely followed spot market sales in setting its market price, that the 'free' market was a leader, and that OPEC councils were codifiers of the accomplished—and that this relationship worked during rapid price increases as well as more recently with price reductions.

Back to the chronological account. When Shah Mohammed Reza Pahlavi was leaving Tehran on 16 January 1979 for 'an extended stay abroad' (in reality for an exile soon to end in his death in Cairo), OPEC's official sale price averaged $13.62 per barrel; in November when the American Embassy was taken over by Khomeini's fanatical followers the price stood at $24.20, a month later at $25.65.

With America humiliated and seemingly powerless, with Iranian output steadily falling and with Western oil demand continuously rising it seemed only a matter of time before we would see a $50 and then a $100 barrel. In January 1980 nobody knew that the turning point would be reached soon and that an incredible weakening of the OPEC was on the way. For the whole year the price rise continued, from the average of $28.15 per barrel in January to $32.95 in December, and the trend carried over into 1981 which opened at $34.86: in two years OPEC's average official sale prices went up 2.5 times and in January 1981 were 13.5 times higher than in January 1973. Prices on the spot market were easily topping $40 a barrel—but the official average sale price had only three cents of rise left.

In March 1981, at $34.89 per barrel, the peak was reached; the next month saw the three cents coming off, then a few cents again and the year ended with a mean of $34.50; the next year shaved the mean to $33.63, and by early 1983 OPEC had to bow to the now inexorable downward pressure and legitimize widespread price cutting by many of its members by setting a new official marker price of $28.74 per barrel in March 1983. The average sale price remained just a shade above that value for the rest of 1983 ($29.31 per barrel) and during 1984 it continued sliding slowly to $28.46 by the end of the year. In early 1985 yet another of OPEC's discordant meetings in Geneva agreed to further price cuts through a 'realignment of differentials' among various kinds of oils, a move which translated into a further 30–40 cent decline for an average exported barrel. Finally, in September 1985 the Saudi's decided to regain their lost market share and doubled their output to more than four million barrels per day. The international price held for a few months and its collapse came only in 1986: in January it dropped below $20, in February below $15, in early April briefly below $10 per barrel to settle afterwards around $15.

Comments, analyses and prognoses accompanying the first dozen years of what can be justifiably called a new energy era have reached enormous numbers, but what I find most fascinating in retrospect is how irrational have been not only many presentations of instant experts but also not a few 'mainstream' perceptions, how even the most prestigious consensus appraisals fell victim to the impressions of the day or the trend lines of a past year, how rare were any correct forecasts. Out of this longish array of misperceptions and erroneous predictions I will select just three high-profile topics for a closer look: I will make a brief inquiry into OPEC's rise, successes, failures, and impacts, then I will turn to the concerns about running out of oil, and close with a review of energy-economy relationships in the changed circumstances.

2.3.1 *Living with OPEC*

> We must bear the medicine on account of its usefulness.
>
> Arabic proverb (J. L. Burckhardt collection)

Before 1973 the thirteen year-old organization was largely ignored, not only by newspapers and TV commentators but, more importantly, even by multinational oil companies, the latter fact going some way towards explaining the accumulated frustration and astonishingly rapid gains of OPEC negotiators; after expensive OPEC oil stalled Western economies, and temporary gasoline shortages, essentially unrelated to OPEC's moves, maddened panicky motorists (who insisted on topping-up their tanks after every day of commuting), the organization was reviled in an incessant stream of invectives and accusations which ascribed to it almost demoniacal powers (we being the defenceless *innocenti*) and foretold either its unstoppable rise to global domination or the much-deserved collapse of the greedy cartel.

Ten years later the passions were spent, the realities not fitting any of the earlier preconceptions. The world of the mid-1980s in general, and the rich Western nations in particular, is not prostrated in front of the omnipotent OPEC since the organization's powers have been much weakened and its revenues have kept on plummeting. On the other hand, OPEC has not collapsed and the prices it sets are now charged by all countries exporting oil and none of those nations (including the United Kingdom, Mexico, USSR or China) wishes to see this precious revenue base slipping.

What real and lasting changes has OPEC brought to the global energy scene? Has it achieved its major objectives or was its heady rise just an inevitable, and for the rest of the world painful, episode before its current, comforting decline? And has it really been laid so low that it will not be able to reassert itself in the fashion of 1973-4 or 1979-81? Answers to these questions are made quite difficult owing to the obvious heterogeneity of motives within the organization, and they also depend heavily on the yardsticks used.

What has failed completely has been the much publicized attempt of the dominant Arab members of the organization to use oil exports as a weapon against the pro-Israeli West in general, and against the United States in particular. To be reminded of the tenor of the original demands, here is how Mana Saeed al-Otaiba, oil minister of Abu Dhabi who liked to pose for photographers with a drawn sword poised high above his head, spoke on 18 October 1973 when he announced with characteristic Arab rhetoric the use of 'petroleum as a weapon in the battle for the first time'. After stating that 'Arab oil is not more precious than Arab blood' he said it would give 'much pleasure' to 'cut off supplies from the USA . . . to turn the scale of this scared war. . . . We shall not deviate from this decision until the USA alters its hostile position towards the Arab nation. In this war we must be victorious. We shall accept no substitute for victory.'

Al-Otaiba's studies in Baghdad and in Nasser's Egypt certainly had a lot to do with his eager pan-Arab stand but the substance of his oration was fully in accord with OAPEC goals. More than a decade later Israel is much more deeply entrenched in Judea and Samaria, it has officially incorporated the rest of Jerusalem, and although it did give back the Sinai peninsula to Egypt, this restoration had nothing to do with the 'oil weapon' and was not at all welcomed by fellow Arab states. Four different American administrations have come and gone since the Yom Kippur war but American support for Israel remains solid. Clearly, OAPEC has not been able to change to its liking the greatest political irritant on its doorstep: the 'sacred war' shows no signs of ending and the 'oil weapon' has joined a long list of Arab clichés.

What must be seen as the outfit's biggest success is its very survival. OPEC was always a most unusual alliance, a peculiar assemblage of religious traditionalists (Saudi Arabia) and revolutionary zealots (Libya), conservative feudal fiefdoms with smallish populations (United Arab Emirates) and huge unruly republics run by soldiers (Nigeria), of highly urbanized (Kuwait) and rural (Indonesia) societies, of very rich and very poor (both in inter- and intra-country comparisons).

The Arabic veneer given to the organization in common Western perceptions was inevitable—Arab members of OPEC supplied nearly 60 per cent of OPEC's total production in 1973, 73 per cent in 1980, and Saudi Arabia has been such a dominant member that OPEC has been seen by some as nothing else but a cover for the royal family's devious plans. Both the Shah and Ayatollah Khomeini would have resented such an interpretation: the first considered himself the foremost architect of OPEC's new militancy, the other's dislike of non-Shia Muslims has not been strong enough to stop him sending his volatile ministers to Vienna or Geneva to square things with the Saudis—and eventually to patch up yet another shaky compromise.

Descriptions of OPEC's infighting would occupy volumes and most of the information leaked from closed-door meetings or given afterwards, usually by Saudi Arabia's Minister of Petroleum and Mineral Resources, Shaikh Ahmad Zaki Yamani, has always been heavily rhetorical, self-serving, and propagandistic. If all such pronouncements were taken seriously and contrasted with realities then OPEC's achievements would not look so glorious. But if an ultimate proof of OPEC's resilience is needed then the Iran–Iraq war provides one: Iran's continuing presence in the organization is perhaps the best example of OPEC's far from perfect but repeatedly surprising cohesion.

Discounting on the one hand, and ignoring the assigned production quota on the other, had been common enough but not so flagrant and prevalent as to cause a precipitous collapse of the official price structure before the Saudi break-away in late 1985. Saudi capability to extend or to cut the organization's total output by about eight million barrels a day was undoubtedly a critical factor here—in a tight market the Saudis pumped as much as 10.3 mbd and by the summer of 1985 their output slumped to just over two mbd.

However, until September 1985 the Saudis were using their balancing act not to help the world at large as claimed by some commentators but to protect their country's future earnings: letting the price rise too high would lead to levels unsupportable in the longer run (as happened with the $34 marker price after February 1983). Producing at full tilt in an overflowing market would lead to sharp price declines so that even a higher output would not bring in the monies required to run the country's ambitious development plans (as happened in 1986 after the Saudi's abandoned their output restraint in a desperate bid to regain at least some of their lost market share). Pre-1985 Saudi *de facto* price-fixing within OPEC by means of their production swings thus helped everybody in the organization, including the discounters and quota-breakers (whether Nigeria or Iran) who did not have such a power over the market price.

Ali Jaidah put it precisely: 'The gains from sticking together are substantial; the losses that would be incurred after a collapse are simply enormous.' And so even after the Saudis boosted their production in the fall of 1985 they did not go anywhere near their full capacity and attempts to patch up a new OPEC-wide policy are still continuing. Recognition of the necessity of some eventual cooperation transcends the enormous differences among the member countries and is responsible for what must be listed as OPEC's fundamental, fulfilled achievement: the organization, buried countless times since 1974 by confident American energy administrators, talk-show energy experts, and sneering editorial writers is still around. Its power to set the global oil price has been much weakened but not irrevocably eliminated. Also, there is little doubt that most of the Middle Eastern oil sold at $15 (1986) per barrel is still far above any level justifiable by costs of exploration, development, and production enlarged by generous but only slightly extraordinary profits.

This power to set oil prices was, of course, the principal aim of OPEC long before its acronym became a bogy of Western governments and so in this respect the organization succeeded between 1973 and 1985 certainly beyond its most audacious hopes of the 1960s—and it may get another rewarding chance. Moreover, by taking away all the concessions and compensating the operating companies just for the fixed assets (undoubtedly a huge bargain) OPEC achieved something that it did not originally anticipate to get for decades: full nationalization of the industry. More than that, national oil companies of producing countries have been increasingly entering into direct oil marketing, bypassing the previously indispensable multinationals, as well as into transportation, refining, and petrochemicals.

This is precisely what the early OPEC organizers wanted back in the 1960s: not only what they considered a just revenue for their non-renewable resource but access to the latest technology, and genuine barter of oil for economic modernization. Even with tens of billions of dollars wasted on and diverted to Swiss villas, private Boeings, New York real estate, advanced jet fighters, collapsed nuclear power plants, or PLO acounts, the earnings have left plenty for what have been some of the most impressive construction efforts of the post-World War II years and for spectacular rises in standards of life in OPEC's less populous Arab states.

The desert turned into wide, tree-lined boulevards, air-conditioned housing areas, expensively irrigated crop fields, huge automated refineries, universities, airports, hotels, ports—all of uncommon dimensions and of the latest design—built from scratch, the acquisition of a modern infrastructure which other countries took generations to assemble concentrated into a decade. Clearly, all undeniable benefits: earnings may be sharply down now but the decade as a whole has to be seen as one of unprecedented gains, an indubitable sign of OPEC's success.

But the obverse of the gleaming facade is surprisingly disquieting. In many places the facade itself is so obviously marred: living with OPEC's ups and downs has been surprisingly hard on the member states. For OPEC's smaller but populous producers—Indonesia (the world's fifth largest nation with some 175 million people in 1985) and Nigeria (Africa's most populous country with 100 million in 1985)—rising oil exports were the main barrier separating them from a slide into greater indebtedness and economic stagnation. As soon as the oil earnings started to slide in 1981 their current account balance went into the red by billions of dollars. Indonesia, with a relatively diversified export base (80 per cent from oil and liquefied natural gas) and higher foreign reserves has been doing better than Nigeria (98 per cent of exports from oil). The latest Nigerian military take-over came as a direct response to the country's collapsing economy.

With per capita GNP much below $1000, life expectancy at birth below 50 years, and literacy around 50 per cent there is little that is enviable among the socio-economic indicators of these two countries whose oil earnings bought some industrial and urban improvements but in Nigeria it also led to an astounding neglect of agriculture and to increased dependence on food imports (Indonesia, in contrast, was self-sufficient in rice by 1985). What is surprising is the extent to which similar difficulties and other trials affected the richer states of the organization.

The Iranian story needs little recounting. One of the world's boldest modernization experiments, planned across decades rather than years and involving everything from supersonic jet fighters to more than 20 GW in nuclear power plants, ended in the rule of medieval imams and in mass slaughter of teenagers in the marshes and deserts on the Iraqi border, a tragedy notable even in this century of genocides. The other warring party, Iraq, would be economically even more prostrate if it were not propped up by increasingly scarce transfers from other Arab Gulf states.

The Algerian economy has made some steady gains but two-thirds of the country's population are still illiterate and life expectancy at birth is just above 50 years. The low levels of both of these critical quality-of-life indicators in virtually all Arab OPEC countries (literacy just around 25 per cent in Saudi Arabia, Qatar and Iraq, life expectancy below 55 years) are a perfect illustration of how real development is far from just being a function of money. Many Asian countries with per capita GNPs only one-tenth of the Arab *nouveau riche* do at least as well or even much better on both accounts!

Another good indicator of how far even the richest OPEC countries are from economic emancipation is their continuing dependence on skilled expatriate labour in running their affairs. Perhaps the best example of this dependence on foreign manpower is the case of Muammar Ghaddafi's Libya. While the leader of the Jamahirya publicly despises just about anybody abroad and tries (both openly and clandestinely) to hurt Western interests by actions ranging from assassinations in European capitals to invasions of neighbouring African countries, he goes out of his way to proclaim and to assure the safety of thousands of foreigners running Libya's oil (and also other) activities.

Saudi reliance on foreign expertise is similarly large and the kingdom, although relatively stable, has had its share of worries related to rapid economic development within such a traditionalist society. The fundamentalist attack on the Grand Mosque of Mecca in December 1979 was an event much more serious than the Saudis admitted, and their ambivalent policies *vis-à-vis* both the United States and other Middle Eastern countries, as well as their peculiar rôle in OPEC (professing to be the staunchest advocates of orderly, gradual price changes although their actions critically contributed to sharp price rises both in 1973–4 and in 1979–80), attest well to how uncomfortably they wear the responsibility attendant on their central position in the global oil market and on their riches. While it is hardly another Iran in the making, Saudi post-1973 experience has been definitely the proverbial mixed blessing.

By far the worst difficulties now affecting every OPEC state are their falling earnings. By 1983 only the small Arab countries (Kuwait and the United Arab Emirates) and Gabon did not have a negative current account balance; for the Organization as a whole the balance peaked at just over $100 billion in 1980, slid to $44 billion in 1981, turned negative ($−10.5 billion) in 1982, and deepened to more than $−50 billion in 1985. Figure 2.9, which charts this development, also shows how OPEC's spending trend has not lagged behind the organization's earnings— and hence how the rich countries, the leading exporters, benefited from these newly wealthy importers.

Yet just the opposite was the surpassing concern just after the first round of price increases when it was feared that within a decade or so all the liquid wealth of industrial civilization would have to be surrendered and all the real estate would have to be mortgaged just to keep all homes warm and lit, meals cooked, basic

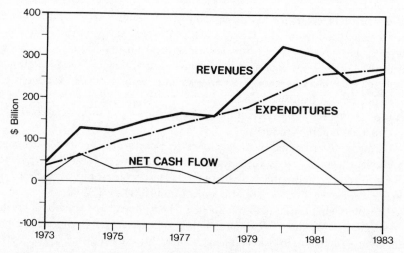

Fig. 2.9 A roller-coaster decade for OPEC: revenues kept rising but so did expenditures and the net cash flow went form very low to high (1974) to very low (1978) to very high (1980) to red after 1981.

industries running and, occasionally, perhaps even cars driven. Monies involved in what was repeatedly stressed to be the greatest transfer of wealth in history were causing feelings of bewilderment and desperation. In 1970 total government oil revenue of OPEC members was just short of $8 billion; a year later it almost doubled and then, as charted in Fig. 2.9, it took off to surpass $100 billion by 1974 and $300 billion by 1979.

What burden these OPEC earnings represented for the global economy is best shown by expressing the oil purchases as shares of the gross world product (GWP). Before 1973 oil purchases accounted for less than 0.5 per cent of GWP—while today they add up to about 5 per cent, a ten-fold difference. The effects of this huge shift in the world economy have been plain to see and unpleasant to experience: diversion of income from consumption and investment, reduced consumer spending, sluggish economic growth, and slipping current account balances were real and they did hurt. However, the largest importers whose demand has always accounted for the bulk of OPEC earnings have been also in the best position to supply OPEC countries with their much needed imports, both resource-intensive (above all food) and capital-intensive (products ranging from drilling equipment to computers, from irrigation devices to fighter jets).

The rapidity with which the Big Seven was able to boost its exports to OPEC has been extraordinary: after the first round of oil price increases OPEC sold to these industrialized countries three times as much as it bought from them, but by 1978 the Big Seven's exports/imports ratio with OPEC rose to 0.66—and after it was knocked down by the second round of price rises in 1979–80 it rose again smartly to 0.67 by 1982. In fact, between 1973 and 1983 the Big Seven exports to OPEC rose faster (seven-fold) than the group's oil imports and its relative trade deficit with OPEC was smaller in 1983 than it was before 1973.

The Big Seven's aggregate annual trade deficit with OPEC has been below $50 billion since 1982, equivalent to less than 1 per cent of the group's GNP—or to less than one-fifth of their combined expenditures on the military. Clearly, living with OPEC has not been all that crippling for the industrialized West! Moreover, the Big Seven, as well as smaller West European nations, have also benefited from the huge investments of surplus OPEC revenues in financial markets, real estate and, to a lesser degree, also directly in some manufacturing.

OPEC's foreign investments have been especially helpful to the United States by providing 'painless' funds for the growing borrowing by the Treasury, as well as for corporate financing. Investment of surplus earnings in the West motivates the investors, especially the biggest ones (Saudis, Kuwaitis, people from the Emirates) to maintain the long-term prosperity of the industrialized economies—and it goes a long way to explain the moderate attitude these countries have taken towards oil price-fixing: the proverbial avoidance of killing the goose that lays the golden eggs applies here so well.

Those who have really suffered in consequence of OPEC's price increases have been the poor oil-importing countries, ranging from the most destitute African nations to the upper middle-income economies such as Argentina or Brazil. With a few exceptions (Indian coal, Brazilian hydro–electricity) these nations did not have any significant production of indigenous energy resources and hence imported liquid fuels, although consumed in relatively small amounts compared to the levels in rich nations, have come to play a large role in their commercial energy supply.

The dimensions of this problem are staggering: fewer than 20 of nearly 100 poor non-OPEC countries are currently self-sufficient in crude oil at their low consumption levels, and only 10 of these are able to export more than 50 000 b/d (Mexico, Oman, China, Egypt, and Brunei each ship over 20 000 b/d). The remaining 80 poor countries are now buying about 4.5 mbd of crude oil. Although most of this fuel is destined for cities and industries (and also for the military), and the rural areas in most of these nations continue to rely almost solely or predominantly on local biomass energies, the cost of these imports is extremely burdensome in relation to their export earnings.

Energy imports of industrialized Western countries before the 1973–4 oil price rises equalled about 15 per cent of their total merchandise exports and then rose to around 25 per cent in 1982, but for the world's 40 poorest countries this ratio rose from about 10 per cent in the early 1970s to 45 per cent a decade later! In the early 1980s half or even more than half of many poor countries' exports went to buy oil— yet the amount of money available was most often no greater than in the early 1970s, and the huge expenditure preempted purchases of much needed modern technologies. This squeeze is well illustrated by comparing the pre-1973 crude oil import burden for Brazil and three large Asian nations with recent expenditures (Table 2.2).

True, during the early 1980s Japan also spent about half of its export earnings on oil purchases but it then upgraded the value of the fuel through sophisticated manufacturing, fuelled with it the world's third largest economy, and supported with it an affluent standard of life. In contrast, the poor world's burdensome oil imports barely kept those countries at the per capita consumption levels of the early 1970s while the value of exports was hard to maintain even with rising volumes

Table 2.2 Crude oil import burden of four leading industrializing economies

Country	Crude oil imports (thousand b/d)		Crude oil import costs (million $)		Total exports (million $)		Crude oil import cost as share of exports	
	1972	1982	1972	1982	1972	1982	1972	1982
Brazil	515	790	470	9 700	3 991	20 180	12	48
India	313	340	285	4 100	2 415	7 300	12	56
Philippines	188	200	170	2 400	1 029	4 970	17	48
Thailand	151	220	135	2 700	1 081	6 940	12	39

when the prices of agricultural commodities (be it cocoa, cotton, palm oil, or sugar) and metals (the items figuring high on the export list of all low-income economies) have been at best stationary but predominantly falling ever since the late 1970s.

And while the poor oil-importing countries would greatly benefit from a prolonged period of cheaper oil—price averaging $15 rather than $27 would save annually almost $5 billion for Brazil, more than $1 billion for the Philippines at the 1985 consumption levels—the same shift would now be ruinous to OPEC's Nigeria (which would lose about $5 billion on the same decline) and also to such poor non-OPEC producers as Mexico (whose losses would be roughly the same), Egypt, China, or Malaysia, and would be felt painfully by treasuries in London, Oslo, Ottawa, and Washington.

Simple judgements of OPEC's post-1973 impact are thus impossible. On the whole, the rich world should ultimately benefit from the experience: focusing attention on the criticality of a previously ignored energy supply and forcing both households and industries to do more useful work with less fuel—if nothing else these two changes must be seen as big gains. Other consequences for rich countries have been much more painful. Perhaps the most difficult challenge has been the rapid decline of primary metal industries led by the huge retreat in steelmaking.

This downturn would have come sooner or later but large jumps in energy prices (driven, of course, by OPEC's new fixings) accelerated the closing down of old, inefficient producers in this highly energy-intensive industry. Slow growth or outright stagnation in global car and ship production, and general economic slowdown, further deepened the trend the end of which is still not in sight. Even Japan, with the world's most modern furnaces and mills has been producing well below its capacity while the EEC countries closed 15 per cent of their enterprises before the end of 1985, and the American story is much the same.

OPEC's price rise impulse may be thus seen as a principal reason for the emergence of a leaner and more efficient steel industry but the social disruption has been considerable. Similarly, America's car manufacturing is again booming and highly profitable—but it will never employ again all the workers let go in the years of difficult adjustments. Clearly, this role of OPEC's as an accelerator of structural shifts and efficiency improvements does not fit easily into simple gain or loss categories.

But if OPEC's price-fixing helped to push many less efficient energy-intensive operations out of existence the new high prices also opened rich opportunities elsewhere: looking for oil and gas in the most difficult environments—deep offshore, in small fields, in formations deeper than 5 km—became suddenly appealing and profitable. Development of bituminous coal and lignite reserves got a significant new stimulus, as did the construction of hydroelectric projects (including small hydros previously considered too costly). Innovative alternatives, ranging from small wind generators to central solar power plants, started their slow reach for meaningful shares of the market, and energy conservation efforts, extending from proper house design to meticulous energy audits of industrial processes and involving solutions as simple as stuffing the attic with fibreglass and as complex as modernizing steel mills, brought impressive savings. Opportunities for vastly increased exports to OPEC countries provided the rich nations with yet another important economic stimulus boosting the production of items as diverse as automatic telephone exchanges, refinery equipment, lamb meat, and offroad trucks.

In the case of poor oil importers it is harder to find the sunny side. Perhaps the only, and decidedly painful, benefit came as a consequence of the second round of steep price rises. After coping surprisingly well with the first rise, poor oil importers were stunned by the 1979–80 jump, and the greatly increased oil import burden was definitely one of the key contributors to the great unravelling of global debt in the early 1980s. Although the view may be considered callous, I would maintain that the sooner the pyramid of excessive borrowing starts falling apart the better: it is preferable to have the mess of repayment troubles, rescheduling manoeuvres and default threats now than a decade later and, unwittingly, OPEC's actions speeded up this hurtful time of reckoning.

Maybe the strangest thing of all is that OPEC's unanticipated, and in retrospect almost incredibly rapid transformation from the feared price-setter of the 1970s to the struggling residual supplier of the 1980s has weakened the organization almost beyond belief, yet it has not taken away its power of deciding the outcome of the next supply crunch.

As shown in Fig. 2.10 global crude oil production in the early 1980s was just about the same as a decade earlier, but OPEC's extraction share fell from about three-fifths to less than one-third and the organization's total exports declined by more than 50 per cent between 1973 and 1983 (Table 2.3 provides all details).

Moreover, the world's largest importers of crude oil became much less dependent on the organization's exports. In September 1973 OPEC oil accounted for nearly 70 per cent of American imports, 95 per cent of West German purchases, and 92 per cent of Japanese and French shipments; in September 1984 these shares were sharply down in the USA (to 38 per cent), West Germany (to 54 per cent), and France (65 per cent), and even Japan reduced its dependence on OPEC to below 80 per cent of its imports.

OPEC thus became the last resort supplier and its output and sales plummeted faster than the global decline in crude oil demand (see the gradient of lines in Fig. 2.10). However, in a fast rising market OPEC has the best capacity to fill the need by increasing its output much faster than the renewed global growth of oil demand: the organization has a large unused production capacity and its Middle Eastern members do control just over one-half of the world's oil reserves which are bound to play a critical role in long-term global supply.

When the oil glut passes, OPEC will be ready, once again, to exercise its power as the world turns to the capacities which are idling at present to fill, or at least to narrow, the demand gap. But this, the third, price jump episode will be determined, as were the two preceding ones, by what the Western economies will do. OPEC's seemingly unshakeable grip on the Western economies resulted from our indolence, mismanagement, downright stupidity, and lack of collective will—it was

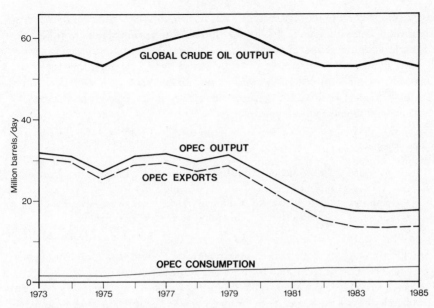

Fig. 2.10 OPEC's declining global influence. In a decade its output fell from about three-fifths to a mere one-third of global production as its exports were more than halved.

Table 2.3 OPEC's declining importance (unless otherwise noted all figures are in million barrels per day)

Year	Global crude oil production	OPEC production	OPEC's share (%)	OPEC's domestic consumption	OPEC's exports	OPEC's official average sales price (current $)
1973	55.75	30.96	55.5	1.51	29.45	3.39
1974	55.04	30.73	55.0	1.61	29.12	11.29
1975	52.99	27.14	51.2	1.66	25.48	11.02
1976	57.29	30.66	53.5	1.94	28.72	11.77
1977	59.48	31.16	52.4	2.17	28.99	12.88
1978	60.13	29.81	49.6	2.30	27.51	12.93
1979	62.58	30.93	49.4	2.52	28.41	18.67
1980	59.47	26.94	45.3	2.78	24.16	30.87
1981	55.82	22.68	40.6	2.96	19.72	34.50
1982	53.05	18.89	35.6	3.25	15.64	33.63
1983	53.36	17.48	32.8	3.46	14.02	29.31
1984	54.39	16.66	30.6	3.55	13.11	29.00
1985	53.48	15.25	28.5	3.50	11.75	27.90

not an inevitablility of modern history. They happened to have much of the existing reserves of conventional crude—but they cannot control most of the world's as yet unexplored territory and, above all, *they could never dictate our energy consumption.*

The probabilities and timing of any future pricing crunch are, naturally, dependent on the global magnitude of recoverable crude oil reserves, and on intensity of energy demand of rich economies, both ideal subjects for some matter-of-fact demythologizing observations.

2.3.2 *Depletion Gothic*

> Depletion ... the reduction of natural resources at a faster rate than replenishment
>
> Gothic ...   referring to a literary style in which a bleak and dismal setting ... and a feeling of doom and decay are characteristic.
>
> *New Webster's Dictionary*

Strictly speaking, fossil fuels are not nonrenewable resources: they are continually formed at countless sites around the planet—but their replenishment rates are many orders of magnitude slower than has been the pace of their extraction. This great mismatch of rates must eventually bring a complete exhaustion of all producible deposits, that is those resources which are accessible and whose extraction can be carried out with the technologies at our command, and at a cost guaranteeing economic profitability. The part of resources (be it fossil fuels or other minerals) satisfying all these three requirements is known as reserves, and the extent keeps changing as a combined result of extraction, new discoveries and development.

Investment, incentives, and technology transform resources into reserves which can be well appraised and quantified. The situation is analogous to having a big, completely dark cave whose dimensions and contents can be perceived and examined only as far as our light and reach can go. With better technologies we can 'see' further (modern geophysical exploration can uncover oil- and gas-bearing structures below a depth of 6 km or even deep under ocean waters), and we can also reach there.

Our wildcatting land rigs can now drill routinely to depths greater than 5 km, and offshore exploration has progressed from simple jack-up rigs whose legs were anchored in shallow near-shore waters to drilling far out of sight of land in more than 1.5 km of water, and the record depth, set in Wilmington Canyon Block 587 in 1983, is just over 2 km in open ocean 160 km from the Delaware shore. Although deepwater production technology has yet to catch up with these rapid advances in exploratory drilling, new designs—ranging from giant fixed production platforms (Exxon plans one in 400 m of water in the Santa Ynez unit in offshore California for the late 1980s) to guyed towers and tanker-based systems—are moving in and production from discoveries in most of the ocean waters is a clear possibility within one or two decades (demand for this inevitably costly oil will determine the timing).

And, of course, on land our reach has been extended to virtually any known environment. Helicopters transport the parts for lightweight rigs to the otherwise inaccessible rain forests of Peru or Papua-New Guinea, crude oil is flowing through the pipelines laid from the wells deep in Libyan deserts as well as from the polar regions of Alaska's North Slope or from the impenetrable swamps of Western Siberia. But the most fascinating thing about this extending reach is not the admirable overcoming of obstacles considered to be insurmountable just a generation ago: rather, to return to the earlier analogy, it is the size of the cave, the magnitude of global oil resources, which astonishes most.

Estimates of the total oil-in-place, or of the ultimately recoverable reserves (a fraction of the former, as standard oil recovery gets out only about one-third of the oil present in the parent rock formation), have risen fabulously since the beginning of this century—and even when leaving the oldest values aside and concentrating on post-World War II appraisals one sees an unmistakeably impressive upward trend. During the mid-1940s estimates of ultimate global crude oil reserves by Duce, Pogue, Weeks, and Pratt ranged between 55 and 82 billion tonnes. A decade later values in the range of roughly 140–200 billion tonnes prevailed, and by the time of OPEC's first round of price increases the estimates ranged mostly from levels close to 300 to values in excess of 400 billion tonnes.

While OPEC was quintupling the price, and OAPEC was embargoing the United States, the trend in Fig. 2.11 was pointing toward estimates of more than 500 billion tonnes of ultimately recoverable crude oil reserves by the year 2000,

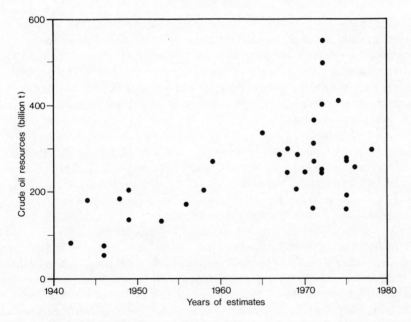

Fig. 2.11 How much oil is there? An historical review of crude oil resource estimates shows an encouragingly rising trend as well as no reduction in accompanying uncertainty.

definitely a calming prospect. Instead—a spate of predictions forecasting an early end to the oil era, an inevitable peaking of world production as the known reserves become exhausted, and then a rapid output slide. A generation or three decades at best, we were repeatedly told, and hence we should hurry to bring the alternatives in. Why this failure of nerve in the face of far from disconcerting evidence?

Arguing against the continued ascent of estimates in Fig. 2.11 made sense on several grounds. High values offered by overly optimistic estimators pull the trend up and it is the duty of realistic scientists to discount such figures and hence to depress the rise. Caution and avoidance of unpleasant surprises are the only valid supports for this stance—scientific evidence is still elusive enough to allow equally well an argument for leaving out the lowest estimates. But even with all values in excess of 400 billion tonnes left out the trend is still rising and the spread of the 'realistic' estimates still spans about a two-fold range. More serious was the argument concerning giant and supergiant fields: these structures have been of dramatic importance for global oil production and prospects as they contain a disproportionately large share of known reserves.

But undoubtedly the most potent push towards a pessimistic outlook came from observing the shrinking difference between the new crude oil discoveries and the rising global consumption of the fuel. This variable, commonly expressed as the reserve production ratio stood at 35 years in the early 1970s and it fell to 28 years by 1979: theoretically, if no new discoveries were to be made in the decades ahead, the world, consuming crude oil at the 1979 clip, would run out of oil by the year 2007.

Nobody expected the new discoveries to cease but since they were getting smaller (in spite of our search in forbidding environments) while the demand kept growing, broad consensus opinion maintained that the end of an era was near. A few decades and we should bump into the furthermost recesses of the cave. The CIA's influential 1979 assessment of world oil prospects put a seal of respectability on the sensationalist prognosis: 'Although the world is not running out of oil, current consumption is *greatly* exceeding new discoveries of oil. If this trend continues, as most experts expect it will, output *must* fall within the decade ahead.'

Even more ominously, another CIA study projected that before the mid-1980s Soviet oil production, the world's largest, would start falling and that the Soviet Union, previously a large oil exporter, would have to seek additional fuel on the shrinking world market. Not only an oil-scarce future then—but one increasingly likely to lead to an armed superpower confrontation in the Middle East, or worse. Enough of fanciful constructs, let us turn to realities.

On 1 January 1974—the day OPEC's new price passed the $10 mark—*The Oil and Gas Journal*'s review of global crude oil reserves (the most prestigious and the most reliable account of this kind prepared annually by this journal for decades) showed a total of roughly 628 billion barrels (about 85 billion tonnes). During the following ten years worldwide oil consumption reached an unprecedented total of 208 billion barrels, yet on 1 January 1984 global crude oil reserves stood at 669 billion barrels, 6.5 per cent above the 1974 levels. This means that new discoveries during those ten years added up to almost exactly 250 billion barrels,

20 per cent more than the period's total consumption, and that the reserve/ production ratio in 1983 stood at 35 years, higher than it stood during 28 out of 38 post–World War II years for which data on the ratio are available (Fig. 2.12).

One can also put it this way: ever since the beginning of massive global crude oil consumption in the early 1950s enough new oil has been discovered each year to keep the reserve/production ratio at not less than 25, and usually at about 30 years, a most remarkable achievement in view of the fact that the annual output is now

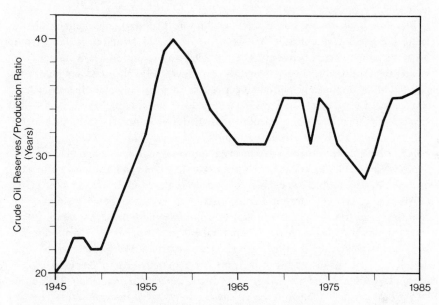

Fig. 2.12 Crude oil reserves/production ratio derived from *The Oil and Gas Journal* annual compilations has displayed a surprising stability from the mid-1950s with fluctuations confined between 28 and 36 years.

roughly four times larger than at the beginning of this period. To push the ratio higher does not make economic sense: why invest more in discovering additional oil (that is in making new reserves out of resources) when enough is at hand for a span sufficient to justify the developmental expenditures and to get satisfactory investment returns?

Many uninformed instant energy experts interpreted the reserve/production ratio as 'running-out' deadlines, but in 1984 we were as many years away from this possibility (which could result only by combining complete cessation of new exploration with steady consumption) as we were in 1970. And glancing further back one notes that 1984 was to be the year we were to 'run out' according to the 1954 reserve/production ratio of 30 years: yet three decades later global reserves and annual output are about four times larger, and the year of the theoretical reckoning has advanced again, now to the second decade of the next century!

These great reserve gains have not been uniform. The crude oil reserves of several large established industrialized producers—USSR, USA, and Canada—have declined but not all old hydrocarbon provinces are waning. In spite of sharply increased production between 1973 and 1981, and with only modest exploratory drilling, Saudi Arabia's reserves are now at least 25 per cent higher than they were in 1973. Venezuela is a comfortable 75 per cent above the 1973 level, and Iraq is at least 35 per cent ahead.

New crude oil discoveries as well as additions in long-known (but until the 1970s not so promising) basins have been spectacular. Mexican reserves rose from 3.6 to 48 billion barrels between 1973 and 1983, a net addition nearly 30 per cent greater than the total United States crude oil reserves in 1973! United Kingdom and Norway added in the same period 50 per cent to their considerable offshore riches, and Malaysia and Brunei listed in 1983 nearly three times as much crude as in 1973. Exploratory, mainly offshore, drilling has made oil producers from such poor African nations as Ghana, Ivory Coast, and Cameroon.

And the reserve situation looks even more comfortable when natural gas deposits are added. This is not mere accounting sleight of hand: natural gas must be included in any complete assessment of hydrocarbon resources as it is a perfect, or nearly perfect, substitute for most crude oil uses, ranging from power generation to countless industrial direct heat applications, and from household heating to petrochemical syntheses. Global reserves of natural gas grew from about 57 trillion m^3 in 1973 to 91 trillion m^3 in 1983, a rise of 60 per cent. More importantly, this expansion has been shared by all hydrocarbon regions: Mexican gas reserves rose sevenfold, Saudi verified deposits grew 2.4 times, the Iranian total climbed by about 80 per cent, as did Canadian reserves, and the huge Soviet reserves, already by far the world's largest by 1973, have doubled! Indeed, in many old hydrocarbon basins it seems that horizons below 4.5 km, untapped before the late 1970s, are fairly saturated with natural gas!

When crude oil and natural gas reserves as well as total global hydrocarbon extraction are converted to common energy units, production of 1.83×10^{20}J in 1973 would have exhausted the reserves of 6.4×10^{21}J in 35 years. In 1982 oil and gas reserves of 8×10^{21}J could accommodate output of 1.85×10^{20}J for 43.2 years, certainly a most comforting and undeniably very significant leap in just one decade! Even when leaving natural gas aside why should not we believe that the global petroleum industry, which has now for some two generations discovered and developed every year enough new crude oil reserves to maintain the reserve/production ratio fluctuating closely around 30, will not keep a 25–35 year supply horizon for generations to come?

The answer of those who have been forecasting an early end of the petroleum era is obvious: we are running out of opportunities to discover new oil on a big scale. To support this claim arrays of United States drilling and discovery statistics will be cited: in spite of record drilling levels new discoveries cannot keep up with extraction and reserves have been continuously declining since 1960; no giant fields get discovered any more, more wildcat wells are dry, much more drilling must be completed for every barrel of new verified reserves. The conclusion should then be inevitable: soon these American patterns will be repeated in other major oil-producing regions. Cheap, easily accessible oil will be exhausted, and it will be more economical to turn to alternative fuels rather than to develop outrageously costly oil.

Once again, we come back to face the estimates about the ultimate size of the cave, and no argument has been more persuasive in refuting the forecasts of the imminent end of the oil era than the demonstration of the drilling gap between the United States and the rest of the world by Bernardo Grossling of the United States Geological Service. This proof is best expressed graphically: the world's principal petroliferous areas (sedimentary basins onshore and offshore on the continental shelf, that is to a water depth of 200 m) are shown in proportionately correct rectangles and each full black square represents 50 000 wells (both exploratory and development) drilled while smaller numbers of wells are approximated by appropriate portions (Fig. 2.13).

The graph's principal message is immediately impressive: compared to the United States the rest of the world has been as yet little explored for oil and the opportunities in Latin American, African, and Asian countries are especially large. Of roughly 3.5 million oil wells drilled since the beginning of the petroleum industry in 1859 in Pennsylvania, over two-thirds were sunk in the 48 continental states of the United States, and, in turn, most of these were in the southern 'oil

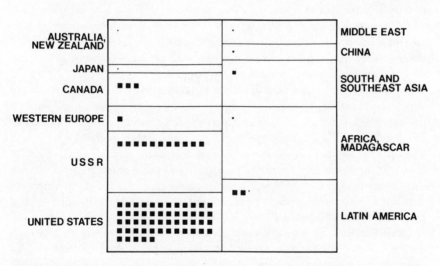

Fig. 2.13 Bernardo Grossling's impressive appraisal of global oil prospects: most of the world's petroliferous areas are yet to be explored intensively (each black square represents 50 000 exploratory and production wells drilled before 1985).

patch' (Oklahoma, Texas, Louisiana). And of the approximately 700 000 exploratory wells drilled between the 1860s and the early 1980s, 95 per cent were sunk in
industrialized nations, nearly three-quarters in the United States alone.

All of this leads Grossling to believe that 'it is within the realm of the possible
that the magnitude of the conventional petroleum resources might turn out to be
more than two or three times larger than conservative prevailing views'. And,
perhaps even more significantly, it is even more probable that the real cost of most
of this yet-to-be discovered oil will be well below the OPEC price or the marginal
cost of oil exploration and development in the United States! Grossling's highest
estimate of the total amount of recoverable conventional reserves is 5.96 trillion
barrels (813 billion tonnes), about 2.7 times higher than the mean of 298 billion
tonnes calculated from eleven other estimates published during the 1970s and
representing a conservative expert consensus.

Other dissenters, most notably Peter Odell of Erasmus University in Rotterdam
whose refreshingly upbeat writings on the future of oil have been commendably
consistent since the early 1960s, have argued for a relaxed outlook. But, even should
the ultimately recoverable reserves be much smaller than indicated by the highest
available estimates, Odell believes the end of the oil era would still be many
comfortable decades away—a not so surprising prediction once one considers the
most likely course of consumption rates.

The reserve-to-production ratio, the most suggestive indicator of 'running out',
is, of course, determined no less by its numerator than by its denominator—and the
1970s have shown the enormous possibilities of lowering that figure for virtually all
developed nations, the crude oil's main consumers. For several years after the 1973
price rise it seemed that the trends of historical increases of crude oil consumption
in many industrialized countries could not be broken by any price increases:
OPEC's prices doubled, tripled, quadrupled, and more—yet the imports either kept
on rising or, at best, remained close to the peak 1973 level. In 1979 the total
shipments for the Big Seven were down only by 5 per cent compared to 1973, an
inexplicably small response to such a large price rise.

Then the response came—precipitously and on a grand scale. The Iranian round
of price increases helped to bring on the severest global economic slowdown since
the early 1930s and resulted in an unexpectedly large decline in crude oil demand.
The Big Seven crude oil imports fell by 15 per cent in 1980, another 13 per cent in
1981, and yet another 15 per cent in 1982. A mere three years after panicky 1979
the leading industrialized countries were buying 37 per cent less crude oil abroad.
The decline is only slightly lower when the imports of refined oil products are
included: the Big Seven crude and product purchases fell from 24.31 mbd in 1979
to 17.14 in 1982, a 30 per cent decrease.

Obviously, these deep demand reductions could not continue for long: in 1983
the Big Seven oil consumption was only 1.3 per cent below the 1982 level, and the
real long-term question is what will be the future growth rates of global oil
consumption in comparison to the historical record. Between 1945 and 1973
worldwide oil consumption grew by almost exactly 7.5 per cent a year (i.e. doubling

every nine years and four months); between 1973 and 1979, the year of the highest oil use, the annual growth fell to just 1.9 per cent—and since 1979 the consumption has been falling (the four year decline, 1979–83, averaged 4.1 per cent annually).

Is this all just a short-term aberration or a lasting change in the long-term trend? I believe it is the latter, and the affirmation can be made with plenty of confidence largely owing to the disappearance of a unique combination of important socio-economic conditions characterizing the pre-1973 period. Low and still falling prices of crude oil (in real terms) were an obvious encouragement to use the fuel indiscriminately and wastefully; industrialized nations, the largest energy consumers, used this cheap oil to switch their previously largely coal-fuelled economies to a high dependence on liquid fuel in the most massive and concurrently the most rapid energy-base transition in history, and the effects of this transformation were further strengthened by its coincidence with the period's high population growth, growing consumerism, widespread suburbanization, and massive diffusion of car ownership. And while European countries and North America fuelled in the nineteenth century their initial, most energy-intensive, industrialization by coal, virtually all newly independent poor countries based their economic development plans from the 1950s onward almost solely on inexpensive, largely imported oil.

To indicate the extent and the rapidity of these processes one should recall that between 1945 and 1973 crude oil jumped from about 30 to 45 per cent of global commercial primary energy consumption, that coal's share fell from 45 to less than 20 per cent during the same period, and that private car ownership rose from just around 30 to 230 million vehicles. A repetition of this concatenation is most unlikely. Every country (except for a few OPEC nations) has been trying to curtail oil consumption through conservation and development of alternatives (coal, natural gas, hydroelectricity). Western economies, with their slow-growing populations have little need for expansion of energy-intensive production, and the poor countries realize that development plans, energized by loose borrowing abroad and growing crude oil imports, are not sustainable. The inevitable conclusion is that when the post-1979 decline in oil demand is reversed new growth rates will never again average 7.5 per cent annually in the long run.

Arguments about how high the new global rate will be would fill easily another chapter of this book. Having in mind the recent rather steep decline in consumption and the general expectation of a nearly stationary oil market during the rest of the 1980s, Odell and Rosing's carefully documented finding that there is a 90 per cent probability that the annual average increase in oil consumption will not surpass 2.5 per cent can be taken as a very realistic example of the possible outlook.

Assuming that the ultimately recoverable reserves of conventional crude oil are just two trillion barrels (273 billion tonnes), and that the life-cycle of the production must, more or less closely, resemble a normal curve (otherwise a record output would be followed by a precipitous collapse, an economically ruinous proposition) the global oil output could continue rising until about the year 2010 when it would peak at nearly 40 billion barrels a year (about twice the current level), and the subsequent decline would bring production to the 1945 level sometime around the year 2070.

Should three trillion barrels be the recoverable base (a figure most analysts would probably agree to be the best conservative value) the peak would come between 2015 and 2020 at close to 45 million barrels a year, and in 2045 the global oil industry could be still producing roughly the same amount of crude oil as today! These simple calculations show persuasively that the end of the oil era is not imminent—and further considerable extensions are a matter of solid, fairly conservative assumptions rather than of irresponsible wishful thinking.

Ultimately recoverable resources of unconventional oil (in tar sands and oil shales) can enlarge the total oil reserves by at least one trillion barrels—or more when the more expensive recovery of even larger resources of such oil becomes justified by higher demand and eased by better technology. Higher recovery rates can eventually wring out an additional 10–30 per cent from the known fields. And, of course, consumption growth rates lower than in the foregoing example would extend the ascent phase of the production cycle curve and lengthen its tail.

In *The Future of Oil* Peter Odell and Ken Rosing assign a 50 per cent probability to a scenario consisting of a 1.9 per cent growth rate, 5.8 trillion barrels of conventional and unconventional recoverable oil, and a maximum annual addition of 75 billion barrels to reserves. The peak consumption then comes only in the year 2046 at 3.5 times the 1980 flow, and a century from now the then declining output would be still above our current consumption.

Details result from slight changes of assumptions, and endless curves can be computer-generated, but it is the basic proposition which matters: there is no reason why even conservatively considered recoverable resources of conventional oil, coupled with moderated growth rates, should not sustain a gentle expansion of oil production for another 25–35 years and assure the existence of an important oil industry for at least another century. A more optimistic outlook, well-supportable by geological and economic arguments, can see the oil industry growing, if need be, for about another two to three generations and being still vigorous in the early decades of the twenty-second century.

2.3.3 *Energy and economic growth*

> 'Just how is that significant, Mr. Mason?'
>
> Earl Stanley Gardner, *The Case of the Postponed Murder* (1970)

Remembering the years of minuscule GNP growth (and frequently of decline) and rising unemployment, it may seem that the drastically lowered oil imports of the early 1980s were simply a result of incapacitated economies, painfully paying with Western economic decline for the demise of OPEC's extortionism, hardly a desirable swap. Michael Posner called this 'energy conservation by the most expensive method of all' and estimated that in the United Kingdom a saving of 1 per cent of oil consumption to forgo 1 per cent of GNP is like paying for the oil at a rate of $(1979) 2000/t. When one looks at real falls in GNP and oil consumption declines during the 1979–82 both the British and the American conservation penalties are very similar and come actually to some $(1980) 200 per barrel (or nearly $1500/t), much less than Posner's value but still almost six times larger than the market crude oil price prevailing at that time.

Fortunately, this dreadful impression is only marginally correct. Economic slowdown certainly shaved off an important part of oil imports (current analyses disagree how much precisely) but the untenability of the sacrifice theory becomes obvious as soon as one compares GNP and energy consumption changes during the 1973–82 decade. When expressed in constant US (1982)$ the combined GNPs of the Big Seven rose from $5.37 trillion in 1973 to $6.41 trillion in 1982, a real growth just short of 20 per cent. Yet in the year when their economies were putting out 20 per cent *more* goods and services than in 1973 these countries were buying abroad 26 per cent *less* crude oil and refined products (and 30 per cent less than in 1979, the peak import year).

More remarkably, the Big Seven did not replace those 7.2 million barrels a day of imported oil cut between 1979 and 1982 by increased domestic extraction. True, after 1973, the United Kingdom emerged as a major crude oil producer (from 2000 barrels a day in 1973 to two million barrels a day in 1982) but this increase was largely negated by American and Canadian production declines so that the net gain amounted to no more than 900 000 barrels a day during the decade and just 322 000 barrels a day between 1979 and 1982. This leaves still over 6.85 million barrels a day to be accounted for but this difference was only partially filled by other energy sources.

In 1979 the total primary energy consumption in the Big Seven reached a record equivalent to 67.52 million barrels a day—but in 1982 it fell by about 5.5 million barrels, while during the same period crude oil and product imports declined by 7.2 million barrels a day. This means that less than a quarter of the oil import drop between 1979 and 1982 was apparently substituted by other energies. And the most dramatic comparison of all: the 1982 annual primary energy consumption of some 62 million barrels a day was virtually identical with the 1973 total. Although using the same amount of energy as a decade before the Big Seven turned out—in real, inflation-adjusted terms—20 per cent more goods and services!

How did we do it? Not through any grand designs. Following the 1973–4 shake-up countless strategies were laid out to emasculate OPEC but their basic thrusts were almost invariably dubious—and soon proved unworkable. American reactions, critical in determining OPEC's future might, were exemplarily wrong. First, President Nixon's Project Independence called for the achievement of complete self-reliance, a politically appealing but economically hardly a most sensible goal. Then President Carter, after soul-searching retreats in Camp David, preached to the country dressed in energy-conserving cardigan and implored it to save for the sake of patriotic feelings—not a very rational thing to do with controlled prices of oil and natural gas being, in relative terms, a lower share of average disposable incomes than a generation ago!

Concurrently, he tried to launch a gargantuan synthetic fuel industry converting coals and shales into gases and liquid fuels—but this move, too, soon foundered, not in the least because OPEC's prices, high as they had risen, were still lower than the recurrently underestimated costs of developing these expensive synthetic alternatives (I shall return to these plans in some detail in section 2.4.3).

Where grand governmental designs failed, millions of small changes, adjustments, and savings undertaken by industries, commercial establishments, and households added up to impressive oil consumption cuts. In the United States industrial use of energy in 1982 was one-third less than would have been predicted by the long-term trend established before 1973, a huge change in a mere decade.

Of course, one cannot have a lasting economic growth with falling energy consumption (a combination equal to *perpetuum mobile* contrivances), but if the incentives to conserve are right we will continue to be surprised by how much more we can do with much less. Even a cursory glance will show that opportunities in recycling (for example, a tonne of recycled steel costs a mere 25 GJ compared to 45–65 GJ for *de novo* metal), substitution of more efficient convertors (the traditional, and still dominant, incandescent light bulb delivers between 10 and 20 lumens/W compared to 55–80 lumens/W for fluorescent lights), and general use of readily available low energy solutions (such as super-insulated houses requiring no heating furnaces at least as far north as Iowa or southern Ontario) still remain not just large but truly enormous.

Falling energy use and rising GNP is a relationship unprecedented in modern economic history when changes of these two variables have always marched not only in the same direction but mostly in a surprisingly close lock step commonly expressed in terms of energy/GNP elasticities. The measure is simple, easy to calculate and hence quite popular: the numerator is the relative change in energy consumption during a given period, usually one year (although longer periods can be also considered), the denominator is the change in economic product.

For decades one could keep coming across two key conclusions about these elasticities. First, in rich countries, with their infrastructures in place and with growing stress on services and information-processing activities, energy consumption increases at rates lower than those of economic growth, resulting in elasticities less than 1.0. Conversely, the second major conclusion held that nations in earlier stages of industrialization, which require large energy inputs not immediately equalled by corresponding increases in economic output, have elasticities well above 1.0.

Reality, as usual, is a bit less orderly and is best illustrated by historical comparisons for a few nations. By far the longest trend available is for the United Kingdom and it shows a secular decline—but one composed of waves of surprisingly large amplitude (Fig. 2.14). The most notable rise between 1830 and 1850 can be attributed largely to a rapid diffusion of steam-power in textile factories, and to the onset of the energy-intensive railway age with its sudden need not only for fuel but even more for new iron-making capacities to supply track and rolling stock.

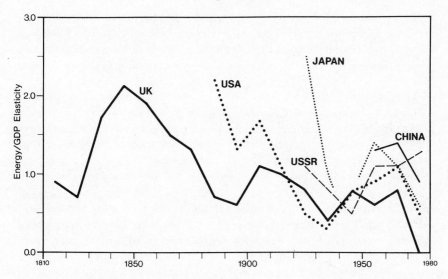

Fig. 2.14 Energy/GDP elasticities calculated for the longest available spans for the world's top five energy consumers. A general declining trend is unmistakable.

Greatly increased efficiencies which took place in iron smelting during the 1870s and 1880s were the principal cause of the subsequent decline. The country's participation in the two world wars of this century and the economic depression of the 1930s make it difficult to discern 'normal' long-term trends. Most recently, a small but clear increase of the energy coefficient during the 1960s, and its rapid fall after 1973, are noteworthy. In broad outlines American experience is remarkably similar (Fig. 2.14) but the post-1930s rise of the measure is considerably stronger (best explained by the mass adoption of the automobile, suburbanization, and the spread of energy-intensive industries such as aluminium smelting and petro-chemicals), and the decline of the 1970s, although proceeding at a similar rate, does not reach as far as the British reduction.

Japanese data are broken by World War II but the precipitous pre-war decline is based on impeccable evidence, as is the rapid rise of the late 1940s and 1950s (Japan's switch to imported crude oil from domestic coal as the key energy source), and continuous decline since the early 1960s.

The Soviet case is quite peculiar: keeping the limitations of available national product statistics in mind, one must still conclude that the Soviet energy elasticities have been growing not only during the 1950s (as were the American) but even during the 1970s—yet another sign of inefficiency and stagnation permeating the country's economy.

China, the world's most populous poor nation, had a predictably high energy/ GDP elasticity during its First Five Year Plan (1952-7), which was basically a

Stalinist programme heavily tilted towards heavy industry, and the coefficient kept on rising during the 1960s. Then its moderation during the early 1970s was followed by a rapid decline resulting from the adoption of more sensible policies (greater stress on light industries and services), by an increasingly pragmatic post-Mao leadership. The coefficient may rise again as some of the recent gains were clearly just one-off affairs (closing of outdated enterprises, consolidating production in larger units), but the Chinese experience is still a very good illustration of how access to modern, efficient technologies can lower impressively the energy/GDP elasticities even for a nation which is still in the relatively early stages of its industrialization.

Even more notably, this Chinese experience is a perfect example of the dangers inherent in the frequent use of elasticity coefficients in long-range forecasting of energy requirements. Energy writings of the 1960s and 1970s abound with examples of this appealingly simple practice: take the anticipated rates of economic growth and multiply them by elasticity coefficients identical to those of the recent past for shorter term forecasts, or somewhat lower than the past experience for longer term outlooks. Even a quick look at Fig. 2.14 will confirm that setting out elasticities for the next few decades is a perilous enterprise and that, not surprisingly, this popular method of energy-reward forecasting does not work very well.

In the early 1980s one could conclude that what was happening in almost all rich countries was nothing less than the decoupling of the normal link between economic expansion and energy use. Forget the low energy coefficients—in 1980 and 1981 the direction of the two variables was opposite and voices started to be heard telling us that this is not a miracle, not a temporary deviation brought about by the sharp price increases and change in economic direction, but the beginning of a new trend. The potential for energy conservation in all rich nations was seen to be so huge that the inevitable requirements for new energy to run new factories and processes could be covered—not by extracting the fuels from the ground but from our buildings, cars, and industrial processes. Conservation would provide for all the current needs with progressively less energy and these savings could go to new uses.

In 1982 the decline of energy consumption among the Big Seven continued—but economic production was down too, although only by 0.6 per cent. Economic recovery came only in 1983 when the Big Seven increased its gross economic product by 2.5 per cent in real terms—but its energy consumption declined for the fourth year by about 3.5 per cent! Still, this period of primary energy consumption decline has been too short and has been too much a part of the worst Western economic downturn for a generation to offer any confident long-term forecasts. With only half a dozen years of sustained (although not necessarily high) growth behind us—that is only by 1990 if things go well—we will see how realistic were the expectations of those optimists who postulated the growth of Western economies without expanding primary energy consumption.

But even if the economic growth of the Western nations during the late 1980s and 1990s is once again linked with growing energy consumption there is a very high probability that the energy intensities of this rising production will not be increasing. Energy elasticities may tell how much energy is needed for new economic growth but they do not reveal how energy-intensive a particular economy is already (that is how many units of energy are, on the average, contained in a unit of national product). Intuitively, it may seem that highly efficient, modern economies will have both low elasticities and low energy intensities while in poor countries both indicators will be rather high. Reality is, again, much more unruly.

Comparisons of the Western industrialized economies—whose uniform GNP accounting, easy currency interconvertibility and reliable energy statistics make such exercises most meaningful—show that in the early 1980s most of the large West European countries needed close to 20 MJ for each US(1982)$, but that the United States and Canada consumed close to 30 MJ for each GNP dollar, and Japan only about 12 MJ, a surprisingly large—more than twofold—spread between the best (Japan at a mere 12.3 MJ) and the worst (Canada at 27.2 MJ) performer.

Several obvious variables explain the spread but usually only when they are considered *en bloc*: it is amazing how many of the most obvious explanations do not work! Those which work best have to do with final uses and conversion efficiencies. For example, in the early 1980s there were 2.6 times as many private cars in the United States as in Japan but much more frequent, and longer commuting in the United States, greater leisure use of vehicles and their larger average size and higher specific fuel consumption translated into nearly twelve-fold higher gasoline purchases. Less pronounced but still important is the difference in electricity consumption. In the early 1980s Americans were using 2.1 times more electricity per capita than Japan and as only about one-fifth of it came from hydro-stations and nuclear power plants—compared with almost one-third in Japan—Americans had to convert, per capita, more than three times as much fossil fuel as the Japanese.

As for the conversion efficiencies, burning liquid fuels in cars wastes 75–80 per cent of the initial energy, while burning them in industrial boilers wastes only 20 per cent. In the early 1980s the Japanese split their use of refined fuels roughly between driving and running their factories—while American gasoline and diesel fuel consumption surpassed industrial refined fuel uses more than four-fold. Similarly, using a kerosene heater (the Japanese burn in this way nearly 15 million tonnes of this fuel a year) yields about 60 per cent of useful heat, while using the fuel to generate electricity results in useful energy not surpassing 35 per cent.

When the shares of transportation consumption in the total use of liquid fuels and those of non-industrial consumption of electricity—the two leading indicators of the most inefficient conversions—are plotted for the United States, Canada, UK, Sweden, France, and Japan for the latter half of the 1970s the resulting ranking (Fig. 2.15) is a perfect replication of energy/output ratios charted for the six nations in Fig. 2.16. So even without performing sophisticated statistical manipulations (such as multivariate factorial analysis) one can see the crucial importance of what might be suggestively labelled as affluent energy uses—driving around for fun, flying for vacations, air-conditioning houses in summer, heating them with electricity in winter.

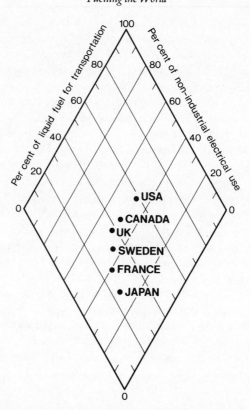

Fig. 2.15 The two variables explaining most of the difference in national energy intensities are the share of liquid fuels used in transportation and the percentage of non-industrial electricity uses. Ranking of the countries in this graph is a perfect replica of their 1980 ranking in the next figure.

Americans and Canadians do these things more than anybody else and hence every dollar of their national production comes much dearer in energy terms than in European countries, and in Japan, a country which in spite of the repeated admiration it elicits from foreigners still does not provide its average citizen with the everyday affluence enjoyed by leading Western European nations. This can be further illustrated by comparing the historical development of per capita energy consumption (Fig. 2.17): Japan now consumes less than a third of North American use and it lags 40–80 per cent behind Western Europe. Even when Japan's higher overall energy conversion efficiency is taken into account (close to 60 per cent compared to 40–50 per cent in other rich nations), the country's flow of useful energy is still lower in per capita terms than in Canada or England.

The trade-off thus appears to be well-established: more mobility, easier

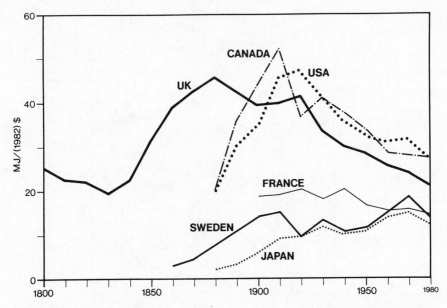

Fig. 2.16 A century, or more, of national energy intensities calculated in constant dollars. North American and British declines during this century are notable for the similarity of their slope while the low-coal economies (Sweden, Japan) always had an efficiency advantage.

opportunities to roam (more restlessness?), greater household comforts (more frivolous consumption?)—in exchange for more widespread conversion inefficiencies and higher energy content in national economic output. And yet the recent North American energy output ratios are the best ones the two countries have had in this century. Reliable long-term gross domestic product and energy consumption statistics show a remarkably similar historical pattern for the United States and Canada—and one resembling the earlier British pattern both in the rate and magnitude of its ascent during the late nineteenth an the early twentieth century, and its descent, roughly since World War I (Fig. 2.16).

I went to the trouble to assemble these historical patterns of energy ratios and per capita usage (backing up in each case as far as standardized national product data and reliable energy and trade statistics permitted) to provide a solid basis for some generalizations about the past and for speculations about the future. The choice of contrasting traditionally heavy users of fossil fuels with more efficient consumers is limited by data availability as well as by consideration of major territorial changes (Germany and Russia would not do) but the six countries represented in Figs. 2.16 and 2.17 are, fortunately, an almost perfect set allowing me to offer several important generalizations.

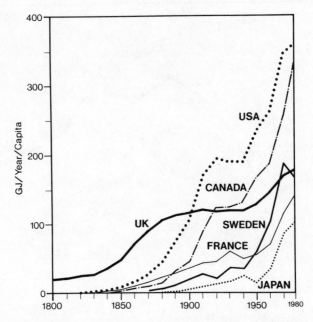

Fig. 2.17 Per capita commercial energy rise from the onset of industrialization: except for the United Kingdom all national curves between 1850 and 1970 were impressively exponential. Differences between countries with comparable incomes remain large.

First, nineteenth century industrialization imposed very similar energy requirements on those countries where coal resources were abundant enough to fuel not only the metallurgical industries, long-distance transport, and large-scale manufacturing but also to be used copiously in households. Initial diffusion of inefficient iron and steel making, railway transportation, factory boilers, and household coal stoves doubled, and even more than doubled, the pre-industrialization energy output ratios. British, American, and Canadian experience resulted in an amazingly similar trend.

Second, the period of the most inefficient performance did not last for long (in the United States and Canada it was especially short, almost certainly an advantage of their later start and faster adoption of efficiency-raising innovations), and it has been followed in these high energy users by an era of steadily declining energy ratios. Excepting temporary setbacks, this welcome slide has been going for over a century in the United Kingdom, for more than 70 years in Canada, and for over 60 years in the United States—and it has brought the current energy/output ratios close to the levels prevailing before industrialization took off 100–150 years ago.

Third, countries with poor or poorer coal resources had to follow a more frugal course. Their pre-industrialization energy output ratios also rose several-fold but the increases were slower and, glancing at Fig. 2.17, their per capita consumption never took off at rates characteristic of coal-rich nations, and until the large-scale introduction of hydrocarbons in the 1950s it remained at surprisingly low levels.

Fourth, when crude oil imports greatly increased the per capita consumption of these low energy users after 1950, their energy output ratios also went up—but the conservation measures of the 1970s set them downward.

Every economic trend will shift eventually but the decline of the energy/output ratio, now shared by all rich industrialized nations, is most unlikely to be reversed soon: it is quite possible that the continuation of current rates of decrease may not bring a levelling-off period for another two decades, and this probability is especially strong in the North American case where the huge per capita consumption levels could be greatly reduced without in any way undermining material affluence and the high quality of life.

But how far can these reductions go? This leads to perhaps the most fascinating question in energy studies, albeit one extremely difficult to answer: what is the optimum (or the minimum) per capita energy consumption needed for (or compatible with) what could be labelled in the simplest, although highly 'loaded' way, the good life? In pursuing efficient energy use, lower consumption elasticities and better energy output ratios are not the goals—they are just indicators of a process whose real aim is to satisfy basic human needs and to support a higher quality of life in a less costly and environmentally less disruptive way. Economists, ever transfixed by theoretical computer models, have paid little attention to quantifying these real-life needs, and the whole task may be viewed as an intractably subjective one.

However, objections about subjectivity in defining the good life can be met largely by ponting out that any rational definition of the good life must include such measurable variables as adequate health care, nutrition, housing, and education, and that the ingredients which are difficult or impossible to quantify (personal freedoms, choice of leisure activities, opportunities open to creative efforts) do not seem to carry high energy costs.

Unless one insists on stock-car racing, dragsters, dune buggies, speedboating, and other such high-powered pleasures as the only worthwhile ways to spend free time then virtually all other common leisure activities come at no extra energy cost (from reading to table games), or only at minimal expenditures of additional muscular energy (dozens of sports and outdoor activities), or at negligible increase of fossil fuel or electricity consumption (from listening to a stereo to building furniture). As for the personal freedoms, one cannot miss a clear link between very low energy use (a good indicator of poverty) and proclivity to dictatorial regimes in countries of Africa, Asia, and Latin America, but in industrialized nations the one-party regimes of the USSR and its East German and Czechoslovakian satellites use more energy per capita to keep afloat their dictatorial and economically inefficient regimes than the West European democracies do. Moreover, all the essential freedoms and democratic institutions we cherish were introduced by our ancestors at times when their energy consumption was only a fraction of ours.

Clearly, a strong case can be made for quantifying energy requirements essential for the good life on the basis of health and educational variables alone: freedom of speech, or faith, or artistic opportunities can flourish in societies with relatively modest energy flows while adequate housing, health care, and nutrition, the essential requirements for greater longevity, and easy access to all levels of education, a precondition to intellectual fulfilment, require much higher energy consumption. Consequently, it would appear that the Physical Quality of Life Index (PQLI), developed by Morris Morris in the mid-1970s and composed of longevity, infant mortality, and literacy, is an ideal measure to correlate with energy use.

Indeed, Philip Palmedo plotted the index against average annual per capita energy use for all of the world's nations and concluded that energy additions beyond 1200–1400 kg of coal equivalent, that is 35–40 GJ per capita a year, bring only marginal improvements; consequently, annual consumption around 40 GJ may be interpreted as a level of satisfaction of basic human needs. I agree that gains in infant mortality and expected longevity at birth do decline after reaching annual primary energy consumption of 40 GJ per capita, but the inclusion of literacy in the PQLI is the measure's great weakness and I feel strongly that a better indicator is needed to capture more satisfactorily the potential for intellectual development: the mere act of slowly reading a paper or writing a simple letter, enough to qualify a person as literate, clearly fails to express this critical ingredient of what I have called the good life.

The percentage of young adults (20–4 years old) enrolled in establishments of higher learning is a much finer indicator. In the early 1980s there were about 35 countries where this share surpassed 20 per cent and in only nine cases did this occur in nations using annually less than 30 GJ of fossil fuels and primary electricity per capita. However, eight of these nine nations—Lebanon, Mongolia, Panama, Peru, Philippines, Ecuador, Jordan and Thailand—must be disqualified owing to their still high infant mortalities, so only Costa Rica is left as a sole exception. Of the remainder, only five countries (Argentina, Cuba, Israel, South Korea, and Yugoslavia) have been sending at least one fifth of their young adults to universities while their annual per capita energy use has remained below 70 GJ. All other nations surpassing the 20 per cent mark consume at least 70 GJ a year per capita, three-quarters of them using over 150 GJ.

Without resorting to extensive multivariate analyses one can thus conclude that to provide both the physical foundations and intellectual opportunities for the good life at least something of the order of 70 GJ a year per capita are needed. This figure is about four times larger than the current average consumptoin for the world's poor countries but only a third of the rich world's average and just 15 per cent higher than the global mean. A perfect redistribution of the current global energy consumption, coupled with a modest output increase, would thus provide every inhabitant of this planet with as much energy as was the Swedish average in the early 1950s, the French mean around 1960, and the Japanese per capita consumption of the late 1960s.

This stunning comparison demonstrates that an impressively high worldwide standard of life (how many people in Asia, Africa, and Latin America would look down at the possibility of sharing the amenities and opportunities of Göteborgers in 1954, or Osakans in 1968?) could be achieved with virtually unchanged global energy consumption—but the possibility is predicated by reductions of average consumption in all rich nations, to the level of the 1950s and the early 1960s in the case of Western Europe (excepting the United Kingdom), and to the mean of the 1890s for the United States.

Although opportunities for energy conservation and efficiency gains remain impressively large throughout the Western economies, and are especially rich in North America, consumption declines of such magnitude are most unlikely. Still, the question should provoke us: would French or Japanese society of the 1960s be such an unpalatable model of energy consumption that we should not even contemplate going back to those levels, especially when better conversions would yield more useful heat and work from the same amount of fuels and electricity?

Given the infrastructural inertia, the consumerist nature of industrialized societies, and the still relatively low energy prices it is very unlikely that we will see truly huge demand reductions during the remainder of this century. Some reductions of high consumption will energize part of the new demand in rich countries, but an extension of the benefits flowing from higher energy use to greater numbers of people in poor parts of the world will not come from redistribution of the existing output. Higher output from conventional sources and new conversion will have to fill the need—and a closer look at the most touted supply alternatives is thus a must.

2.4 Looking for alternatives

> The tide went out
> and the rock was revealed.
> Chinese saying

Even before the first round of OPEC price rises brought widespread apprehension about the security of future energy deliveries, there was no shortage of suggestions on how to cope with what was generally perceived as a serious, long-range supply crisis. OPEC's actions intensified this search for solutions and the spectrum of proposals broadened to include everything from armed invasions to esoteric conversion technologies. An annotated listing of major possibilities will convey best the breadth and the inventiveness of proposals which have spanned hard engineering and soft chimeras and offered choices based on impeccable physics as well as on wishful thinking.

The most straightforward way to deal with a shortage of Middle Eastern crude oil seemed, to some, simply to take over the oilfields. Several detailed studies and contingency plans were prepared and the general consensus was that while the United States could easily defeat OPEC resistance in any locale and seize desired oilfields, preserving the installations intact and holding on to them for a long time would be a much more uncertain proposition; its worst risks would be such a depletion of strategic reserves that the country would have little material and men left for contingencies elsewhere—and the possibility of Soviet intervention. Although repeatedly discussed at the highest Washington levels this option had little chance of seeing a green light. Consensus was overwhelmingly for a supply expansion through tapping of the many alternative resources.

None of these expansions looked more obvious than a return to coal—and coal's big comeback was one of the major recurrent themes of post-1973 global energy assessment. Abundance of the resource (even the most conservative accounting of extractable reserves gives more than three centuries of supply at the early 1980s consumption level), its relatively low cost (in 1985, depending on the country, anywhere between 25 and 65 per cent cheaper than oil per unit of energy content), and its domestic availability in virtually all major industrialized nations most affected by costly oil imports, seemed to present an irresistible combination of opportunities.

Yet the mid-1980s find the coal industry far short of the expectations which had the Western economies, as late as in 1980, doubling their coal consumption by the year 2000; in 1982 the International Energy Agency forecast was cut to growth of 85 per cent, in 1984 to 60 per cent and that, too, appears overly optimistic. In a way, there is little new in such unfulfilled expectations: a post-1945 comparison of various British outlooks and actual performance (Fig. 2.18) could have served as a cautionary example.

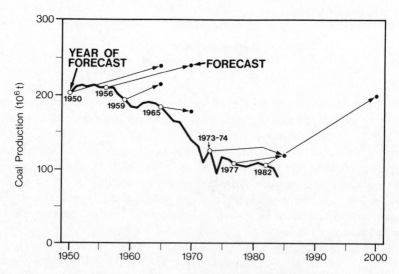

Fig. 2.18 Coal's unfulfilled promise. Realities always lag behind plans, a worldwide phenomenon here illustrated by the repeated failures of British forecasts.

True, mid-1980s global coal output was about 25 per cent above the 1973 level but, besides slipping forecasts, the industry had to face large over capacities in the three large Western producers (USA, Canada, and Australia), closures of many new coal mines started in the late 1970s, prospects of more cheap coal from huge new developments in Colombia and China, and declining demand by such traditionally heavy users as the iron and steel industry. Fluidized bed combustion (see section 5.5) may help to expand coal's use in industrial boilers, but for the rest of this century coal will remain an ageing pretender rather than the much-touted king of future energy supply.

Still, coal has done much better than nuclear electricity generation. This technology, which was in the early 1970s just entering large-scale commercial application, was extolled as the perfect near-term alternative to fossil fuel based conversion and many grand promises were tied to the early commercialization of fast breeder reactors, and later to what was seen by many as the ultimate solution, controlled nuclear fusion. Fusion's future was the first to dim with the growing realization of the enormous engineering difficulties waiting once the practicality of the process had been proven in the laboratory. America's fast breeder programme never got off the ground, and the continuing French and Russian efforts do not signal a mass breeder economy around the corner. Most notably, prospects for fission generation, based largely on pressurized water reactors, whose ascendancy seemed unstoppable in 1973, have turned into an American-led retreat, a phenomenon worthy of consideration in a separate section (2.4.1).

The nuclear retreat has been closely bound up with the rise of renewables which were worshipped especially if dispensed in small doses. A heady mixture of generation revolt, liberal activism, Schumacherian preaching, environmental concerns, nostalgic ruralism, back-of-the envelope calculations, and American faith in salvation through new gadgets produced a simplistic faith in passive solar heaters on every roof, clean electricity from photovoltaic cells (whose main attribute, as any diligent reader of solar literature knows, is their steadily falling cost so that they will eventually break the first law of ecology—the one about there not being any free lunches), in equally clean and captivatingly melodious power from the swooshing of assorted windblades, in benignly sweet-smelling fuel alcohols oozing from cozy farm stills fed with spoiled grain and fuelled with dried straw or stover.

Bolder plans were also not lacking as some renewable option enthusiasts—impatient to usher in their magic on truly continental, even planetary, scales—opted for grandiose schemes. Choice items in this category include fleets of huge solar satellites in geostationary orbits beaming microwaves to giant Earth-based receivers (detailed economic assessments of this high-tech launch have been done in the absence of a reliable carrier to haul the gadgets up there), giant kelp plantations which, to avoid the high costs of mooring the plants, would move with wind and wave-driven propulsors across the oceans and whose biomass production would provide enough energy for over 100 (at the United States consumption level) and food for at least 1000 people from each square kilometre of 'arable surface water', and gigantic offshore arrays of windmills towering hundreds of metres above stormy oceans and feeding electricity via mammoth undersea cables to coastal cities and industries.

In between the alluringly small and the stunningly large—much else: ocean thermal energy conversion projects providing tropical locations with cheap electricity; short-rotation intensive cultivation plantations of fast-growing trees fuelling thermal stations or methanol factories; energy from ocean waves and tides, from solar ponds and sugar cane, from melting the glaciers of Greenland, from high-flying kites (the pull on the rope to be converted at the ground into rotational energy; during calm spells balloons would hold the kites in our skies—altogether quite an enchanting way to deliver energy), from pig manure, and from galls. Galls? Indeed, an otherwise serious senior European energy expert touted giant genetically-engineered galls producing hydrogen (at a rate of $1 \ W/m^2$) which would be gathered by collection pipelines running among trees and which would make all large world regions energy independent!

Hardly anything which moves, grows, or has a thermal gradient escaped the enthusiastic appropriation of renewable energizers and often made one wonder why we had ever bothered with dirty, awkward, and dangerous fossil fuel conversions when surrounded by all these clean and perpetual plenties which required only a bit of R&D and commercialization effort to put away forever worries about energy supplies. But, as noted, the great gains of the first post-1973 decade came from conservation (from the simple stuffing of attics with fibreglass to the sophisticated optimization of energy-intensive industrial processes) rather than from windmills, OTECs, photovoltaics, or fuel alcohols—and in section 2.4.2 I will comment on this discrepancy between the lures and realities of the renewables.

But before turning to the dimmed allures of the two great and opposing salvations—nuclear electricity generation and renewable conversion—a few paragraphs about fancier uses of poorer fossil fuels. This alternative is as obvious as it is sensible: if we are running out of natural, high-quality liquid fuels why not use the truly enormous deposits of low-quality solid fuels and wring the liquids out of them? Why not extract—after mining or, more elegantly, *in situ*—liquid fuels from America's oil shales and turn the Green River Formation in Colorado, Utah, and Wyoming into a new Middle East (after all, commercially exploitable reserves of oil in American oil shales have been put as high as 400 billion barrels, easily an equivalent of Middle Eastern reserves). Why not extract Alberta tar sands whose oil content is no less impressive? And why not open giant surface coal mines in the rich seams of the Great Plains and then liquefy or gasify this cheap coal in huge conversion centres?

The appeal of this 'solution' is multiple: substitution of a major part of costly and insecure crude oil imports by domestically produced liquid fuel (as deposits of oil-shales, sands, tars, and coal are much more widespread and abundant than those of crude oil, many more countries besides the United States and Canada have the resource base to allow such a shift), the spending of money on development at home instead of transferring it abroad, the use of existing pipelines to ship the final products around, and reliance on technologies which need many performance improvements but which are basically well understood.

Pre-1973 research and production experience in the recovery of oil from solids, and in the synthesis of liquid and gaseous fuels, was dusted off and new programmes massively expanded the quest for commercialization. The culmination of this effort was a massive Canadian plan jointly supported by the government and leading oil companies, aiming at oil supply self-sufficiency through the large-scale extraction and processing of Alberta tar sands, and an even bigger American scheme, President Carter's synthetic extravaganza budgeted at $(1979) 88 billion, billed as a 'Manhattan Project' of the 1980s and designed to deliver at least 500 000 barrels of oil equivalent per day in 1987, and two million barrels per day by 1992. When fully operational some 40 huge liquefaction facilities would have supplied around one-tenth of the country's oil demand.

Both programmes have dissolved in the heat of excessive costs and environmental burdens. As for liquids derived from coals, highly uncertain paper estimates put their costs at $(1980) 35-60 per barrel but actual costs would have been certainly even higher. Huge water requirements, land destruction, and release of hazardous air pollutants made the projects even more controversial and there should be little regret about their abandonment.

This listing still has not exhausted all alternatives paraded after 1973: notably missing is hydrogen promoted by many as an ideal synthetic chemical fuel of the future. Others have seen methane as the preferred energy carrier. Then there are magneto-hydrodynamic enthusiasts, advocates of photobiological energy production, diggers of peat in the world's countless bogs, and engineers on the planetary scale damming the Bering Strait or the Strait of Gibraltar and generating electricity from sea level differences. But we shall leave aside all these proposals—some fascinating and sound but hopelessly too expensive for decades to come, others no less fascinating but much less rational—and focus on just nuclear generation and major renewable conversions, the two alternatives which gained such prominence after 1973.

2.4.1 *Rise and retreat of nuclear power*

> The vigorous development of nuclear power is not a matter of choice, but of necessity.
>
> Hans A. Bethe (1977)

The rise has been undoubtedly spectacular. Only fourteen years after the first sustained chain reaction came the first commercial nuclear reactors delivering electricity to utility lines: British Calder Hall (202 MW) in 1956, American Shippingport (141 MW) in 1957. During the next decade the first big wave of new orders in the United States, the leader in fission engineering, pushed the average nuclear power plant size from just 300 MW in 1962 to 1150 MW in 1972, and more than a dozen industrialized countries, as well as half a dozen poorer Asian and Latin American countries, started national nuclear programmes. By the time of

OPEC's first price rise in October 1973 more than 400 reactors were in operation, under construction, or in various planning stages worldwide.

This coincidence seemed to be most fortuitous: rapidly maturing nuclear fission clearly held a great potential for the gradual displacement of fossil-fuelled power generation, especially as the more efficient fast breeder reactors were expected to enter commerial service on a large scale during the 1980s. But the 1980s came to be a decade of withered expectations and shrunken forecasts. Not that the global nuclear industry disappeared: in October 1983, ten years after OPEC's first price shock, there were nearly 300 reactors with about 175 GW operating worldwide, producing some 9 per cent of all electricity or roughly 3 per cent of total primary energy output.

In many countries comparable shares were much higher: in 1985 fission provided about 65 per cent of all electricity in France, 60 per cent in Belgium, 50 per cent in Sweden, 26 per cent in Japan, 18 per cent in Britain, and 17 per cent in the United States. Moreover, in the spring of 1986 150 other reactors were either ordered or already under construction, bringing the total global commitment to over 500 reactors and 383 GW. If all this capacity would generate with a high (say 80 per cent) load factor, the annual output would eventually be about 2.7×10^{15} PWh—and to produce the same amount of electricity from fossil fuels would require some 600 million tonnes of crude oil or around one billion tonnes of bituminous coal, equivalent, respectively, to roughly 20 and 25 per cent of early 1980s global extraction of these key fossil fuels.

But these impressive figures are just one face of the reality: they are the outcome of strong commitments made during the 1970s which contrast so sharply with the 1980s, when the utilities have been not only abandoning the planned projects, but also forsaking those which have already cost them billions of dollars and which are nearly completed. Cancellations have been a worldwide phenomenon but the United States nuclear industry has been the one in the greatest retreat with some ninety plant orders cancelled since 1975.

This American decline is of critical importance as the country is the world's largest producer of nuclear electricity with 27 per cent of all reactors and 35 per cent of all installed capacity. The supersonic transport (SST) analogy fits here rather well: when the Americans abandoned SST in the early 1970s the French-British Concorde and the Soviet Tupolev kept on flying but American withdrawal from the race guaranteed a downhill course for SSTs; with Americans virtually out of the nuclear game can the French, the British, and the Soviets sustain the global expansion? A short historical review will retrace some of the key phases of this rapid expansion–contraction cycle and survey the principal reasons of the recent collapse.

When the First International Conference on the Peaceful Uses of Atomic Energy met in Geneva in 1955 no fission reactor was delivering electricity to a commercial network, and all cost estimates were based just on paper studies which showed a range between $(1955) 183 and 450/kW, with $350 perhaps the most likely value. Expectations were that normal engineering progress (technological improvements, simplified design) and economies of scale (multi-unit plants, standardization of designs) would bring the typical capital cost to \$180–\$250/kW (for comparison, a large coal-fired station cost \$130–40 in the mid-1950s) and generation costs to 4–5 mills/kWh within a decade.

These expectations were reached and surpassed by the late 1960s and the lowered costs led to the first large wave of new capacity orders. While only twelve nuclear powers plants were ordered before 1965, American utilities concluded firm contracts for 83 reactors during 1965–9. Typical cost assumptions in 1970 were for capital cost of \$(1970) 170/kW (without transmission), compared to \$135/kW for coal-fired, and \$115/kW for oil-fired stations.

With electricity demand going up at a sustained rate of 7 per cent annually, with increasing experience in reactor construction, with the first signs of significant upward movement of long-stable fossil fuel prices, and with OPEC's first stirrings, the early 1970s seemed to be the beginning of a real take-off for the nuclear industry. American utilities ordered 108 new reactors during 1970–3, with a record of 38 units in 1973, the year of OPEC's first huge price boost.

Not surprisingly, a bright future seemed to be assured: fission, in an increasingly electrified economy, was clearly destined to be *the* next major source of primary energy. How providential it seemed that such a source was now commercially mature enough to aid the oil-importing countries, rich or poor, in overcoming the burdens of massive oil price rises. Just a few vignettes of that time when nobody anticipated that a long (fatal?) decline was just a few years ahead.

Glenn J. Seaborg, Chairman of the United States Atomic Energy Commission (USAEC) opened the Fourth International Conference on the Peaceful Uses of Atomic Energy in 1971 by predicting that by the end of this century nuclear energy would bring 'unimagined benefits that will directly improve the quality of life for the greater part of the six to seven billion people who will then inhabit the earth', and had 'little' doubt that half of the US total of more than 1500 million kilowatts would be nuclear, with a similar pattern of growth taking place in other major regions of the world. And Seaborg was confident that this 'phenomenal' success would further diversify: he foresaw positive energy production from fusion as almost certain by 1980 and before the end of the century he had nuclear-propelled spaceships ferrying men to Mars, nuclear-powered tankers plying the seas, and nuclear explosives employed on a widespread scale to improve the recovery of underground resources.

USAEC's 1974 forecast of nuclear power capacities put the nationwide total at 260 GW by 1985 and at 1200 GW by the year 2000, when the reactors were foreseen as generating at least two-thirds of all United States electricity. Yet, at the same time, the nuclear industry's record started to slip in several worrisome ways. Many plants which were to be finished in the mid-1970s were still far from completion, their costs doubling, then tripling compared to the original estimates, their safety questioned by highly vocal anti-nuclear lobbies. Several key figures will illustrate the retreat from the high promise year.

The average time to construct a nuclear power station rose to more than 10 years. At the beginning of the 1970s it took about 50 months to build a nuclear generating unit, a decade later 130 months were needed, representing an average slippage of 48 months behind original schedules. During the 1970s construction time was thus increasing at a stunning rate of seven months per year! Labour requirements expanded at similarly rapid rates. Ramesh Budwani's detailed analyses of the American power plant construction industry during the 1970s show that, in comparison with the early 1970s, nuclear units of the early 1980s needed 3.7 times more man-hours both for the design and for actual construction, and that average peak manpower requirements grew 2.1 times. Requirements for skilled on-site work (craft man-hours) were increasing by an average of 13.3 per cent a year, and on-site non-manual man-hour inputs grew from 13.8 to 22.2 per cent.

The reasons for these schedule slippages and escalating labour needs are complex but none appears to be more important than constant regulatory changes whose flow started to rise in the early 1970s and grew exponentially for a decade. By 1978 new regulatory or statutory requirements imposed on the nuclear industry averaged thirty a month, a veritable barrage of uncertainty making impossible any fixed standard designs and reasonable cost estimates. The Three Mile Island accident on 28 March 1979 only strengthened this crippling trend and the changes required in the late stages of construction carried particularly high costs and delays.

Many of these changes, ranging from better seismic protection to plant security measures, were undoubtedly essential as the large scale construction programme of the early 1970s started with relatively little accumulated experience with big reactors. Other requirements reflected more the public pressures and distrust of the industry and to this day have no counterparts in other countries with sizeable nuclear programmes (France, UK, USSR). But these regulatory requirements, no matter if seen as harrassing or merely annoying, would not have had such a deep effect if the projects underway had been fewer and if a few standardized designs had been used. Utilities expected demand for electricity to keep on rising at 'historic' rates (around 7 per cent a year), and new nuclear orders were seen as absolutely necessary to satisfy this demand.

Yet American utilities committed themselves too fast to too many reactors (from a few hundred MW to nearly 200 000 MW during the ten years between the early 1960s and the early 1970s), ignored standardization of design, construction, and quality maintenance procedures (reactors coming in any sizes between 450 and 1345 MW, more than a dozen companies designing reactor systems, more than a score of main contractors building them, over three scores of utilities operating them), and they did not at all anticipate the declining rates of electricity demand (7 per cent during the 1960s, 4.5 per cent during 1970–5, 2.5 per cent during 1975–80, 0.5 per cent in 1981, −2.5 per cent in 1982, 3 per cent in 1983).

Construction delays and higher material requirements resulting from new regulations came coincidentally at the time of rising interest rates and double-digit inflation and this combination pushed cost overruns to unprecedented highs. In the early 1970s units entering service could cost well below $200/kW. For example, a detailed account of expenditures by Commonwealth Edison Company, America's largest user of nuclear power at that time, showed that large reactors completed between 1970–4 cost \$147–280/kW (compared to \$113–231 kW for similar size coal-fired units), and generated electricity nearly 40 per cent more cheaply than coal-fired stations.

Units entering operation in 1974–6 cost \$400–500/kW; by 1980 the mean reached \$1000/kW, the \$2000 level was breached in 1982, and had reached \$3000/kW in 1984. Coal-fired power plants got costlier as well, but even those equipped with highly efficient flue gas desulphurization will cost in the near future about 40 per cent less than an average nuclear station. The verdict thus seems simple: after three decades of commercialization, nuclear generation failed the market test in the country where its prospects should have been by far the brightest. The lesson follows: to avoid further economic burdens this costly experiment, this unfortunate delay on the road to cheap and abundant energy, should be abandoned. The past three decades have not been an opening of a new era—just a plunge into a *cul-de-sac* of economic, environmental, and political disputes.

But the conflicts surrounding the development of nuclear power have been too long-drawn, too passionate, and too complex to let these (now nearly universally accepted) conclusions go unchallenged. The failure of initially excessive promises should not be mistaken for the inherent failure of the technology; runaway capital costs for new American nuclear stations should not be interpreted as the inevitable attribute of this electricity generation process; continuing uncertainties regarding the operational safety, long-term disposal of radioactive wastes, and security implications of fission generation must not be seen as irremoveable curses making our attempts at peaceful, safe utilization to be at best foolish Faustian bargains, at worst horrifying genocidal flings.

The uncertainty of the longer term outlook is perhaps best characterized by quoting the key result of a study published by the Congressional Office of Technology Assessment in February 1984: during the decade of 1984–94 anywhere between 0 and 400 new large nuclear power plants will need to be ordered to satisfy the demand, a range from a complete moratorium to a burgeoning industry.

Still, even with a complete and lasting moratorium on new nuclear generation, there would be the operating reactors to watch for signs of problems which might lead to their early retirement and, starting in the first decade of the next century when the first large commercial reactors will be forty years old, a difficult challenge of decommissioning. This will be done most likely by delayed dismantling, by finding some useful function for containment buildings while allowing the radioactivity to decay for fifty years or so. Robert Blumberg of the Oak Ridge National Laboratory estimated that in the United States decommissioning might be a \$(1982) 500 million/year industry from the year 2008 onwards.

The great nuclear retreat offers a perfect opportunity for analysing all the critical blunders and wrong turns: above all, promising too much too soon and then trying to scale the gadgets too fast to meet commercial expectations; creating the industry standards *en route* and thus burdening the march with unpredictable detours, backtracking, and delays; not doing enough to dispel the enormous distrust of the technology, a process much better accomplished by meticulous quality control and (in comparison with other industries) by excessive safeguards—rather than by theoretical calculations (one Three Mile Island accident and, even more so, one Chernobyl disaster, weighs endlessly more in public perception than stacks of risk-probability studies).

In fact, long-term safety considerations rather than current high capital costs will decide the fate of fission generation. French experience proves well enough that nuclear power plant construction does not have to slip seven months every year and end up costing four or six times the original estimate. And even if the capital costs would not be lower than those of a fossil-fuelled station this alone would not be the undoing of fission generation. Fredric C. Olds pointed out the strange paradox of environmental anti-nuclear activists decrying nuclear generation's runaway costs, and condemning electricity from fission at $1000/kW as un-economical, while enthusiastically promoting photovoltaics, costing several tens of thousands of dollars per killowatt, writing enchanted reports about solar satellites ($1 million/kW), and recommending fuel alcohol from corn whose production entails a net loss of energy.

For those who would bury nuclear generation, recent high costs are just another convenient argument—but not the key one. Fission at a tenth of the current cost would not make them more enamoured of the technology which they consider dangerously unsafe and socially pernicious. These perceptions are wrong but strong. In a survey to rank thirty sources of risk in the United States both the members of the League of Women Voters and college graduates, two of the most politically and electorally active blocks of population, put nuclear power in first place—while the actual annual contribution of fission generation to the number of deaths is best put about 50 per cent below the side-effects of contraceptives (of which both groups certainly approve), an order of magnitude below bicycling (which those enlightened citizens increasingly take up as a modish recreation), and three orders of magnitude less than smoking (more women than men now smoke in the USA).

The scientific consensus has been reasonably clear. Low level radiation arising from routine operation of nuclear power plants is lost amidst other natural (cosmic, substrate, building materials, food) and anthropogenic sources (diagnostic X-rays above all), making the lasting dispute between the proponents of cumulative and threshold effects of little practical impact: even if cumulative the hazard is too small to be detectable (see also Fig. 4.4). However, the risk of accidental releases, especially those caused or compounded by human errors, is a much more uncertain realm. In the West even the worst cases so far have not led to anything truly threatening (at the Three Mile Island plant, the maximum dose theoretically receivable by outsiders was 70 millirem, about a third of the natural background radiation at the site). In contrast, the Chernobyl accident in April 1986 was, whatever its eventual casualty toll (from delayed radiation effects) and economic costs, of a clearly catastrophic and totally unacceptable magnitude. There is also no denying that such risks will grow with more and larger stations in operation.

On this account alone a go-slow strategy makes sense: once we have accumulated close to 10 000 operation years with large reactors (some time in the first decade of the next century) we will be in an incomparably better position to assess the accidental risks which the informed consensus has been consistently putting at comfortably low levels but whose possibility, so frighteningly realized by the Chernobyl explosion, has been a mainstay of anti-nuclear arguments. Yet even an exemplary future safety record may not turn the tide of unfavourable public perception: the Hiroshima syndrome coupled with the continuing superpower nuclear arms race, possibilities of terrorist acts, and diversions of the spent fuel for bomb production in many nations (i.e. dangers of nuclear weapons proliferation) make fission generation irrevocably the most politicized and the most emotional segment of energy debates.

The current retreat of nuclear generation is a complex function of the industry's obvious blunders (ranging from rash developments and serious design flaws to slack management and the tendency to over expand), the public's perception of the technology as unusually dangerous (even before Chernobyl and Three Mile Island, as a result of environmental consciousness diffusing since the late 1960s, also owing to fission's irrevocable link to military applications, and because the fission processes and effects are understood only by a slim section of the public), as well as a result of the general slowdown of energy consumption growth throughout the rich world.

The rich world should not stress large-scale nuclear generation, not because its costs are hopelessly high or because its assorted risks are plainly unacceptable. If things are done right neither is true (after all, Chernobyl reactors do not even have containment buildings!); even the ultimate problem—safe, long-range disposal of high-level radioactive wastes—will have acceptable technical solutions. We should go slow because the history, complexity, and passions determining nuclear generation development call for measured advance, one stressing safety, security, and public acceptance.

In the poor countries the go-slow imperatives are even stronger. Nuclear power in Asia or Latin America should not be dismissed *a priori* as hopelessly inappropriate. Careful studies have shown that per capita GNP and assorted quality of life indices are not the most effective indicators of fission's suitability. A country may be relatively poor but if the overall size of its economy, and the magnitude and growth rates of its electricity demand are large, and fossil fuels or hydro-energy capacities are scarce, introduction of nuclear generation may still be an economical way out of supply shortages.

But, considering the costs and possible complications, only the nations with vibrant economies and an adequate pool of skilled and managerial labour should attempt the construction, and the process should be incremental and deliberately cautious. Laying out grandiose plans where the country's finances are wobbly and where no previous nuclear experience exists, or trying to build new plants at rates matching those of some rich nations is irresponsible and unattainable: Brazilian and Iranian examples are clear proofs. Western efforts to promote nuclear generation in any poor nation whose dictatorial rulers show interest have been definitely very unwise: only a few select poor countries could benefit, and only from programmes which are even more modest than careful national efforts in the rich world.

The fact that in most of the democratic countries people are more willing to tolerate the higher costs, proven damages, and environmental uncertainties arising from traditional fossil-fuelled generation, or from renewable technologies, may be easily perceived as groundless but it cannot be dismissed as irrational: nuclear fission *is* a very unique way of generating electricity. Neither the American 'market failure' nor the Swedish construction moratorium, nor the Chernobyl disaster are the irrevocable signs of failure. Nor is the continuing French push a clear proof of a successfully charted road.

We have been too impatient, too resolute in demanding absolute proofs where probability assessments can be sharpened—but not wholly eliminated—only by generations of everyday experience. Rapid diffusion of nuclear generation is not a necessity making the difference between a comfortable and a distressed future for the industrialized nations during the next two decades. By that time we shall know much better how (in)tolerable it is to live with hundreds of large power reactors—and whether we should build thousands of them during the century ahead.

The current shift is a retreat—not a demise—and such a pause may be actually the best thing to ensure that nuclear power has a future: clearly, it is a valuable option the discarding of which would not ease the dangers of nuclear war, but whose substantial future improvements (higher efficiencies, enhanced safety, satisfactory waste disposal) could, especially when combined with a longer operation experience, offer an appealing alternative to fossil-fuelled electricity generation. Still, the coming of the second nuclear era is not a foregone conclusion. The complex interplay of future energy demand, fossil-fuel-based generation, technological advances, and diffusion of new conversion processes will determine the timing and the extent of any nuclear renaissance. And the fortunes of renewable energy sources will be a key indicator to watch.

2.4.2 *Lures of the renewables*

> Long experience has taught fatal consequence may follow if I get too addicted to a theory.
>
> Charlie Chan in Earl Derr Biggers' *The Chinese Parrot* (1926)

In societies prone to embrace uncritically any appealing panaceas the sermonizing of renewable energy enthusiasts was given very generous attention and regarded surprisingly often with awed reverence. Even among otherwise sensible engineers, whose lifetimes had been spent on perfecting designs and performances of technologies so they would fit the cost and reliability requirements of the real world, not a few were captured. When even some of the people who knew best the huge gap between conceptual designs, working prototypes, and rapidly diffusing commercial devices started to speak in tongues of a renewable tomorrow, how could the public media and popular scientific accounts be blamed for seeing every house heated by solar energy, every kilowatt-hour produced by photovoltaics, wind turbines, or ocean thermal conversion, every car running on alcohol?

Yet the appeal is not difficult to understand. In a world either rapidly running out of fossil fuels, or threatened by crippling environmental pollution following their further extended use, conversion to nuclear energy would be an even more dangerous proposition. With all the escapes closed only one solution remains, a massive adoption of renewable sources and technologies. The most powerful part of the appeal certainly rested on the renewability of the flows to be harnessed—out there all the time, free to be captured with devices ranging from the simple to the ingenious, and with appealingly low costs.

Actually, the renewable *inamorati* invested their solution with a longish array of other, far from modest attributes. Nobody spelled out these better than Amory Lovins, the most publicized guru of the new nirvana. He assured the nervous post-1973 world that harnessing of renewable energies, which 'are always there whether we use them or not' and which represent an enormous potential, would not only be 'elegant' but also economical, that the possiblity of small-scale, decentralized operations would especially benefit the poor by contributing 'promptly and dramatically to world equity and order', that these designs 'for democracy from the ground up' would be safe and ecologically inoffensive, spreading the virtues of community resilience, self-sufficiency, and a sustainable future, a gentle balm of soft energy technologies healing the distressed planet.

Specific claims made by the uncritical proponents of renewable energy sources read like wishing lists in fairly-tales: William Jewell wrote that 'development of marginal lands for biomass production of fuels could result in generation of the total U.S. energy requirement'; Bent Sørensen foresaw that residues from biofuels could replace all chemical fertilizers within 50 years; Donald Klass had a single 100 000 square mile desert area in the Southwest USA supplying two-thirds of America's energy needs!

What a stunning combination of appealing features unalloyed by any draw-backs, literally oozing the confidence that *the* solution to global energy problems was at hand—what a crafty assemblage of generalities, a skilful manipulation of realities betraying more of a messianic zeal than a critical understanding of the complexities and limitations surrounding the use of renewable energies. I am still astonished that such rubbish could have been published in scientific journals when a few minutes with a small calculator (and, of course, appreciation of basic biospheric realities) are enough to unmask the irrationality of such claims. For example, to turn Klass' scheme into reality such a gargantuan energy plantation would require daily eight trains made up of 100 50-tonne hopper cars to bring the requisite nitrogen, phosphorus, and potassium fertilizers and it would consume nearly 30 per cent of all surface runoff from the 48 contiguous United States!

But let us look at some of those superficially so persuasive general arguments in favour of the renewables—and the availability of renewable energies is a natural place to start with. There is perhaps no finer refutation of the claim that renewable energies are 'always there' than a closer look at the availablity of solar radiation in the poor countries whose 'fortunate' location in the 'sun-drenched' tropics is supposed to be very advantageous, not only for direct solar conversions but also for the cultivation of high-yielding food and energy crops.

True, the area between the two tropics contains regions with the planet's highest insolation, such as Southern Egypt and west-central Saudi Arabia, but it also contains the extensive inter-tropical convergence zone (moist tropics), and the large monsoonal regions where annual insolation is surprisingly low. And so most of the Brazilian Amazon, or all of southern Nigeria receive less radiation than Georgia or Kansas, and virtually all the very densely populated, poor places from Southern China through Vietnam to Malaysia, Sumatra, and Kalimantan have solar inflows comparable to those in northern France and southern England, the latter two being hardly the regions usually described as 'sun drenched'. Naturally, this does not make Manaus, Lagos, Guangzhou (Canton), or Singapore unsuitable for collecting solar radiation by appropriate (i.e. flat-plate rather than concentrating) devices but, as in Philadelphia or London, it makes using direct radiation a good deal less practical, efficient, and reliable than in Tucson or Aswan.

The crop productivity myth, perpetually fostered by the rich, lush appearance of natural tropical vegetation, should not be around any more as it has been exposed many times in the past. In the dry tropics productive farming is predominantly confined to irrigated land and already severe water limitations will ensure the lasting priority of food and feed crops over biomass harvested for energy. In the humid tropics yields of annual crops are depressed in comparison with temperate latitudes owing mainly to the previously mentioned lower insolation (during the summer growing season Amazonian, Zairian, Indonesian, or South Chinese ricefields receive 10–15 per cent less radiation than even the Canadian Prairies at about 50°N) and to the relatively high minimum night temperatures promoting respiration losses.

The effects of reduced potential photosynthesis are further aggravated by water deficits (differences between the potential evapotranspiration and effective rainfall in many parts of the tropics are comparable to those of the temperate desert regions), often low fertility soils, and favourable conditions for diffusion of numerous pests and diseases. This unalterable climatic impoverishment of the tropics will always put the region at a disadvantage in growing highly efficiently all but a few specialized crops. Counter-intuitive as it may be, even humid tropics cannot compete in growing food with temperate zones, and it then makes hardly any sense for countries which cannot feed themselves to turn large blocks of their farmland into sugar-cane or cassava fields to produce fuel alcohols and to sink even deeper into dependence on food imports. Significantly enough, most of the populous tropical nations located fully or partially in the moist or monsoonal tropics are already large, and increasing importers of food!

In capturing the solar flow directly or indirectly through food and energy crops, many poor nations are thus already facing fundamental environmental limits which are dispelling erroneous notions of the tropics as an especially propitious region for an easy and swift establishment of modern renewable energy economies.

The infatuation with the enormous potential is due to an almost always overlooked difference in essential categories: what the uncritical proponents of renewable energy sources were almost exclusively citing were, in fact, *resource* estimates rather than *reserve* assessments. Biomass energy appraisal serves as an excellent example of this fallacy. Unlike the obviously intermittent solar-radiation flows, biomass appears to be *always* available, conveniently 'storable' in trees and crops and it has been assigned by renewable energy planners the essential roles of supplying liquid transportation fuels (alcohols), and generating electricity in small, decentralized (and hence supposedly less vulnerable) power plants.

Yet the impressively huge figures for the total national or regional availability of wood wastes or crop residues are analogous to our estimates of total oil *in situ* and these sets of resource estimates share a large range of uncertainty and a fundamental limitation: whatever the total resource might be, only a fraction of it will be recoverable. While, in the case of crude oil, the recoverable reserves cluster around only one-third of the resources *in situ*, waste biomass recovery ratios may be still smaller.

A detailed feasibility study on the co-generation of electricity and steam by using wood waste as fuel in timber-rich Lewis county in Washington illustrated perfectly this discrepancy. The potential of logged fuel, sawdust shavings, chips, yard wastes, and all kinds of logging residues was found to be worth 115 MWe—but the effects of fluctuating annual forest cut, and lumber and plywood markets, alternative (and almost invariably more lucrative) uses for chips, sawdust, and merchantable cull logs, and prohibitive transportation costs for diffuse, low-density residues from remote sites, cut the feasible size of the facility to just 25 MWe. Extractable wood waste 'reserves' were in this case only about one-fifth of the appraised resource.

A similar situation arises with the removal of crop residues. Again, a simplified estimate (usually based on very unreliable grain-to-straw ratios) will show a large potential energy resource, and theoretical calculations can be used to advise a farmer how much straw or stover he can remove from his fields without adversely affecting soil tilth and fertility, and without exposing the land to excessive erosion. Yet such general recommendations may prove recklessly excessive in a year when the harvest is followed by a snowless, windy winter and dry, and no less windy, spring. During the August or September harvest we have, of course, no way to predict the weather for six or nine months ahead, a simple reality forgotten by renewable energy proponents—but rarely lost on the farmers who are, after all, playing a risk-minimizing game.

Consequently a careful appraisal of straw availability in the heart of Kansas wheat country (Pratt county) had to conclude that concerns about wind erosion would limit a *partial* removal of residues to only some 20 per cent of wheat area, and that fields would have to be modelled individually to arrive at the allowable 'extraction' rates. Farmers' perceptions were even more conservative: 60 per cent of the farmers queried in the county were unwilling to have *any* residues removed from their fields, and 40 per cent would allow a partial removal from only a part of their land.

These two examples, coming as they do from the areas having North America's richest concentrations of, respectively, woody and crop waste biomass readily convertible to energy, cannot be overemphasized. They are striking illustrations of the ever-present disparity between grand estimates of resources *in situ* and recoverable reserves in a particular locality, and they also illustrate that renewable options are not automatically elegant, ecologically inoffensive, and sustainable. As we get down from generalities about renewable, inexhaustible, and environmentally benign resources to a close, pinpoint accounting of reliably available waste biomass whose extraction will be compatible with long-term preservation of soils, nutrient cycles, and water balances, the availabilities will shrink quite substantially.

Similar limitations would apply to harvesting the whole trees specifically cultivated for energy in projected fuel plantations. As Harold Young, the originator of the whole-tree utilization concept, aptly remarks, we can build a power plant and fuel it with short rotation trees in three years but scientific information regarding long-term effects on forest nutrients can easily take ten times as long to gather. Yet before we possess such, usually highly site-specific, information we might be over-extracting the resource, degrading and eventually ruining its environmental foundations, and making it nonrenewable.

Preoccupation with smallness and decentralization has been an outright obsession with advocates of 'soft renewability' but in the real world there are inherent and predictable, as well as hidden and surprising, advantages and drawbacks to scales small *and* large: judging a technology solely by its scale is neither rational nor useful. Small scale goes too often hand-in-hand with mismanagement, inefficiencies, high energy waste, and uncompetitive costs—as best illustrated by the prodigious Chinese experience with small-scale industrialization between 1958 and 1978.

The Chinese, the world's leading promoters of self-sufficient smallness, concluded that the approach was not only wasteful but that it was clearly insufficient to provide a foundation for the industrial advancement which is so badly needed by all poor nations. In this respect power output considerations become critical. Most operational technologies based on renewable resources can be, when properly maintained and run, very helpful to a rural household or a small village supply—but they cannot support such basic, modern, energy-efficient industries as iron and steel-making, nitrogen fertilizer synthesis, and cement production.

'Renewable' ironmaking provides the best example of scale mismatch. Charcoal would be the only practical renewable substitute for metallurgical coke and no less than 2.5 m^3 of it are needed to smelt a tonne of pig iron. Wood requirements for charcoal vary with tree species and kilning procedure but, by coincidence, 2.5 m^3 should be taken as the minimum. This means that securing enough charcoal for the current global annual output of 500 million t of pig iron would claim some 3.25 billion m^3 of wood—and even if this wood could be grown in intensive (fertilized, irrigated) fast-maturing tree plantations yielding around 5 t per hectare, it would take an equivalent of nearly 30 per cent of the world's farmland, clearly an impossibility.

This is just one example of an irremoveable mismatch between high power densities of industrial processes (which require energy supplies at rates of 10^2–10^7 W/m^2) and large cities (which consume energy at 10^1–10^2 W/m^2) on the one hand, and low power densities of renewable energy supplies on the other. Biomass fuels and electricity from wind cannot be produced with power densities higher than 10^{-1} W/m^2; hydro, tidal, geothermal, and solar energies can be converted into heat or electricity with power densities of 10^0–10^1 W/m^2—but still need storage to overcome random flows. In contrast, fossil fuels can be extracted at densities of 10^3 W/m^2, matching much more easily the needs of industrial civilization without exorbitant land requirements.

Poor countries need it all: two hot meals a day, higher crop yields, widespread industrialization. Obviously, going 'from the ground up' by means of small decentralized energy serices would not, costs aside, be without many local benefits, but a prompt and dramatic contribution to world equity would hardly follow. After all, small-scale decentralized energy sources are still an everyday affair for the poor world's peasants, and while replacing today's inefficient rural stoves or open fireplaces with solar cookers and biogas digesters (providing local environmental and resource constraints would allow their operation) would be an important qualitative gain, it would still be only a partial one as growing populations, rapid urbanization, and higher food production cannot be managed without reliance on large-scale industrial processes.

About the economics of renewable resources little has to be said: claims of embarrassingly cheap energy deliveries have been based on a very limited body of everyday, long-term experience, and often the values represent just dubious guesstimates which only a few decades of cumulative experience can dispel. Until such time, most of the cost claims resemble nothing more than the cost appraisals of nuclear generation proffered during the 1950s.

Curiously, 'soft' energy proponents share even more with their counter-parts in the nuclear camp: both groups of proselytizers are bound, though none would admit this, by a misplaced and mistaken faith in technology as *the* solution for complex problems of energy supply. Zalmay Khalilzad and Cheryl Benard noted how the claims made on behalf of the renewable energy technologies are amazing carbon copies of earlier claims made on behalf of fission generation. After a generation or two of buffeting in the real world these faultless utopias get a worn-out look!

Many more cutting comments have been made about the presumptuousness, fallacies, and fundamental flaws of soft energy schemes. Harry Perry and Sally Streiter pointed out that there is no evidence about the irreconcilable exclusivity of 'hard' and 'soft' paths which formed a basic premise of Lovins' argument and asked a simple key, question: 'How certain are we that the soft technologies can be successfully developed for widespread commercial use?' Alvin Weinberg argued that the use of renewable sources may not be the best choice to satisfy important trade-offs of energy use and time savings, and that other considerations (environmental, economic, reliability) militate against simple-minded stress on maximum thermodynamic efficiency.

Perhaps the most distressing characteristic displayed by the pushers of soft energy was the intellectual poverty of their grand designs, their impatient dismissal of all criticism, their arrogant insistence on the infallible orthodoxy of their normative visions. There is little doubt about the origins and the real message of soft energy dogma: the roots are in the muddled revolts of young Americans in the late 1960s and the early 1970s, the goal is a social transformation rather than simply a provision of energy. The latter fact explains the widespread appeal of soft energy sources among zealous would-be reformers of Western ways.

Michael Stiefel's question cuts to the core: 'What will happen when soft paths become superstition instead of heresy?' I am afraid that in the way it was put out the 'soft' sermon was superstitious right from the start. That we need all workable, reliable, economical alternatives to diversify our current energy production is clear; we need renewables—not as a means to redeem our fossil-fuelled or nuclear sins, but as part of a complex effort respecting that enormous heterogeneity of natural endowment, environmental, and economic conditions, and available human skills and including all worthwhile conversions on any scale. To worship a single, simple, infallible precept is not only to discard excellent available alternatives and pass up the opportunities for improving other approaches but, most unfortunately, to foreclose tomorrow's choices: a soft thinking, indeed.

2.5 Are there any sensible forecasts?

> Alice laughed. 'There is no use trying,' she said: 'One *can't* believe impossible things.'
> 'I daresay you haven't had much practice,' said the Queen. 'When I was your age, I always did it for half-an-hour a day. . .'
>
> Lewis Carroll, *Through the Looking Glass* (1871)

As the title suggests, the shortage is not absolute: the 1970s saw more long-term energy forecasts than during the whole previous century of fossil-fuelled civilization. But what a sorry assemblage most of this effort is: a junkyard of endless computer printouts, some nicely bound books, and plenty of boldly embossed binders, a repository of a great many millions of dollars bestowed by private agencies, big companies, and anxious governments on the select groups of experts whose multivariate alternative models were originally awaited with such expectations.

What went wrong? This question cuts deeper than just into the flabby body of energy forecasts: the whole global forecasting vogue of the 1970s, handfuls of men with computers scaring us or soothing us, much like the fairy-tale purveyors of yesterday. I shall return to this phenomenon later in this book—here are just a few guidelines to any would-be forecaster, illustrated appropriately by choice energy forecasting examples, before offering, yes, one sensible look at the likely global energy use in the year 2000.

The first warning is so obvious that its very mention should be a source of genuine embarrassment—but I shall state it without such feelings because energy forecasters appear to be prone to selecting techniques whose assumptions suddenly fall apart: the unexpected happens. Energy affairs of the 1970s, so linked to and so much influenced by political considerations offer many outstanding examples.

What forecast prepared in 1970, when few people paid attention to the 10-year old OPEC which was selling its oil at US $1.85 a barrel, could have foreseen that before the decade had ended crude oil prices would have risen seventeen-fold in current terms, but that this staggering rise would be paralleled by increased, rather than by declining, oil consumption! How unpredictable was the 1973 Yom Kippur breaking of the Bar Lev line, an overt cause of the Arab oil embargo, and how even more surprising were Sadat's and Begin's toasts a mere four years later in Jerusalem.

How unexpected was the election of an ambitious peanut farmer to become the first Southern President of the United States (coincidentally during the month when Iranian crude oil flow reached the record of nearly 7 million barrels a day), a President who just a year later toasted the Shah Mohamed Reza Pahlavi as America's pillar in the Middle Eastern morass (with no other leader did Carter feel 'a closer personal friendship' than with the Shah while the Iranian flow was still at 5.5 million barrels a day), but who a mere fifteen months later would not let the exiled, ailing Shah remain in the United States. The deposed sovereign's short stay incurred the Imam's curse anyway and led to the take-over of the United States embassy in Tehran (Iranian flow down to a shade above 3 million barrels a day), to 14 months of American humiliation, and to Carter's forlorn exit (Iranian crude just a bit above 2 million barrels a day, and crude oil price nearly triple the 1976 value).

What is astonishing is that, even after the first oil price jump, one of the most laborious long-range forecasting studies tied all of its outcomes to rather rigid price assumptions for the next quarter century! The Workshop on Alternative Energy Strategies (WAES), conducted between 1974 and 1977, was a high-level, inter-national, inter-governmental, multi-volume effort including some of the world's top economists, engineers, and planners which based all of its scenarios of the future world energy situation until the year 2000 on three equally unrealistic oil price assumptions: a constant price of US $11.50 per barrel; a price rising to US $17.25, and a price falling to US $7.66. The second round of oil price increases devastated the whole effort just two years after its much acclaimed publication!

The second reminder concerns the infatuation with new sources and new conversions, an affliction common to the most determined promoters of untested, complex, highly advanced technologies as well as to the preachers of renewable simplicity. Consistently over-optimistic forecasts of national and global contributions of nuclear generation, exaggerated hopes for recovery of crude oil from tars and shales, and unrealistic programmes for synthetic fuels from coals have already been noted earlier in this chapter. Lovins' forecast of global energy consumption by the year 2000 is perhaps the best summary example of overdoing things from the other extreme.

His uncritical belief in the omnipotence of small decentralized energy conversions led him to forecast global energy needs of just 223 EJ by the end of the century, a reduction of about 20 per cent compared to the early 1980s level. But by the year 2000 populations of poor Asian, African, and Latin American countries will increase to about five billion and, even should their per capita consumption remain at the current grossly inadequate level, an additional 30 EJ would be required to cover the increase. So to keep the total at 220 EJ, the rich nations would have to reduce their current use by about 100 EJ—a most unlikely drop in just fifteen years!

The third warning to all long-term forecasters is perhaps the most difficult one to heed: divorce yourself from any prevailing moods at the time the forecast is made. Again, the post-1973 experience offers fine examples of ignoring this maxim. The outlook was so different when the mullahs had just taken over Iran, and crude oil prices doubled in a year from an already high base of $ (1979) 17—or a mere three years later when OPEC was producing at just 50 per cent of capacity, energy prices were declining, and imports were falling.

As the world crude oil market shifted from continuous increases to rapid output decline, followed by inevitable lowering of the OPEC price, the feeling took hold that this new situation was a result of fundamental structural readjustments, and that this situation would continue for many years, if not decades—and this lowering of prospects for global oil consumption led to generally lower total energy forecasts. Some of these declines were impressively rapid, as best illustrated by a series of OECD primary energy demand projections for the year 2000: 8.9 billion tonnes of oil equivalent (toe) were forecast in 1977; just a year later the total decreased to 6.91 billion toe, and in 1982 it fell to 5.5 billion toe.

But perhaps the best illustration of the strong influence exerted by perceptions of the day is a comparison of all important forecasts of global energy consumption in the year 2000 made between 1976 and 1983 in the OECD countries charted in Fig. 2.19. The trend is impressively downward. But may not the widespread belief that the drop is due to fundamental structural changes in Western economies which will hold consumption down for years, perhaps even decades, be shattered during the late 1980s?

Another good example of the reigning short-term trend translated into a grotesque long-term forecast is the completely unrealistic importance assigned to synthetic fuels in the elaborate global energy forecast prepared at the International Institute for Applied Systems Analysis. This study was assembled during the late 1970s when growing crude oil imports and tightening global oil supply led to heightened interest in synthetic fuels from coals, a trend culminating in the Carter administration's decision to opt for a massive coal liquefaction and gasification programme in the wake of the Iranian upheaval.

The IIASA study assumed that coal liquefaction, together with oil production from non-conventional sources (shales and tar sands), would be the key mode of closing the huge gap between its postulated demand and available conventional oil after the year 2000. According to the study's high scenario, coal liquefaction was to provide 211 EJ by the year 2030, that is an equivalent of about 70 per cent of global

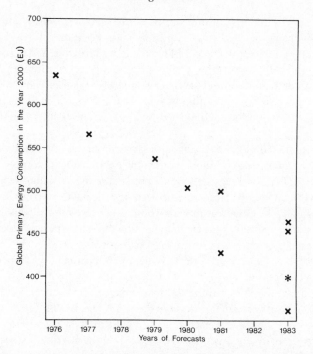

Fig. 2.19 The effect of prevailing moods on long-term forecasts. As the perceptions of an energy 'crisis' eased global consumption forecasts plummeted. The star marks my best estimate whose derivation is described in the text.

energy use in the early 1980s, and even in low scenario 107 EJ, or almost exactly an equivalent of the worldwide crude oil production in 1983, were to be supplied by an industry which is as yet virtually non-existent.

My last point here is a plea for unglamorous, common-sense approaches to energy forecasting. Smitten by the fashion wave of ever more involved and further forward-looking global models, most energy forecasters seem to spend all their time on devising structures of such interlocking complexity, demanding the determination of future values for such finicky variables, that reading their scenarios makes one wonder if all of it is not just a big put-on at public expense.

Once more, the IIASA's super-study can serve as a paragon of this mania; for a person who wants to see a grand display of sybilline powers a plunge into MEDEE-2 assumptions (remember, computer models must always have fancy names) will be most satisfying. Have you ever wondered what will be the average number of persons riding a bus in Latin America in the year 2000? What will be the share of steel production in non-electric furnaces in Middle East and Northern Africa in 2015? Or how about the average demolition rate of dwellings during 2025–30 in IIASA's region III, an admirably organic unit comprising Western Europe, Japan, Australia, New Zealand, South Africa, and Israel?

MEDEE-2 will give you, prophetically, all these answers—then it incorporates them into its equations and comes up in its high scenario with 530 EJ of global consumption by the year 2000, 429 EJ for the low alternative. Yet much the same or, I strongly believe, superior results can be obtained without pretentious computer models, and without seven years and 880 pages of glamourized trials. The key to any sensible forecasting is to select variables which are least likely to undergo rapid fluctuations and then to appraise a few of their most plausible courses.

One of these simple and relatively least error-prone ways is to rationalize plausible per capita consumption figures. Of course, such a forecast is fundamentally dependent on long-term global population dynamics but when looking just at the year 2000 the differences in projections are tolerably small. Continuation of the average 2 per cent a year growth would bring the world total population to 6.6 billion by the year 2000; gradual moderation to just 1.5 per cent (certainly the best one could expect) would mean a 6.0 billion total. The World Bank's 1984 *World Development Report* opts for 6.3 billion, just a 5 per cent difference from the two plausible extremes.

The simplest case is to assume the maintenance of the current consumption mean of 60 GJ a year per capita: instead of the current nearly 300 EJ at least 360, and as much as 400 EJ, would be needed, with 380 EJ the best single value. But such reasoning is dismally defeatist on at least two grounds: no room is left either for growth of those plainly inadequate consumption rates throughout the poor world, or for further efficiency improvements, above all in the rich countries.

So in a much more realistic scenario the numbers get split and 1.4 billion rich people are assumed to get by with 190 GJ, the same per capita amount of energy as today (that is, they will cover the requirements of their modest population growth by better use of the total they already consume), while the poor 4.6–5.2 billion people double their mean to 40 GJ (this would be an equivalent of a 4 per cent exponential growth). This reasoning brings us to 450–75 EJ.

Looking back one sees that doubling of per capita energy consumption in poor countries before the year 2000 is certainly possible: between 1960 and 1980 per capita use of commercial fuels and primary electricity rose on the average 2.1 times in 21 upper-middle income nations (such as Mexico, Brazil, or Malaysia), 2.3 times in 43 lower middle-income countries (such as Philippines, Nigeria, Turkey) and 1.7 times in the poorest group of 34 nations including China and India. The ambitious development plans of many poor countries would aim at much higher future growth rates but, considering the rate of past improvements, the financial state of most of these countries, and the long lead times in developing fossil fuel resources, the assumption of doubling is the most plausible approach to determine a ceiling value.

On the other hand, it is obvious that in the rich countries the opportunities for energy savings, so clearly demonstrated since the late 1970s, have been far from exhausted. In the OECD nations primary energy consumption continued to rise until 1979 and then it declined by almost exactly 10 per cent in a mere three years (in per capita terms it would be about 11 per cent as just over ten million people were added to the OECD population in those three years).

Using an average annual 3.5 per cent rate of decline would bring the OECD energy use to only 54 per cent of its 1982 per capita value by the year 2000, that is to just about 110 GJ. With some 890 million people by the year 2000 OECD would then consume less than 100 EJ of primary energy. However, continuation of the 1979–82 declines for another two decades is most unlikely. These years of rapid energy consumption decline were also the years of the sharpest decline in industrial growth for decades: in 1980 and 1981 four of the OECD's seven largest economies registered absolute production declines, in 1982 five countries produced 1.6 to 10.8 per cent less than a year before.

That the gains in energy efficiency were real no one can doubt; but that the lower energy use had nothing to do with such substantial production declines is naïve to believe. Moreover, it has been convincingly shown that the easiest energy efficiency and conservation gains come first at relatively low cost, and that subsequent efforts are much more sluggish: to deny this trend is impossible. Consequently, granting half of the calculated decline in per capita energy consumption would still mean an outstanding performance as the OECD economies will certainly grow (albeit certainly more slowly than during the first post-World War II generation) between now and the year 2000: the OECD's total energy use would then be 130 EJ of primary energy.

Although Soviet energy consumption is still increasing—with the recent rates between 1.5 and 3.0 per cent a year—one can safely assume a lowering of future growth rates which is already clearly discernible in all six European Communist countries, some of which have been actually consuming slightly less in the 1980–2 period than in the late 1970s. Given their economic infrastructure with heavy reliance on primary industries I do not think it possible to credit developed Communist nations with future energy savings comparable to the OECD potential. Still, by the year 2000 they might manage with 10 per cent less energy per capita than today, that is with about 160 GJ per capita. With 470 million people in the year 2000 they would then use about 80 EJ of primary energy.

If the poor countries, whose population will reach at least 4.6 billion people by the year 2000, could not raise current average per capita energy use—a depressing but by no means implausible prospect owing to their economic difficulties and continually high population growth—their total primary energy consumption would be as low as 80 EJ, and the lowest conceivable global total (with 80 EJ in the Soviet block and 130 EJ for the OECD) would come to 290 EJ, or virtually the same as in the early 1980s.

These rather simple, common-sense exercises based on population growth, and on the likely changes in per capita consumption, bracket the global primary energy consumption by the end of the century between roughly 300 and 475 EJ, or basically between the maintenance of the current level and an increase of up to about two-thirds. I have a great deal of confidence in this interval forecast as it is based on fundamental considerations affecting energy use.

Mature industrial societies are most likely to have increases in their per capita use of energy but, at the same time, assumptions about the declines of energy intensity should not be governed by the most recent records whose rates would be extremely difficult to sustain in the long run. Analogously, the Soviet bloc nations have considerable conservation potential but their chronically poor economic management and heavy industry bias will not enable savings comparable with the OECD performance. And the extremes for the poor nations are exceedingly unlikely to sink below the currently so inadequate levels, or to be doubled in less than two decades.

If forced to choose a single value I would opt for a total close to 400 EJ, most likely approaching it from the low side. This consumption would imply basically no change in the global per capita mean but if this stagnation resulted from a relatively greater drop in the rich world's fuel use there would be room for a modest growth in the poor world's consumption. Efficiency improvement and lowering of energy content in each unit of the world product would be essential to reach this goal which would, the saddest truth in this forecast, still leave most of the people on this planet outside the realm of the 'good life' as defined earlier in this chapter.

As for the breakdown of this aggregate in terms of sources I do not envisage any fundamental shifts. Of the three leading coal-mining nations—the United States, USSR, and China—the first two will not, almost certainly, reach their long-term output objectives repeatedly announced during the 1970s, and it is easy to envisage only moderate increases or even the stagnation which marked the most recent past.

On the other hand, China, where coal supplies 70 per cent of primary commercial energy, is expanding its output rather vigorously—as it must. To cope with an extra quarter billion people during the next generation, and to raise the low living standards of today, the Chinese will need, even with widespread energy conservation, considerably higher fuels output and their rich resources present no limit in this respect. Other major coal producing nations, such as Poland, South Africa, India, and Czechoslovakia, where coal still accounts for a large share of primary energy use, may not expand their output much but they will not reduce it substantially either. Consequently, we will not definitely see doubling of global coal output by the year 2000 prophesied by many until recently, but a gradual expansion of coal extraction is certain.

Global crude oil output had been falling between 1979 and 1983—4.6 per cent in 1980, 6.2 per cent in 1981, 4.6 per cent in 1982, 0.8 per cent in 1983—and the mood of the moment was that it will do so for years (if not decades). One must repeat that the consensus feelings at the time immediately preceding that fall were that the demand for oil would keep rising so rapidly that global shortages would develop by the mid-1980s and that even the Soviets, the world's top producer, would start looking for imports by the mid-1980s. This contrast of the consensus of 1977-9 with that of 1982-4, the rapidity of the unanticipated change, speaks volumes about the impossibility of forecasting such particulars even less than five years ahead.

As the OECD energy savings and economic slow-down have been largely responsible for these rapid falls in oil demand I just have to repeat what I stressed earlier: such strong trends cannot last very long. The continuation of an average annual 5 per cent decline would bring the global output to 40 per cent of its current, already depressed, level by the year 2000, a clear impossibility. Indeed, in 1984, the worldwide crude oil flow rose again (by 2.3 per cent), only to dip slightly once more in 1985. The Soviets, defying many Western predictions, have managed modest annual increases to their huge output (now 22 per cent of the global total) every year until 1983, but their output dropped slightly for the second successive year in 1985. Still, they are looking ahead for further minor growth.

The Chinese are using (so far with little reward) the skills of many Western and Japanese companies to open up what is generally believed to be one of the last most promising easily workable offshore oil provinces—the South China Sea. With lower crude oil prices exploratory drilling around the world has been well below the peak levels of the mid-to late-1970s, but the multibillion efforts in thousands of far-flung locations continue. And, a fact so often forgotten in the West where savings opportunities are so obviously large, potential unsatisfied demand throughout the poor world remains huge, and lower prices and brighter economic performance may easily lead to much higher imports once again and even by rich nations. In sum, to relegate crude oil to a 'has been' resource category is an inexcusable error of mistaking some surprisingly rapid rates of short-term changes for irremovable long-term trends.

Similarly, natural gas has assured long-term prospects for gradual expansion, as does hydroelectricity generation with huge new plants under construction from China to Honduras. As noted earlier in this chapter, the nuclear future in North America looks at best very uncertain, at worst decidedly unpromising but, once again, these perspectives cannot be applied universally when the USSR, China, Japan, and many European nations are going ahead with numerous projects.

Consequently, the established energy sources and technologies will continue to dominate the markets, especially as it is virtually certain that the relatively rapid growth rates in energy consumption which marked the three decades between 1950 and 1980 will not be replicated during the next generation. As a result there will be no pressing necessity for an accelerated development of either non-conventional fossil fuel sources and technologies (oil shales, tar sands, coal gasifica-tion), or for a large-scale introduction of several far from commercialized renewable sources (central solar electricity generation, medium-to large-scale wind generators, liquid fuels from biomass) previously seen by many enthusiasts as absolutely essential to cover the energy needs of the next few decades.

Currently under-used extraction and conversion capacities can deliver about 320 EJ of fossil fuels and primary electricity so the mere maintenance of this level could satisfy the minimum forecast for the year 2000. Going to 400 EJ annually would necessitate annual growth of a mere 1.25 per cent between 1983 and 2000, an expansion which would present little technological challenge and would pose no resource constraints. Current resource breakdown of the primary supply (roughly one third coal, three-fifths hydrocarbons, the rest hydro and nuclear electricity) could remain very similar. Only the highest case, requiring an annual growth rate of over 2 per cent between 1983 and 2000, would be most likely to call for important interfuel shifts, but the fossil fuels would dominate it as much as the previous cases. Allowing 10 per cent, or 50 EJ, for all renewable energy contributions appears to me to be a liberal estimate.

This would mean that the current annual combustion of fossil fuels may, in the case of extreme expansion, increase by about 50 per cent within two decades—or that it may remain virtually static, with 20 per cent growth being my most likely estimate. Clearly, two fundamental properties of complex systems are being increasingly felt in global energy development: the impossibility of repeated high-growth spells with very large absolute outputs, and the inertia of costly infrastructures. The first one excludes any possibility of global energy consumption doubling, or near-doubling, by the year 2000; the second one limits any fast changes to the existing output-use patterns.

Wolf Häfele (leader of the IIASA energy study) and Amory Lovins, the two extremists of the global energy debate sharing a belief in a rather quick technical fix, albeit from different directions, are wrong. At the beginning of the next century we shall be living neither in the world of burgeoning synthetic fuels and fast breeder reactors with nearly doubled energy consumption, nor in one powered by tiny gizmos on the roofs and in the backyards and using impressively much less energy than now. Gadgets of both categories will be around but—it is an unglamorous forecast—the most, or the least, surprising fact will be how the everyday energy supply realities of the year 2005 will resemble those of 1985.

3

Feeding a Planet

The servants who belong to thee
Come with the dinner things;
They are bringing beer of every kind,
With all manner of bread,
Flowers of yesterday and of today,
And all kinds of refreshing fruit.

From Turin papyrus (323–330 BC)

For those of means—be it a beautiful girl in the ancient Egyptian poem whose servants hastened to please her and her lover under a sycamore tree laden with ruby fruit, or her more recent (and perhaps less enchanting) counterparts at the royal and princely courts and in rich merchant houses—food was always there, not just enough of something but plenty of astonishing choices. And so while Chinese peasants were eating monotonous millet and rice gruels for millenia, their rulers were indulging in such exquisite combinations as steamed eels with eight treasures, stewed turtle with chrysanthemum, or breast of smoked duck with swallows' nests. Now the Chinese peasants eat better but few of the 800 million have ever seen the everyday meals of Jiang Qing, Mao's fourth wife and would-be empress, who enjoyed crabmeat dishes and different broths 'with floating exotic seafood and fungi cut to resemble wilted flowers . . .'.

There is nothing new under the sun: indulgence for the rich, bare subsistence, frequent malnutrition, recurrent famine for the poor. But a coincidence of developments during the 1950s and 1960s did turn the global food prospects in an unmistakeably optimistic direction. After World II basic inputs into modern farming—machinery, fuel, fertilizers, pesticides—became for the first time available not only in large quantity but also at gradually declining real prices, and their diffusion did not stop at the borders of rich nations as poor countries, taking advantage of newly developed rice and wheat cultivars (these being yet another unprecedented advance of the post-1945 period) started to use increasing amounts of farming chemicals, refined fuels, and electricity for field machines and irrigation.

The results of this new industrialized farming, higher energy subsidies, and high-yielding seeds were impressive: between 1950 and 1970 the global harvest of cereals doubled, as did the production of oilseeds; harvests of pulses grew 2.4 times, sugar output was up 2.1 times, and among the principal foodstuffs only tuber production ended just shy of a two-fold increase (Fig. 3.1). During the same time the world's population grew by 50 per cent which meant that the average per capita output of food rose by no less than 33 per cent—an impressive and unprecedented change in just one generation.

Naturally, continental, national, regional disparities persevered, malnutrition was not eliminated, and brief famines did return on several occasions in various poor countries—but the course was so evidently upwards that it was foreseen that the 1970s and beyond would become yet another instalment of the success story: output would go up everywhere (although obviously at different rates) and, should bad weather cause national or regional shortages, abundant grain reserves would be brought in to cover the emergency needs.

Then in the early 1970s another set of coincidences appeared to shatter these expectations. Drought in Africa's Sahel ravaged the crops and herds from Senegal to Ethiopia, poor harvests were gathered elsewhere in Africa (Tanzania, Zambia, Rwanda, Burundi) and in Asia (Nepal, Burma, Bangladesh) where India's 1972-3

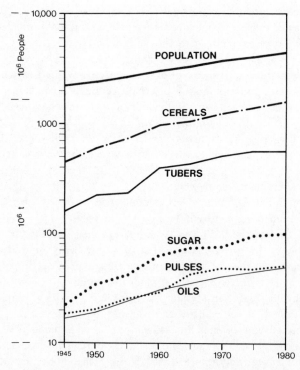

Fig. 3.1 Since 1945 growth rates of worldwide production of all principal foodstuffs have been running ahead of the natural population increase. Tubers and pulses are the only exceptions, victims of the general preference for grains, sugar and oils.

grain output fell 11 per cent in comparison with the 1970–1 record. The Soviet grain crop was just 161 million tonnes in 1972, 13 million less than in 1971, and about 20 million below the long-term trend. Russians, who rode-out the previous shortfalls (in 1963 and 1965) by reducing grain feeding to livestock, decided to make up the entire 1972 deficit by imports: during the 1970–1 crop year they were a net exporter of 7.5 million tonnes of grain, a year later they were a net importer of just 1 million tonnes while in 1972–3 they bought 19.6 million tonnes.

These jolts brought the carryover grain stocks in the United States, Canada, and Australia (the three top exporters) from 92.1 million tonnes in 1972–3 to 57.9 million tonnes a year later, a 20-year low which was promptly lowered to 45.7 million tonnes in the 1973–4 crop year. Markets responded predictably with corn prices up about 2.5 times, wheat prices peaking in early 1974 at more than 4 times the 1972 level, and soybeans going up in 1973 to almost 4 times their 1970 cost.

At a time when the world had only started to cope with the global consequences of the first round of OPEC's drastic crude oil price increases a new crisis was born. Headlines such as 'The world food crisis', 'The coming world struggle for food' and 'The empty breadbasket' became commonplace as photographs of Sahelian villagers collecting grain strewn over the barren ground from burst sacks dropped by aeroplanes and of emaciated children with swollen tummies filled the pages of Western journals and TV screens. In the second week of November 1974, as *Time*, the world's most widely read weekly, carried a special section on the new global crisis, the first World Food Conference, sponsored by the United Nations, gathered in Rome.

Analogies to gloomy energy prospects were unmistakeable: forecasters agreed that global grain reserves would remain low and food prices high, and rising in the years ahead when global food scarcity could become chronic. With the return of rains to the Sahel, recovery of Indian harvests, more orderly arrangements for huge Soviet grain purchases, and increased American outputs the acute phase of global food concerns passed during the late 1970s, but the long-term outlook remained decidedly pessimistic as hundreds of papers and scores of books spread, more or less pointedly, Malthusian tales of woe.

One might choose many examples but Lester R. Brown's writings, always so persistent in their gloomy tones, have been certainly among the best embodiments of the worried outlook, of recurrent bad news. I will summarize his main conclusions and quote bits from the food chapter of *Building a Sustainable Society*, a book which came out in 1981 and which in so many ways epitomized the often faulty consensus views of the 1970s about energy, food and environment.

Brown's portrayal of the global food situation was anguished on three main accounts. First, he saw the middle of the twentieth century as 'a watershed in the evolution of world agriculture' because, from 1950 on, most food output increases came from rising yields rather than from expanding the cultivated area, a change which made modern farming with its synthetic fertilizers, machines, and irrigation increasingly energy intensive. In logical extension, the shrinking relative availability of arable land will lead to yet higher energy intensity with food production 'claiming a growing share of the world's scarce energy resources by the year 2000.'

Second, while the world had traditionally two great reserves providing 'a cushion against any imaginable disaster'—grain carry-over stocks and fallow American cropland which could start producing within a year—it entered the 1980s' with only the first protection: 'For the first time in a generation, there is no cropland idle under U.S. farm programs. The loss of this reserve, which provided security for the entire world, may be permanent.'

The third major worry was the dramatic transformation of inter-continental food flows as 'more than 100 countries rely on North American grain. The worldwide movement of countries from export to import status is a much travelled one-way street. The reasons vary, but the tide is strong. No country has gone against it since World War II. ... Not only are more countries joining the legions of importers, but the degree of reliance on outside supplies is growing.'

These three undeniable changes would seem to be an irresistible combination of deteriorating trends but one must keep in mind the two often forgotten simplicities which I tried to illustrate with many examples of the tangled energy story: no trends last forever, and closer examination of their reasons is a must before judging if they are truly worrisome. Global crude oil prices could not continue to rise and the reasons for perceiving the omnipotence of the price-setting organization were Western misunderstandings rather than OPEC realities. Analogously, standard interpretations of the world food troubles, persuasive as they sound (no less than did the talk of $100 barrel by 1985 in 1980) require much re-evaluation.

The first argument, about a growing share of scarce energy resources going into energy-intensive farming, is dubious on two grounds. The preceding chapter tried to demonstrate that the resources are not so scarce, and in the fifth chapter a more detailed outline of the energy costs of modern farming will show that the burden is surprisingly small. Just a brief illustration here: during the 1970s, production of grain corn, by far the largest American crop grown in a totally mechanized system with heavy fertilization, widespread irrigation, and almost universal drying, required between one-fifth and one-quarter of the country's total agricultural energy consumption.

In turn, American farming consumed no more than about 3 per cent of the nationwide energy use so that grain corn accounted at most for a mere 0.75 per cent of the country's aggregate energy consumption, less than the energy needed to produce the nation's throwaway steel and aluminium beer and soft drink cans! (Yet *The Futurist* wrote in 1974 that Brown 'refuses to own an automobile and uses public transportation, *so that more energy can go into food production*'. The italics are mine, no comments required.)

Even this single example illustrates abundantly how relatively low are the demands made by modern farming on a country's energy supply in comparison with numerous wasteful uses which contribute little or nothing at all to economic and social well-being. Even in countries much poorer than the United States, intensive farming claims a fraction of the energy which could be easily conserved elsewhere in the nation's economy by better management and technical improvements of inefficient industries. And, as will be shown later (section 5.3.1), considerable energy savings can also be made in just about every current farming practice, further decreasing the undoubted but far from uncomfortable energy dependency.

Brown's second worry about the loss of American cropland left idle which 'may be permanent' is a perfect illustration of the powerful spell cast by strong trends. Indeed, after idling an average of about 20 million ha during the 1960s, American farmers returned all of this land into production between 1973 and 1978 and then, after a slight temporary drop in 1979, they added more, to surpass all the previous cropland records in this century (Fig. 3.2). But this rapid expansion combined with good weather to produce huge grain surpluses and the record high was immediately followed by an even faster return to unprecedented idling: in 1983 the United States government through its payment in kind programme induced farmers not to grow crops on about 33 million ha, or almost one-fourth of the cultivated area, so the country had less land under production than at any time during this century (Fig. 3.2)!

In 1976, at the height of the 'food crisis', Brown calculated an index of world food security by adding reserve stocks of grain and grain equivalent of fallow United States cropland, and then expressing these total reserves as a share of global grain consumption: in 1973 this fraction fell to a mere 10 per cent (from the high of 26 per cent in 1961 and the mean of 18 per cent during the late 1960s) as the stocks stood at 105 million tonnes and fallow cropland could contribute just about 20 million tonnes.

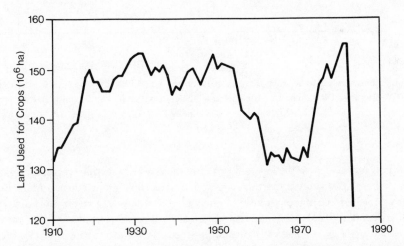

Fig. 3.2 American cropland in productive use went from the lows of the late 1960s to the record highs of the early 1980s. These were immediately followed by a precipitous drop in 1983, a development disproving the predictions about the permanent loss of this food reserve.

A decade later, after the 1983 harvest, the world had stocks of 120 million tonnes of wheat alone (as a share of demand, the highest level in 30 years), American surplus of grains reached 130 million tonnes, Chinese and Indians had problems with finding storage for their surpluses, and the idle American cropland could have produced at least another 120 million tonnes. Moreover, all this abundant grain could be bought in 1983 at prices cheaper in real terms than they were in 1973 (Fig. 3.3). So much for the prognoses of chronic world food insecurity, permanently lost farmland reserves, and inexorably rising food prices!

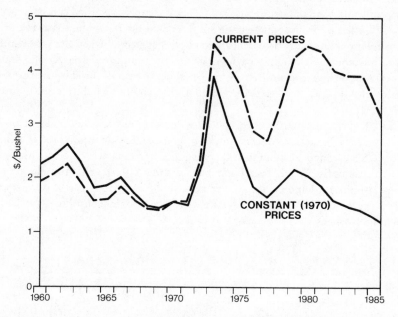

Fig. 3.3 Grain prices took off much like the crude oil in the early 1970s and, in yet another coincidence, they had another sharp increase in the late 1970s. But the slips in the mid-1970s and after 1980 mean that the cost of a bushel of wheat in constant money value was less in 1985 than at any time during the previous 25 years.

This leaves us with the last, and certainly the most disquieting worry, the rising number of countries importing larger shares of their grain consumption. But searching for the reasons of this 'much travelled one-way street' one quickly discovers that this shift has been largely a matter of preference, of affluence, not of necessity, not even a need to eat well. Between 1980–4 about half of all exported grain went (as it did a generation ago, in the early 1960s, when the total cereal imports averaged less than half of the early 1980s level) to rich industrialized countries, with Europe taking about as much as the USSR. All these nations do not import annually about 100 million tonnes of grain to stave off malnutrition and hunger—but to feed it to their animals in order to keep eating unprecedented quantities of meat. Annual per capita averages of meat production have been lately over 130 kg in the Netherlands, nearly 120 kg in East Germany, about 100 kg in France and Czechoslovakia, 80 kg in Bulgaria and Romania, and over 70 kg in West Germany and Poland.

Is it not ridiculous to worry about the world food prospects when half of all grain imports go to maintain, and even to increase the inordinately high carnivorous urges of rich populations which have nothing to do with proper nutrition, health, and longevity? After all, to give just one example, East German life expectation at birth is lower than in Japan where less than 30 kg of meat are eaten each year per capita. And is it not obvious to ask if all these countries would have got into this senseless importing habit if the North American grain had not become so cheaply and so abundantly available in the first place, and if the exporting countries had not gone out of their way to subsidize and to promote sales of those feeds?

Since the 1960s several better-off industrializing countries have been moving relatively fast towards the grain consumption pattern characteristic of the rich nations. Taiwan is the best example: a generation ago the island's farmers were feeding a mere 1 per cent of grain to livestock, by 1982 the share passed 60 per cent. Although the island has long been self-sufficient in rice (in fact it exports it), its new meat tastes require large feed grain imports—but the vibrant economy can afford them. Other richer poor countries (those with annual per capita GDP of more than $1000) have been moving in the same direction, but this still leaves plenty of imports (about a third of the total) by the truly poor ones who buy the grain for food not for feed.

These imports are relatively small—for example in Africa nine-tenths of the people live in countries importing less than 10 per cent of their cereal consumption, and the total annual grain shipments to the continent have been recently just a bit more than Japan alone buys—but they are taking away precious foreign exchange which should be put to better use. But this growing dependence on imports cannot be ascribed either to worsening environmental conditions (droughts, floods, and pest outbreaks have not become uncommonly frequent during the recent decades) or to the near exhaustion of productive potential in the affected countries. Why were African countries producing less food per capita in the early 1980s than a generation ago, at the time they were gaining their political independence? Why during the 1970s, when Asian per capita food production rose by an average of 10 per cent, did the food output in sub-Saharan Africa fall by about the same amount—and then decline by as much again just between 1980 and 1983 (Fig. 3.4)?

Obvious 'natural' explanations cannot explain this huge gap. True, African populations have been growing faster, but even if Asian countries had grown equally fast the continent's average per capita food supply would have still been rising. True, the Sahelian drought of 1972-4, and a new rainless spell between 1982 and 1984, which has again affected the Sahel and much of eastern and southern Africa, had to hurt—but many Asian countries were similarly affected in the mid-1970s and the early 1980s (for example, in early 1980 drought affected one-fifth of

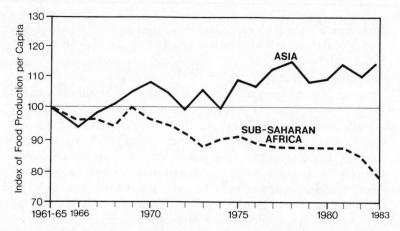

Fig. 3.4 After similar poor performances in the early 1960s, food production totals in Asia and sub-Saharan Africa started to drift apart and by the early 1980s the gap was a stunning 30 points compared with the 1961–65 mean!

China's crop fields) but the consequence was just a temporary downturn on a generally ascending trend, not an almost uniform collapse.

The names of the African countries that have experienced the worst food production declines will give a ready clue to the continent's dismal performance. Between 1970 and 1980 Angola's per capita food output fell by 43 per cent, Ghana's 28, Mozambique's 33, Uganda's 30, Ethiopia's 15—but from an already very low level, and since 1980 it has plummeted by another at least 30 per cent, the direction in which Sudan has been heading since 1982.

The common points are clear: the chronic political instability of African nations, their endless *coups d'état*, civil, border, guerilla, and tribal warfare, and their dismal economic mismanagement reflected in disastrous agricultural policies and dubious developmental goals favouring city-dwellers at the expense of farmers, provide the real explanation of the shocking fact that people in four-fifths of all African countries now eat less than a generation ago, a record of unparalleled decline, and that in about a third of them starvation is on the rise. This brings the second essential question: are ever higher regular imports of staple cereals (less than 3 million tonnes in the early 1960s, over 4 million tonnes in 1970, just over 20 million tonnes in the early 1980s), and other foodstuffs, and increasing emergency aid, viable solutions to this painful degradation?

Food aid for famine areas is essential for obvious humanitarian reasons—but giving food outside emergencies is a dubious policy. It is a clear deterrent of proper management and intensified production. I am writing this in the midst of yet another Ethiopian famine about whose man-made roots there can be no doubt. Since the overthrow of Haile Selassie's medieval autocracy in 1974, the communist colonels have waged wars in the east and south against Somali nationalists, and in the north against separatist movements in Eritrea and Tigre (spending at least one third of the country's budget), set up artificially low prices for all staple grains, and established large state-owned farms which are shoddily run but receive most of the available modern inputs.

Food output, inevitably, started to fall, and the absence of adequate grain stores (discouraged as capitalist hoarding), and a prolonged drought crippled this deliberately weakened system to the point of mass starvation. The solution to this tragedy is not in North American and European food donations but in removing the criminal ruling system. But at least one poor African country of leftist leanings has done relatively well. In Robert Mugabe's Zimbabwe the price of corn, the country's main crop, has been rising faster than consumer prices since independence, loans to small farmers are increasingly available, stockpiling is taken seriously, and food subsidies for city dwellers are being reduced. The results: a 30 per cent rise in farm output in spite of a three-year drought between 1982 and 1984!

On a much grander scale Indian and Chinese farming are the two best examples of what can be done. Both countries have had their harvests repeatedly influenced by serious droughts and floods, China also by recurrent political turmoils, and at times they needed a good deal of outside grain: after poor monsoons in 1965 and 1966, one-fifth of the American wheat crop went to India to prevent famine, and China entered the world grain market following the collapse of the Great Leap Forward in the early 1960s, and became a steady large importer. Fluctuations in domestic grain production, natural and man-made, have been considerable but the recent changes are most encouraging.

India's green revolution, so often criticized for favouring wealthy farmers, appeared stalled during the late 1970s but the early 1980s have shown nice impressive gains. Punjab, with just 2.5 per cent of the country's population produces over one-fifth of all wheat, a feat achieved only by uniformly high performance and not due just to rich farmers (the state's average wheat yields are now approaching 3 t/ha, well ahead of the American mean). More importantly, rice harvests are rising in Orissa, Bihar, West Bengal, and Uttar Pradesh.

Government subsidies helped to push up fertilizer consumption, and higher guaranteed grain prices, the spread of high-yielding varieties, better distribution of commercial seeds, massive give-aways of modernization kits to smallholders (containing high-yielding seeds, fertilizers, and pesticides), and greater availability of extension advice, are finally pulling Indian rice yields up—the 1985 summer crop was a record 58.5 million tonnes, 27 per cent above the 1982–3 level, but the harvests still have a long way to go to catch up with those of Indonesia and China. Indian food grain stocks surpassed 21 millikon tonnes in summer 1984, creating a shortage of storage space—the problem recently encountered to an even higher degree in China, where the *de facto* privatization of farming and greater availability of inputs have led to unbroken years of record harvests between 1981 and 1984—in spite of serious droughts and floods affecting some key grain-growing provinces.

Although the area sown to grains has declined by about 4 per cent since the late 1970s, 1983 yields of wheat were 50 per cent, and those of rice 23 per cent up compared with 1980, and the total 1983 harvest of wheat, rice, and coarse grains was nearly 25 per cent above the 1980 level—and the top Chinese officials took to reassuring worried American and Canadian grain traders that China will still buy some of their corn, wheat, and soybeans—although it does not want any more long-term import deals for large amounts of grain.

Undeniably, world food supply is not a topic for complacent comment but suggestions that the developments since the early 1970s are clear signs of the world rapidly approaching the stage where demands of growing populations of the poor countries on one side, and unceasing demands for higher meat consumption in the rich nations on the other, will combine to surpass the capacity of global grain production are obviously false. Parallels with the global energy situation are self-evident. Higher prices worked wonders for energy conservation and they lowered energy intensities around the world in a most impressive fashion (see section 2.3.3).

Higher prices paid to farmers in poor countries would, together with earnest efforts to increase the availability of basic inputs, produce bigger harvests. Recent Chinese experience is a perfect illustration of the rapid turn around achievable in a very short time. As long as farmers are growing food just for their families and not for the market—a thing to be expected where urban food prices are controlled and subsidized by rigid governments, be they in African countries or in Eastern Europe—their countries will have to keep importing.

One more perfect example: in Communist Europe the contrast between grain-exporting Hungary and grain-importing Czechoslovakia or Poland owes virtually everything to smarter management aiming to produce for the market—not to superior natural conditions. Later in this chapter I will examine the potential for higher grain harvests (in section 3.3.1) but it is certain that, even with just the current know-how, the global yields have decades of increases ahead—providing the right political and managerial decisions are taken and steadily implemented.

Let it be stated quite plainly: vagaries of climate (yields will always fluctuate, but good management will reduce the ranges of such inevitable departures from long-term trends), shortages of arable land (relatively severe in some nations, none elsewhere, and a good deal of abandoned oddlands, and inexcusable and unnecessary encroachments on farmlands everywhere), inadequate farming inputs (as often the unavailability of fertilizers and pesticides where and when they are needed as their absolute shortage, a result of inadequate transportation and distribution infrastructures), and continued high population growth rates (declining impressively in some poor countries, still uncomfortably high in most Latin American and African nations) make the task of improved per capita food supply in poor countries a difficult challenge but not an impossibility.

All the real obstacles are elsewhere: in inept governments, in misplaced expenditures (is it not notable that the world's poorest 30 countries now spend, on the average, a *larger* percentage of their central government budgets on military expenditure than the world's 20 richest industrialized states?), and in too ready reliance on food imports. Perhaps nothing could exemplify these blunders better than Nigeria's squandering of its fabulous crude oil export revenues (more than $(1980) 100 billion) during the 1970s with the military taking about two-fifths of the state budget, corruption of phenomenal proportions, tribal and political strife, the rush into the already overburdened cities, the withering of agriculture, a 10 per cent drop in per capita food output in a decade, and rising food imports.

In the rich Western countries where governments are either bribing farmers not to grow grain on part of their land (the American way of propping up world prices at higher levels), or guaranteeing artificially high prices and then paying the difference between them and world prices for exportable surpluses (the European Economic Community's way of depressing world prices), meat consumption appears to be finally levelling out—but these governmental subsidies will almost certainly continue, although in somewhat reduced ways. The result is a more or less stable demand for grain and continuing large exportable surpluses in the United States, Canada, Australia, and France.

For the last time in this section I will note the analogy with global energy: there is no justification whatsoever for any fears of imminent, near term (less than a decade), or mid-term (20-5 years), global food shortages. Staple grains—the principal (direct or indirect) food energy and protein carriers—will, most likely, be produced in great surpluses by the current large exporters, while the shortages may get even worse in many poor countries, a contrast of affluence and poverty no less pervasive than in the distribution of global energy use.

But as recent Indian and Chinese efforts show, the widening of this gap is not inevitable. A combination of a large unrealized production potential in poor countries, and of no smaller opportunities to lower grain consumption in rich nations, could result in better nutrition everywhere—and still leave comfortable reserves for the future.

Further belabouring of the world food prospects would add little to these conclusions at this stage. I will return to important aspects of this fascinating topic later in this chapter when dealing with prospects for higher cereal yields and with old and new ways of extending our food harvests, then again in the next chapter (discussing the losses of farmland in section 4.2.1, and the preservation of genetic diversity in section 4.2.3), and in chapter five. But I will devote most of this chapter to the consumption rather than to the production of food.

Why? Because practically all recent treatises on the world food 'problem' ignore the huge gaps existing in our knowledge of nutrition and concentrate on learned discourses on how to raise cereal yield by transferring nitrogen fixing genes, or transferring appropriate technology to poor nations. A reader eager to pursue these far-off dreams (in the first instance), and often dubious advice (in the second), will find no shortage of writings on these and kindred topics. Instead, I will make eating a (logical, I think) centrepiece of this chapter, concentrating first on the lasting gaps in our nutritional knowledge, and then on the gap between rich and poor diets.

3.1 Human food needs: certainties and ignorance

> We believe that the energy requirements of man and his balance of intake and
> expenditure are not known.
>
> J. V. G. A. Durnin, O. G. Edholm, D. S. Miller and J. C. Waterlow (1972)

This resolute statement was not written lightly: it appeared in a letter to *Nature* in
1972 and the first undersigned scientist was one of the world's leading authorities
on human energy needs, the last one the country's senior expert on infant nutrition,
all four among the most distinguished British nutritionists. The years that have
elapsed since the letter's publication have not changed its conclusion—although
even a brief scan of the popular, and popular-scientific, press will reveal no shortage
of recommendations of what to eat and how much.

The public eagerly tries out all the latest tips, food industries scramble to meet
the new demands or to stave off the collapse of the old—but the substance behind
the words is invariably meagre and nebulous, or ungeneralizable and overly
specific. Michael V. Tracey summarized this perfectly in his excellent contribution
to *The Encyclopedia of Ignorance*: 'It is clear that, contrary to the general impression,
the emperor of human nutrition has no clothes but at least he knows what clothes
he should have, for his subjects have by now produced an almost complete list of
the raiment, haberdashery and accessories necessary—but they cannot be worn with
conviction while there is continuing dispute concerning the size and cut of every
item in the list.'

Some of these disagreements seem minor on the level of per capita daily intakes
but they make a critical difference on a national level. For example, when it is
related later how successive international expert committees reduced their
recommendations of average energy intakes, and how complex such determinations on a large-scale level may be, it will be appreciated that a difference of
100 kcal a day per capita is easily within the range of existing uncertainty about the
average food requirements of a population; for a mean value of, say, 2400 kcal this
addition of 100 kcal a day represents just 4 per cent—but applied to China's
population of just over one billion this increment would necessitate an extra output
of 10 million tonnes of milled rice.

Other disputes involve differences ranging over an order of magnitude, and the
case of vitamin C, where the recommendations range from less than 100 mg to
several grams, has been perhaps the most publicized example. But at least, as Tracey
notes, the list is almost complete and arguments such as whether vitamin E should
be considered an essential nutrient do not involve a dozen other constituents of
normal diet which all experts agree must be present but whose recommended
intakes, especially for some age-specific quantities, have undergone considerable
changes over time. I will review historical developments of these recommendations
for total energy and protein intake in the next section (3.1.1), and afterwards will
focus on a few major disputes involving lipids and some vitamins. Before doing so,

however, a few observations on the reasons for our surprisingly large ignorance of food needs are in order.

That the science of human nutrition is relatively young is perhaps the least important reason; certainly, compared to other disciplines, and even to nutritional studies of animals, the field is rather fresh—but no more, and much less so, than other realms of bioscience with their spectacular practical advances (from crop genetics to dung microbiology). Comparisons with some two centuries of experimental work on the nutrition of domestic animals reveals the first real difficulty: the lack of easily and precisely definable goals.

In pigs, steers, or broilers the nutritional goals are clear as the performance is judged by the amount of meat gained in a given time (and also for a given cost—to minimize the price). For adult humans, whose body weights should not be increasing any more such a criterion is obviously useless—but it is no less inappropriate even for a growing infant. Although in children the normal criterion of nutritional adequacy is, indeed, the rate of growth (translated in the pediatrician's tables and graphs of expected weights and heights) nobody would claim that it is a maximum rate of weight gain (which may be, after all, largely an accumulation of fat and water), or the fastest tissue formation that is the best long-run course for the child.

Obviously, babies do not have to be finished like piglets in half a year and, while the piglets are bred for uniform performance, with children a great latitude for individuality must remain. For example, the most commonly used percentiles of infant and child growth in North America, derived from the Longitudinal Studies of Child Health and Development at the Harvard School of Public Health, show that for infant girls at the age of one year the weights betweeen the 10th and 90th percentile range from 8.2 to 11.1 kg. A year later the range widens to 10.7–14.3 kg, and by the age of 12 years it is 31.0–54.5 kg, a nearly two-fold spread and clearly too broad a base on which to make nutritional recommendations (Fig. 3.5).

Obviously, people are born with highly individualzed nutritional needs, and these inequalities may further increase with their different eating habits and activity levels. Elsie Widdowson's fundamental question—'Why can one person live on half the calories of another, and yet remain a perfectly efficient physical machine?'—has yet to be answered two generations after its formulation. Consequently, the recommended dietary intakes, usually established by lengthy discussions in expert committees and often representing highly unsatisfactory compromises, are just statistical abstractions and hence they are only rarely applicable to individuals; moreover, in some cases they may be misleading or nearly useless even in their abstract form. Such is, perhaps most prominently, the situation with human protein requirements when old concepts became untenable but rigorous new ones are not yet available.

These uncertainties could be greatly reduced if large-scale and long-term nutritional trials were possible but, in contrast to short-lived laboratory animals where several generations can be reared under rigorously supervised conditions, no controlled scientific experiments in human nutrition have surpassed a tiny fraction

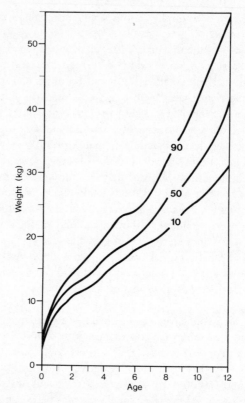

Fig. 3.5 The variability of human bodies is here illustrated by tenth, fiftieth and ninetieth percentiles of weight for American girls. The curves are based on extensive statistics collected by the Harvard School of Public Health.

(0.1 to 1 per cent) of man's normal life expectation. There is no shortage of students to volunteer for a few weeks living on strange diets, but no chance of ever accounting for all nutrients, and all activities, in the lives of a 1000 people over the first two decades of their development.

For this reason alone human nutrition must remain a primarily observational rather than an experimental science—and hence always greatly handicapped in comparison with those studies where the lack of intrinsic individual variability and easy replicability of experiments allow for precise and meaningfully transferable findings. The contrast with the plant, animal, and food sciences, where we can grow wheat with specific potein content, breed animals for lean meat, or prepare repeatedly identical batches of fermenting cultures, is obvious.

Fundamental gaps in our understanding of human nutrition, whose most general goal is maintenance of a healthy and vigorous life, are seen not only in the disagreements and reversals concerning protein needs, clinical implications of cholesterol intake, risks of high lipid consumption, or benefits of ascorbic acid megadoses (all points to be discussed in the next two sections), but perhaps even more basically, and certainly more frustratingly, in our inability to deal effectively with a new major nutritional problem, obesity of large sections of affluent Western populations.

Comparison of per capita food energy availabilities (section 3.2.2) with even the most liberal assumptions of needs (section 3.1.1) shows incontrovertibly that during the last few decades in practically all industrialized nations of Europe (be it Belgium or Bulgaria), in the USSR, United States, Canada, Australia, and New Zealand—countries containing about a fifth of mankind—average food consumption has been well above need, a condition unprecedented in history. As a result obesity has been spreading, but the common-sense, intuitive link between excessive energy intake and storage of body fat has proved to be an untenable generalization.

If all surpluses were turned into adipose tissue obesity would be a prevailing condition throughout the rich world and, conversely, restrictions of food intake should lead to an effective control of weight gain. In reality, prevalence of obesity shows some fascinatingly counter-intuitive correlations with affluence (its higher incidence among relatively poorer segments of Western countries whose excess energy intake may be significantly smaller than for the better-off groups), and the very existence of scores of highly-touted miraculous weight-loss diets is the best proof of the dismal inefficacy of dieting.

More than half a century ago E. F. Terroine noted that we have far more information about the nutritional needs of young pigs than of children, presumably owing to the former creatures' greater economic importance. This harsh and sarcastic judgment could still be endorsed today—in spite of the enormous advances in understanding—for many critical concerns, from nitrogen balances to metal deficiencies. New uncertainties arising from better knowledge are a common property of scientific progress but in the case of human nutrition it appears that we have been moving farther away from an understanding translatable into practical everyday guidelines.

The following sections will survey some of these controversies but it must be kept in mind that they will touch only some essentials of just a few (although perhaps the most prominent) continuing efforts to reach an elusive comprehension of human nutritional requirements.

3.1.1 *Energy and proteins*

> 'What do we need for optimum health, are we being adequately fed?' I suppose the short answer is: 'Since people tend to be living somewhat longer we cannot be doing all that badly.'
>
> Arnold E. Bender at Nestlé Nutrition Research Symposium (1982)

Variables are easy to define but the actual requirements of both individuals and nations are surprisingly difficult to calculate as human food energy needs are determined by age, body weight and composition, by physical activity, by climate, and by individual metabolism. A small group of volunteers can be weighed easily, measured for subcutaneous fat, have the exhaled air analysed for O_2 and CO_2 to find the energy expenditures of particular activities at different temperatures, and have all the food intakes and body wastes carefully noted and assessed; experiments of this kind do not last very long, but with healthy individuals and steady regimens one can make sensible conclusions about the adequacy of a diet and recommend with confidence average daily intakes and their make-up for situations closely resembling those of the experiments.

But how can one say that a nation has enough to eat? Working with averages is unavoidable but undesirable as the means may mislead more than they may illuminate and in several key instances they simply make no sense at all and the whole distribution must be known with fair accuracy to calculate the nutritional requirements: most notably, this is the case with the age and sex distribution of the population. For rich countries these statistics are readily available, but in many poor nations extrapolations must be made from older, unreliable censuses.

The next requirement is even more difficult to satisfy: body size and composition affect both resting (basal) metabolism and activity expenditures, but reasonably reliable anthropometric data at the national level are available only for most of the rich countries. They show much similarity among these countries and a great deal of individual variation, but with repeated measurements the means can be used with confidence. However, application of these averages for Latin American, African, and Asian conditions would be very misleading: a fifteen-year old Canadian boy averages 56 kg, his Nigerian counterpart weighs no more than 46 kg, an average Bangladeshi youngster just 36 kg.

Sensibly, FAO calculations of recommended food consumption link the quantity with adult weight and use appropriate multipliers to modify the child and adolescent intakes: but whilst means of 70 and 58 kg for American men and women may be very close to the actual mean, how well are Indians represented by 55 and 45 kg, and what should be the values for Zaïrians? Perhaps the only certainty is that the international reference man and woman (65 and 55 kg) have a few real-world counterparts in national means which now tend to be higher for the affluent world and still much lower for the poor one.

Finally, of the three principal variables, human activity is usually the most difficult to account for in a satisfactory manner. There is no precise categorization of tasks according to their energy requirements and variations in energy cost of a particular activity performed by different individuals, or even by the same person at different times, may be quite large. For example, Durnin and Passmore's often quoted volume on *Energy, Work and Leisure* which summarizes hundreds of specific energy expenditures lists 3.6 kcal/min as a typical energy cost in manual planting of crop fields, whilst the recent Chinese investigations put this exertion at 6.3 ± 0.6 kcal/min; and even for quiet sitting, energy expenditures for adult 65 kg men will range from as little as 0.87 kcal to 1.94 kcal/min, a 2.2-fold span.

The only practical approach in national calculations is thus to assume moderate activity as the standard exertion and adjust the needs of that part of the population that works more, or less, by applying typical correction multipliers—0.9 for people with light activities (office workers, professional occupations, and housewives in the rich countries fill this category), 1.17 for those very active (such as traditional farmers, forestry workers, miners, steelworkers), and 1.35 for the exceptionally active (in certain farming, forestry and industrial tasks rather than on an eight-hour basis).

Obviously, there are no readily available sources breaking down populations into these activity categories, and compilations based on occupational statistics could be much in error without detailed knowledge of the actual work performed: for example, in unmechanized coal-mining digging the coal with a pick, shovelling it into wagons, and erecting roof supports are all very strenuous activities requiring in excess of 6 kcal/min—but during an eight-hour shift coal-miners may spend 60 per cent of the time on the way to and from the working face, and in pauses and rests, so that the actual average exertion is just a bit over 3 kcal/min, a moderate rather than a heavy work average.

Consequently, the fundamental question of how much energy a nation needs cannot be answered, even in the best statistical circumstances, without using generalized assumptions whose cumulative errors can result in significantly misplaced averages. At a time when we have mapped parts of the Moon down to small boulders and are splicing genes to insert desirable properties into living organisms, the conclusion irritates but stays: we do not know how much food this planet needs in order that its population may lead a healthy and vigorous life. And we cannot specify this consumption level even for the rich nations with all their statistics, surveys and data bases—and not because we have not been trying: after all, the history of energy requirement calculations is now more than a century old. In 1880 Voit put the food needs of a typical German labourer at 3055 kcal; in 1895 Atwater had an average adult male, doing ten hours of moderate work, eating 3500 kcal. Lusk's recommendations in 1918 were for 3300 kcal for an average male working, again moderately, for eight hours, while the League of Nations allowed 2400 kcal for an adult, male or female, living in an ordinary way in a temperate climate, and this minimum was raised by supplements needed for manual work, so that a male working moderately for eight hours would need 3000 kcal.

During the post-World War II period the most influential recommendations have been those of the United States National Research Council (NRC) issued first in 1948 and revised in 1953—for a 79 kg male they ranged from 2400 kcal for sedentary life to 3000 kcal for moderate activity, to 4500 kcal for heavy work—but, with growing concerns about global food supply, the consensus established by four international FAO committees became the most often quoted norm. The First Committee on Calorie Requirements met in 1950 and set an energy expenditure of 3200 kcal for the reference man (25 years old, stable weight of 65 kg, living in a temperate zone, working eight hours, moderately active), and 2300 kcal for a reference woman (also 25 years, 55 kg). In 1957 the Second Committee endorsed these values.

In children energy intakes must be sufficient for both satisfactory growth and the usually high degree of activity. The earliest FAO committees retained the NRC's initial recommendation of an average of 110 kcal/kg for infants from birth up to one year, as well as for children up to 12 years, but the second (1957) reduced the intakes of 13–15-year old boys by 100 kcal to 3100 kcal, and those of 16–19-year old males by 200 kcal to 3600 kcal a day. But it was the third FAO committee, considering both energy and protein requirements in 1973, which brought changes to virtually every age category.

For infants the average was raised almost negligibly to 112 kcal/kg (with extremes of 120 kcal/kg during the first three months, and 105 kcal/kg during the ninth–eleventh month). For boys aged 10–12 the value went up by 100 kcal, while for the girls it decreased by 150 kcal, and for the 13–15-year old children there were reductions of 200 kcal for boys (to 2900) and 100 kcal for girls (to 2500 kcal). By far the most drastic decline affected male adolescents between 16–19 years of age, from 3600 to 3070 kcal, while for the adolescent women the mean was reduced by 100 kcal to 2300 kcal a day.

And the reference adults were redefined to be between 20–39 years of age, healthy, weighing 65 kg (men) and 55 kg (women), working moderately for eight hours, sleeping eight hours, and spending 4–6 hours sitting or moving around in only very light activity, and either walking or engaging in active recreation or household duties during the remaining time. These adults should be able to cover their energy requirements with 3000 and 2200 kcal, again reductions of, respectively, 200 and 100 kcal compared with the 1957 reports.

Practically all small adjustments have been simply the result of better, more representative observations of actual food intake compatible with satisfactory growth, good health, and vigour. The fourth report, issued in 1985, abandoned the reference approach entirely and expressed all energy requirements in terms of basal metabolic rates for given weights and ages, subsequently adjusted for different activity levels. Overall changes in recommended totals were minor but even they can add up to considerable quantities of food once they are used in estimation of energy requirements at the national or population level. Before illustrating this influence, however, I shall sketch the historical development of protein recommendations, a story of similarly notable changes.

Perhaps the best way to present the historical development of findings about optimal protein intake is to plot more than 100 values (for easy comparison all in g/kg/day) published between the early 1880s and the mid-1950s (Fig. 3.6). The ranges, especially at the times of critical growth requirements between 5 and 17 years, are quite considerable and there is hardly any trend in the values: for example, the range is widest for 10–12 year olds but its median value of 2.3 g was advocated by Hasse in 1882—as well as by Eppright in 1954, and the researchers in the intervening years provided figures as low as 1.4 and as high as 2.8 g.

The considerable difference for adults results from the inclusion of Rose and Leverton's minimum intake of 0.3–0.4 g/kg a day of good quality protein and the 1953 dietary recommendation by the United States National Research Council

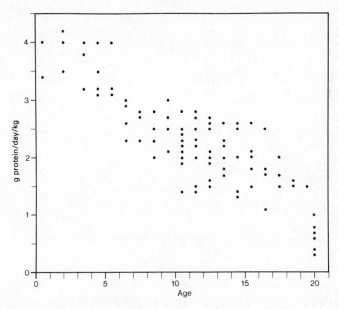

Fig. 3.6 A plot of optimal protein intake recommendations published between 1882 and 1955 illustrates well the substantial uncertainties prevailing at all age levels, and especially during childhood and teenage years.

pegged at 1.0 g. Since the mid-1950s the question of protein requirements has been addressed by several international expert committees working under FAO auspices and the recommendations (often involving a great deal of compromise) of these influential reports represent the reigning truth in this complex matter—with no lack of new re-evaluations.

The first report, issued in 1957 (the meeting itself was held in October 1955) tried to put the recommendations on a rigorous quantitative basis by examining the research on minimal requirements for essential amino acids (that is those compounds which humans must ingest to maintain nitrogen balance), by defining the provisional pattern of reference protein, and by expressing the minimum intakes in terms of this ideal nutrient: the lightest curve in Fig. 3.7 shows these age-dependent intakes for males. Besides assuming consumption of an ideal protein, the recommendations were valid only when no significant losses occurred through incomplete digestion, when diet variation from meal to meal was minimal, and when disease and parasitic infection were absent—altogether a highly idealized situation.

A decade later a new committee's most significant change was to reject the previous setting of reference protein requirements at puberty at no lower than 0.8 g/kg, which was more than twice as much as the recommended adult level: the

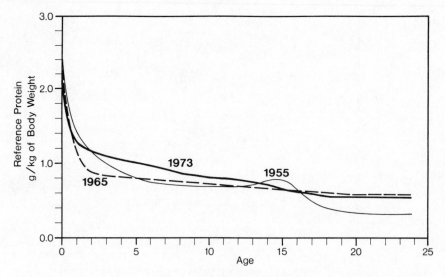

Fig. 3.7 Consensus recommendations of three FAO/WHO expert panels on human protein requirements show significant shifts with time. The fourth report, published after long delays, contains further changes.

group lowered the intakes for 12–16-year olds and raised the adult needs to 0.59 (± 20 per cent), 60 per cent higher than the 1955 level. The 1973 report lowered the adult needs only very slightly, but it put the needs between ages 2–13 higher than the previous two recommendations—and it changed the amino acid scoring pattern (the thickest line in Fig. 3.7).

Still regarding the make-up as provisional it raised the quantities of all but one of the ten essential amino acids so that their total amounted to 2250 mg per g of nitrogen, compared to 2016 mg in the 1955 provisional pattern. This led to some important shifts in chemical scores of common foods: for example, while in 1955 cow's milk scored only 78, in 1973 it was at a nearly perfect 95, but milled rice's score dropped from 72 to 67. Chemical scores and information on observed digestibility of various proteins are essential for estimates of actual dietary protein intakes. These and numerous other considerations mean that in practice an assessment of nutritional adequacy on a population scale is very complicated and that, as already stressed, the results of such calculations can shift significantly with small cumulative changes in assumptions.

An example, dealing first with energy needs, will illustrate these complications. It will be composed of two model countries, the first typifying the situation in a poor Asian nation, with all relevant inputs derived from the 1982 Chinese population census and from other recent Chinese information on nutrition and physical activity; the second basically mirroring the current situation in the United States, a model of a rich nation with readily available detailed statistics.

Table 3.1 outlines all the information necessary to estimate the per capita energy requirements of the Chinese population. Age groups and standard requirements according to weight and age are as recommended by the FAO; age–sex distribution is according to a 10 per cent sample survey of the 1982 census; body weights are averaging 60 kg for adult men and 50 kg for adult women (Chinese Academy of Medical Sciences values); and I have assumed that half of the working pouplation between the years 16–59 are very active while everybody else is just moderately active. These inputs result in an average daily per capita consumption of 2300 kcal—but it does not require big shifts in assumptions to produce different means with significant nationwide implications.

As noted previously, average body weights are difficult to pin down even in rich countries and while there are thousands of measurements of energy expenditures for a great variety of tasks, including recent detailed Chinese figures for peasants and industrial workers, estimates are still unavoidable in setting out those sectors of population which are more, or less, active than the standard moderate level. Measurements of a peasant's energy expenditures by the Institute of Health of the Chinese Academy of Medical Sciences indicate some very heavy physical exertions in ploughing, planting, shovelling, digging, and harvesting, with males using between 3900 and 4900 kcal and females 3400–4400 kcal during a full day of such heavy tasks.

All of these exertions would put these peasants into FAO's 'exceptionally active' category (energy needs at least 1.34 times larger than the moderate level)—but the tasks are seasonal rather than ongoing. Spring digging and summer harvest are the peak exertion periods while in winter peasants may do lighter work; consequently studies of activity distributions show ten hours of sleep in winter and just six hours of work, while during summer harvest the two figures are more than reversed at, respectively, 6.5 and 11.5 hours for full-time labour.

If these periods of the heaviest exertion lasted only 60 days, with 100 relatively restful (moderate exertion, no activity mark-up) days, and the rest of the year the peasants were working at a very active level (activity mark-up of at least 17 per cent), the annual mean for 73 per cent of the employed population would be almost exactly at the 'very active' level. And as heavy labour remains so common in China's mines (where even in the largest enterprise only two-fifths of all extraction is mechanized), in many factories and on construction sites, it might be still somewhat on the conservative side to assume that 60 per cent, rather than one half, of China's labour force is 'very active'. Combined with just 2 kg higher adult body weights this consumption would translate into a daily mean of 2400 kcal/capita.

On the other hand, lowering the average adult weight by just 2 kg, and having just one-third of the labour force in the 'very active' category (an easily defensible assumption with rapidly spreading rural mechanization, and with the commonly slow tempo of labour), brings the total down to 2200 kcal. Another plausible assumption could further extend the range of uncertainty, but the difference of 200 kcal between the two presented, and almost equally defensible, alternatives is already big enough in nationwide terms. For China's farmers to supply the higher

Table 3.1 Detailed calculations of energy requirements of the Chinese population in 1983*

Age (years)	Population share** (per cent)	Energy requirements (kcal/day per capita)		Total (10^{12} kcal/year)
		Standard needs	Very active people	
Both sexes				
<1	2.1	1090		8.6
1–3	5.5	1360		28.0
4–6	5.8	1830		39.7
7–9	7.1	2190		58.2
Males				
10–12	4.1	2600		39.9
13–15	3.9	2680		39.1
16–19	5.2	2820	3300	59.5
20–39	15.1	2760	3200	168.4
40–49	5.0	2620	3100	53.5
50–59	3.9	2480	2900	39.2
60–69	2.4	2210		19.8
70+	1.2	1930		8.7
Females				
10–12	3.8	2350		33.4
13–15	3.7	2260		31.3
16–19	5.0	2100	2500	43.1
20–39	14.1	2000	2300	113.4
40–49	4.5	1900	2200	34.5
50–59	3.5	1800	2100	25.5
60–69	2.5	1600		15.0
70+	1.6	1400		8.3
Total	100.0			867.1***

  * Calculations are done according to FAO recommendations for the 1983 population of 1.025 billion and for average adult weights of 60 kg for men and 50 kg for women. Half of the working population (16–59 years) is assumed to be very active.

  ** Derived from the 10 per cent sample tabulation of China's 1982 population census.

  *** (867.1 × 10^{12} kcal/365 days)/1.025 × 10^9 people − 2,318 kcal/day per capita.

rather than the lower average for the 1985 population of 1.05 billion would necessitate producing food equivalent to about 22 million tonnes of milled rice, or somewhat over 30 million tonnes of paddy (almost 20 per cent of China's mid-1980s rice harvest).

In contrast to China, the question regarding American energy requirements is how low they will be; the fashionable diffusion of jogging, celebrity-promoted dance-aerobics, and the rise of the muscle cults have done more for the sales of sweat-suits, leotards, and assorted fitness paraphernalia than they have for greater food needs. The vogue is still limited to a fraction of the American population, and for an average participant leading an overwhelmingly sedentary life a few weekly spells of exercise will not push his average above the moderate activity mean. Starting with the assumptions of universal moderate exertion and average adult body weights of 70 and 58 kg, and applying them to the latest age–sex distribution of the United States population, results in the mean daily requirement of about 2450 kcal per capita.

But this calculation obviously exaggerates the actual need. Just by strolling down any of the country's huge, cloned suburban shopping malls one gets the impression of an extremely inactive population: everybody came there by car and the numbers of the overweight and obese eating potato chips and licking coloured ice-cream are large indeed. Consulting the surveys one finds the proportions even more shocking. Recent American skinfold measurements show that 15 per cent of adult males and 30 per cent of females are obese (defined by skinfold thickness surpassing the 85 percentile) and, even more notably, every ninth or tenth teenager is already so fat.

Assuming that light, rather than moderate activity is the more appropriate mean of American physical exertion, the average daily requirement falls to about 2250 kcal and the most likely range, 2200–2400 kcal, is thus identical with the best estimates of Chinese needs. Later, in section 3.2.1, we shall see how these calculations square with actual consumption values whose means are no less elusive to establish.

Turning to protein needs, the FAO's recommendations, China's 1982 age–sex distribution and, again, adult weights of 60 and 50 kg, result in an average daily per capita need of 27 g of reference protein. This rather low value is, however, merely a theoretical precept as it assumes a regular intake of proteins with an ideal balance of amino-acids and their perfect digestion. The second condition never exists. Moreover, the latest expert consensus report published jointly in 1985 by the FAO, WHO, and the United Nations University concluded that the mean adult requirement, even for protein sources with nutritional value equal to that of eggs or milk, is approximately 0.6 g/kg/day, about 30 per cent above the earlier recommendation.

For poorer proteins of plant origin net utilization rates (combined measures of digestibility and of the efficiency of utilization of the absorbed amino acids) are considerably lower and the faecal nitrogen losses are more variable, making generalizations even more tenuous. And although previous experiments based on individual protein sources, and on studies of healthy adults concluded otherwise, the latest consensus is that efficiency of nitrogen utilization is also lower at very low protein intakes and with diets of low energy density—conditions comon through-out the poor world.

Net protein utilization also differs with age and may be much affected by various stresses (from minor infections to anxiety) and by respiratory and intestinal diseases so common, especially among the poor world's children. On the other hand, recent studies show that the nitrogen balance may not be the most sensitive criterion of the adequacy of food intake: the relation between nitrogen retention and weight gain is surprisingly variable as children on the same diet differ in the composition of tissue laid down. Consequently, there may be diets with sufficient protein intakes but the weight gains may still be poor if the food does not also have enough energy density (i.e. more fats).

This far from exhaustive survey of qualitative and utilization considerations show how literally useless is the calculated figure of 27 g of reference protein a day: it must be adjusted for quality, digestibility, utilization, losses, energy density, etc., etc., but plausible multipliers can combine to make a much greater final difference. The Chinese case is especially difficult as there is virtually no milk and very few eggs in their typical diet. Plant proteins accounted recently for about four-fifths of the total and they came mainly from rice (50 per cent), wheat and coarse grains (40 per cent), and legumes (most of the remainder). Using average chemical quality scores of 65, 50, and 55 for these three groups of foodstuffs, and weighing their relative contribution, gives a mean of 59 and hence the actual value is most likely to be somewhere between 55 and 65.

Similarly, weighing the chemical scores of Chinese pork, poultry, fish, milk, and egg consumption by relative consumption gives a mean value of about 80 for a combined average of between 60 and 68. Even with perfect utilization efficiency the Chinese would thus need daily between 40 and 45 g of protein in their diets rather than 27 g. Imperfect digestibility alone will raise the total by at least 10 per cent and the same amount would be a good mean to compensate for losses and growth increments arising from the minor stresses of everyday life. Adjustments for low-density diets and infections are impossible to quantify in terms of meaningful averages but if they added up to just another 10 per cent the additional mark-up would be about 30 per cent so that the actual per capita consumption of protein in current Chinese diet should be (rounding the numbers) at least 50–60 g every day.

The overall effect of these adjustments is to about double the level of the calculated requirements for reference protein, and a similar multiplier would be in order for most of the largely vegetarian diets with total energy intakes between 2200 and 2500 kcal—unless there is a widespread tradition of milk drinking (Indians, of course, consume yearly more than ten times as much milk as the Chinese). For poorer diets, especially those below 2000 kcal, reference protein requirements can be easily tripled in reality and the intake may still be insufficient. But where the national food energy mean surpasses 2500 kcal with animal protein supplying at least a third or two-fifths of the total essential amino-acid intake one does not have to bother with conversions: such countries will have at least 70 g of protein available per capita a day (at least 25 g of animal origin), an amount high enough to accommodate, on the average, all normal growth and maintenance needs.

This is as much generalizing as I shall do. Quite obviously, those repeatedly and boldly headlined figures about the multitudes with inadequate protein intake are dubious and misleading, useless as guides for any remedial action. One simply cannot calculate with any sensible accuracy what should be the world's intake of proteins in actual diets. The FAO's influential assumption of a unique level of requirement, implying that everybody consuming less than that amount is undernourished, is plainly wrong. All we can say is that protein–energy malnutrition (the two must be linked) is not a worry in diets of the rich industrialized nations—but putting a number on how many of the roughly 3.5 billion people in the poor world fit the category will remain elusive for decades to come.

While it is easy to recognize all the extreme forms of protein–energy malnutrition (PEM)—mainly variants of the types widely known as kwashiorkor (whose criteria are minimum weight not less than 60 per cent of expected body mass for age, oedema, often hepatomegaly, dermatosis, diarrhoea and mental changes), and nutritional marasmus (weight below 60 per cent of the expected norm), and their numerous intermediate manifestations—which are so striking in small children of the world's poorest nations, it is the milder PEM which is so much more widespread and still so inadequately understood.

Large-scale surveys in the three poor continents showed that moderate forms of PEM are on the average six-ten times more prevalent than the severe ones, affecting easily 20-25 per cent of all children in many countries and regions. The main findings are, of course, weight-for-age and weight-for-height ratios significantly lower than those of well-nourished children. Another undesirable adaptation to energy and protein shortages is a decline in physical activity: reduction of play time and intensity, and growing apathy can set in quite rapidly. Beyond these generalizations numerous studies of PEM offer little material for simple conclusions.

Most notably, low serum albumin levels in kwashiorkor led to a most reasonable conclusion that the condition arises from a worsening deficiency of protein, and since the mid-1950s preoccupation with adequate protein intakes has characterized all FAO and UNO dealings with the poor world's malnutrition. Mark Hegsted of the Harvard School of Public Health, who spent a lifetime studying nutritional needs, characterized this blindness best by writing that 'protein appears to have captured the public imagination, and "protein" and "good nutrition" often appear to be used synonymously, even in the United States where there are abundant protein sources and intakes are generous'.

But this one-sided stress, leading to such arcane efforts as genetical boosting of the protein content of common cereal crops, is clearly wrong. After all, protein needs in infancy, although as much as four times higher per kg of body weight than among adults, are relatively small because the children gain their weight much more slowly than experimental animals whose life-stories nutritional experimenters (and psychologists) love to translate to the human realm. Mother's milk, the perfect infant food, is, to the surprise of many, actually a low-protein food, containing only as much protein as rice and less than wheat or corn. Protein concentration above 7.5 per cent of total energy (expressed as milk or egg protein) does not appear to confer any additional benefits on healthy children. Moreover, children's protein requirements per kg fall rapidly during the first year of life and so, even when the necessary adjustments are made for lower quality and poorer digestibility of plant proteins, it is only in the poorest circumstances with a diet totally without animal proteins where PEM can be explained as primarily due to protein deficiency.

Energy shortage is then the obvious choice, especially in the light of such studies as India's National Institute of Nutrition where pre-school children provided with mainly carbohydrate and fat supplements, containing even less protein than their normal diets, showed improved growth. Yet this hypothesis carries little general weight as well: undersized and clearly malnourished children are not usually ravenously hungry and eager to eat any profferred food. Diseases are the third obvious key factor, as frequent respiratory and intestinal illnesses decrease food intake, increase energy and nitrogen demand, and interfere with growth in many other ways—but, once more, recurrent infections were not found to be a determining factor in some careful PEM studies.

A common-sense explanation, which has to do with logistics rather than with quality, is that the overwhelmingly vegetarian, monotonous diet of weaned children is too bulky to be fed in sufficient amounts during normal feeding periods (i.e. mostly just three times a day), and that more frequent feeding would literally stuff the child with sufficient food. Should this prove to be a large part of the answer, the remedy would lie in changing traditional habits rather than in growing more of the same cereal staple. Although this statement runs against decades of consensus of the international agricultural community with its call for more food to combat hunger, Hegsted's conclusion is the only one the best evidence allows a responsible scientist to offer: 'the information available does not lead to the conclusion that simply increasing food production will wipe out protein–calorie malnutrition.'

Protein-energy malnutrition is a complicated syndrome whose most prevalent manifestations—retarded physical and mental development accompanied by recurrent illnesses—arise from a complex of often self-perpetuating factors commencing long before the birth of a baby. Given the wide variations in individual energy and protein intakes, as well as the adaptation to less than optimal food consumption through reduced physical exertion, recreation, or social participation, many women on marginal diets will maintain their body weights and composition—but it will be low birth-weight babies who will pay the price as they will be more susceptible to diseases, and as the reduced milk output of the mothers will not support them adequately through the critical first months.

Overworked and uneducated mothers then wean their children without any sensible understanding of their nutritional needs, and combinations of unpalatable diets fed in improper ways, low food-energy densities, and low protein qualities and digestibilities from overwhelmingly vegetal foods, and recurrent illnesses lead to malnourishment, general retardation of development, and to the opening of yet another cycle of this grave social disease which cannot be solved simply by greater provision of food.

Many pages could be devoted to just outlining the puzzles characterizing the opposite condition to PEM. Obesity has become an all-too visible social disease of the Western world but I will just touch on some of its health implications when, in the next section, I will examine in some detail intakes of carbohydrates and lipids and their variously claimed effects on human health and premature mortality, a realm encompassing some of the most controversial, and literally deadly, life-sciences disputes of recent years.

3.1.2 *Carbohydrates and lipids*

> Indeed, the lucky grains were sent down to us,
> The black millet, the double-kernelled,
> Millet pink-sprouted and white....
> We pound the grain, we bale it out—
> We boil it all steamy.
> Then with due care, due thought
> We gather southernwood, make offering of fat....
> God on high is very pleased....
>   *Shi jing* (7th century BC) (A. Waley, translation)

Three millenia ago people of the dynasty in northern China made a ritual offering of what was most precious to them—the 'Lord Millet' and lamb's fat. In different cultures, and at different times, the treasures were similar: on the shores of the Mediterranean, bread and olive oil, in Central European villages, dumplings and lard, on the Indian plains, *chapatis* and *ghee*. Always the variations on the same theme: carbohydrates and lipids, the first ones to fill, the others to make the filling more satisfying.

Yet reading some of the modern sermons on healthy nutrition one might think our ancestors utterly reckless. All those cellarfuls of potatoes, sackfuls of flour, crocks of lard and butter, and jars of oils look like nutritional *auto-da-fé*, not only to the devotees of Zen macrobiotic diets, but also to millions of nutrition-conscious consumers who have come to see starches, sugars, and fats as decidedly inferior, if not outright dangerous nutrients in comparison with proteins, vitamins and, yes, indigestible roughage.

But the simpler carbohydrates, sugars, and the complex ones, polysaccharides, are still the principal providers of food energy everywhere. In poor countries they supply on the average three-quarters of daily calorie intake, the global mean of their contribution is now about two-thirds, even in the rich nations they deliver half of all food needs, and a longitudinal look shows that their importance in total global diet is still slowly increasing. This increase, however, is due solely to growing consumption in all those industrializing nations of Latin America, Africa, and Asia where an improved standard of life still translates into higher consumption of starchy staples and, to a much lesser extent, of sugars, as the diets are reaching more satisfactory levels.

For example, the gradual *de facto* privatization of Chinese farming after 1978 which brought great relative increases in consumption of all major foodstuffs, boosted sugar availability from 3.3 kg in1978 to about 5 kg in1983 but during the same time supplies of grains (in unprocessed weight) went up from 250 to 370 kg. In contrast, in all rich countries—be it Sweden or Canada, USSR, or Japan—total carbohydrate intakes have been declining. Individual consumption rates differ very much and decreases are far from uniform but the general trend is unmistakeable (see also section 3.2.2). But the decline hides a critical substitution.

In one of the principal nutritional shifts accompanying industrialization and rising affluence the relative consumption rates of sugars and polysaccharides have changed in what is almost universally seen as an undesirable direction. As will be illustrated in some detail in section 3.2.1, per capita consumption of grains and tubers has declined in all rich nations by as much as 40–50 per cent compared to their traditional, pre-industrialization levels but in energy terms the total carbohydrate intake has changed much less as formerly marginal sugars have risen to one of the top energy sources in affluent diets.

In the United States, where we have reliable annual data since 1909, the per capita availability of all carbohydrates has declined by about 20 per cent in 75 years, but that of sugars has increased by nearly 40 per cent so that they now account for more than half of all carbohydrates, compared to just short of one-third at the beginning of the century. Sugars are now supplying almost exactly one-quarter of all food enegy and, of this share, half comes from refined cane and beet sugar and naturally occurring sugars (mainly in dairy products and fruits), and what are labelled 'other calorie sweeteners' (syrups, honey, molasses) split the other half.

In retrospective, the availability of natural sugars has remained surprisingly constant since the early decades of this century while consumption of refined sugars has been declining since the early 1970s, mainly as a result of strong competition from corn syrup, above all its high-fructose variety which is twice as sweet as regular corn syrup. Moreover, there have been important shifts in the final uses of refined sugars: their direct household consumption declined by more than 40 per cent while intake of store-bought food products rose nearly three-fold as most baking, canning, pickling, and jam-making moved from kitchens to commercial establishments after World War II. This shift, however, did not lead to higher consumption of refined sugar: in fact, its total direct and food product intake was nearly 10 per cent higher in 1910 than in 1980!

Where then does the increased consumption of sugars come from? About a quarter comes from more sweeteners (led by syrups and corn sugar) in processed foods—but the bulk from soft drinks. In the United States the use of refined sugar in beverages rose from just 1.5 kg per capita in 1909 to about 12 kg by the early 1980s when nearly half of all high-fructose corn syrup also ended up in soft drinks. The shift had been proceeding rather slowly until the early 1960s: since then American per capita consumption of soft drinks has more than doubled.

Food consumption surveys in the late 1970s showed an incredible annual average of 384 twelve-ounce containers per capita, a little more than one drink a day for all of the 230 million people, 136 litres of pop a year per person (for comparison, the French drank in the late 1970s 23 litres of soft drinks per capita per year). Teenage American boys (15-18 years) had the highest intakes, averaging an equivalent of nearly two cans a day, about half of all children between 6 and 14 were devoted consumers and, incredibly, 40 per cent of infants below two years and 11 per cent of babies were given the drinks by their parents (while in 1965 a mere 3 per cent of children below one year drank the concoctions).

Not surprisingly, sugars in soft drinks became noticeable suppliers of food energy—about 4 per cent of the total for the population as a whole, up to 6 per cent for older teenagers and young adults (19-22 years old). As the rest of the world, (regardless of its frequent dislikes of American ways) is bent on emulating the country's worst habits, consumption of those sticky, unpleasantly sweet, artificially flavoured and coloured liquids which do anything but quench one's thirst has been rising in predictable (Western Europe) and unpredictable places (from China to Kenya) alike. Bread, rice, and potatoes will recede in more countries so that the new, relative affluence of their inhabitants can replace a surprisingly large part of those abandoned carbohydrates by sweet, bubbly liquids, certainly one of the nutritionally most irrational swaps.

Nutrition science is a realm of chronic disagreements but on this score there is a solid uniformity: total sugar intakes in the rich countries are too high (up to twenty times higher than in most poor nations), and the high shares of refined sugars (in the United States now representing nearly one-sixth of all food energy) are seen as particularly inappropriate. While there is no foundation for recommending their optimum share, it has often been recommended that the intake of refined sugars should be limited to 10 per cent of all food energy, and a mass of popular writings has linked 'excessive' consumption to a multitude of ills. Yet elucidating and quantifying these links has proved surprisingly elusive.

Perhaps most notably, the deep-rooted public belief that the consumption of sugars—as well as of starchy foods—favours the development of obesity has yet to be substantiated. While there is no doubt that obesity is detrimental to well-being and presents a clear health hazard—most prominently, the state has been implicated in the etiology or worsening of hypertension, diabetes, cardiac, renal, gallbladder, and pulmonary respiratory diseases—no reliable evidence implicates any specific nutrient contributing to the excess energy intake and thereby to the genesis of this most serious form of malnutrition in affluent nations.

And the role of carbohydrates is similarly unclear in dietary treatments of obesity. Charles Hollenberg has compared the past control efforts to the exertions of a football team that has lost every game it has ever played. The physical consequences of this spreading condition have been much studied but the mental dimensions are just emerging. According to a 1984 American study, as many as

16 per cent of female high school students (every sixth girl) 'binge eat', 6 per cent frequently vomit after eating, and nearly half describe themselves as being terrified of getting overweight. This fear has become translated into growing evidence of female adolescent *anorexia nervosa*, of which one severe case is now found in every 250 young women between the ages of 15 and 24, and the frequency of less acute forms of this disorder is as high as 1 per cent among private school attendants.

Among the scores of 'revolutionary' diets unfailingly promising instant and effortless weight losses are those which consider carbohydrates as unimportant components that may be safely reduced, those that insist on their drastic reduction or complete elimination, and yet others which actually make carbohydrate consumption a central part of attempts to correct and control energy imbalance (these high-carbohydrate diets have also been claimed to prevent the development of atherosclerosis—although there is no known mode of action through which this could occur).

Experimental evidence to support these contradictory positions is inconclusive but the hazards of low carbohydrate diets are well known and they do not make a nice list: an abnormal accumulation of ketones in the body, fatigue, nausea, calcium depletion, higher serum cholesterol (when higher fat intakes supplant abandoned carbohydrates), kidney failure in predisposed individuals, hyperuricemia and the possibility of gout attacks, and in pregnant women risks to foetus from chronic ketosis. Obviously, the elimination of the principal nutrient which supplied the bulk of energy needs throughout human evolution would not be a sensible approach.

Pinning the blame on carbohydrates in general and sugars in particular has proved no less difficult in the case of coronary heart disease (clinical and epidemiological data do not support a relationship between its development and sucrose intake) and does not serve as an explanatory factor for higher serum lipids (some studies show a rise in serum cholesterol after replacing starch by sucrose, others fail to demonstrate the difference). Also, there is no clear evidence that the consumption of simple sugars is of importance in the development of diabetes and, in some ways the most surprising conclusion, even the link between dental caries and sugars is not as obvious as commonly thought.

Ernest Newbrun's recent thorough review of numerous studies of special population groups, epidemiological surveys, and controlled longitudinal investigations of sugar and its substitutes indicates that frequent or high consumption of sweet foods predisposes to dental decay—but that the relation is far from clear-cut and that most studies supporting the link had serious methodological problems and limitations.

Impressive proofs range from multi-country correlations between sugar consumption and mean numbers of decayed, missing, and filled teeth in 10–12-year old children to very low caries prevalence among persons with hereditary fructose intolerance. But the automatic equating of high sugar intake with high occurrence of caries is not possible. Modes and times of sugar consumption matter—sticky solids eaten between meals have much greater impact than sugar consumed at meals in solutions—and the sequence of eating various foods affects their cariogenic potential and the cariogenicity of individual foods cannot be predicted with any degree of accuracy.

Even the experimental evidence looks weaker once it is realized that the diets of human subjects cannot be controlled over the two years or so needed for clinically detectable lesions to develop, that the diet's sugar content is difficult to assess (owing both to the inadequacies of food consumption data—more on this in section 3.2.2—and to the non-uniform values of the sugar content of various foodstuffs), and that most of the studies were not only short-lived but have not involved enough persons for any meaningful statistical analyses. A qualitative statement that more sugar, especially when eaten more frequently in between-meal snacks, will tend to cause more caries is thus the only unexceptional conclusion: quantification of the relationship remains elusive and undoubtedly it will stay so for a long time to come.

Before leaving carbohydrates I must devote at least a few pages to a story so characteristic of the 1970s and so well fitting the goals of this book, a story which made the consistency of human faeces an object of lively interest to large segments of Western populations, a saga which has enriched many producers of breakfast cereals who, while continuing to take away the outer layers of the grain, put a small part back to the accompaniment of mass advertising campaigns in which they assured the consumers of their ever so serious commitment to healthy nutrition. Yes, you had to guess right, it will be a story of bran or, more appropriately, of dietary fibre!

The story started in 1971 with Painter and Burkitt's paper in the *British Medical Journal* where they suggested that a low intake of fibre causes diverticular disease; next year Trowell, writing in the *American Journal of Clinical Nutrition*, labelled fibre as a protective factor in ischemic heart disease. A veritable explosion of interest followed almost immediately and by 1975 Burkitt and Trowell combined their forces to bring out a bible of the movement, their *Refined Carbohydrate Foods and Disease: Some Implications of Dietary Fibre.*

Some implications indeed! Diets rich in fibre were claimed to protect against an impressive array of common killer diseases much feared throughout Western civilization, illnesses ranging from coronary heart disease and stroke, to diabetes and cancer of the large bowel. As we will see shortly (in the section on vitamins) the claims launched almost concurrently by Linus Pauling on behalf of vitamin C were in several ways startlingly similar and anybody learning about these great 'dis-coveries' through duly sensationalized columns in Sunday newspaper supplements could be forgiven for believing that megadose combinations of bran muffins and vitamin C pills would ensure immortality.

But the cited works published in the early 1970s were just the rekindling and refinement of older observations and hypotheses. Leaving ancient remarks on the desirability of coarse cereals aside, Graham's efforts in the United States and Atkinson's work in England during the last century, the writings of John Harvey Kellogg on cereal roughage in the early 1920s, and a spate of papers evoked by his views, Cleave's publications in the 1950s and 1960s (although ascribing most of the ill effects to sugars rather than to lack of fibres), and Painter's work on diverticular disease in the 1960s represented an almost continuous span of 150 year-old interest in the effects of roughage in man which Burkitt and Trowell so effectively popularized starting in the early 1970s.

While there is no doubt that these popularizers sought to diffuse their views as much as practicable, a position almost always entailing undesirable simplifications, their message was never so simple as to recommend a retention of current diets supplemented with large doses of bran, and their arguments have gradually developed in less deterministic ways. Burkitt himself put it best at a 1982 conference: 'Earlier emphasis was on a deficiency in fibre being causative of a number of diverse diseases . . . that adequate fibre can be protective against these diseases that proved to be more readily acceptable.'

But, he continued 'a more important change of emphasis has been that it is not merely one single component of food that distinguishes diets of communities with high and low prevalences of Western diseases. Diets low in fibre are also low in starch and high in fat and sugar and vice versa. In particular recommending a high-fibre diet almost inevitably implies a low-fat diet.'

This is an appeal for a dietary change, a far cry from a spoonful of bran at breakfast. Moreover, recent extensive studies of dietary fibre have shown that high intakes of it may not always be beneficial as they influence the bio-availability of various nutrients. Reductions of protein and lipid availability appear nutritionally insignificant, absorption of carbohydrates is reduced more significantly, but it is the possiblity of an altered balance of trace minerals which has caused most concern.

Calcium malabsorption follows high intakes of whole grain and bran used to supplement high-protein diets. Serum iron levels can decline just after a few weeks of wheat bran supplements and as iron deficiencies are so common, especially among women (see the next section), regular consumption of bran can further worsen the availability of this essential mineral. Absorption of phosphorus, magnesium, zinc, and copper may also be lowered. In general, supplementary wheat bran, the substance promoted as the best source of fibre, causes most of this interference while consumption of whole highly-fibrous foods appears to have negligible effects.

The disappearance of simple precepts and beliefs in a single mode of action is not surprising. Those carbohydrates generally termed dietary fibre are complex mixtures of lignin and polysaccharides that are not hydrolyzed by the endogenous enzymes in the human digestive tract and, as they range from water-soluble gelatinous, acidic compounds to insoluble, neutral fibrous materials, they will have a variety of effects and, moreover, their interaction with one another, and with other food components, may be quite different from their independent effects (for example, if isolated cellulose is not fermented in the large intestines it does not mean that no cellulose in the plant wall ever breaks down), and laboratory *in vitro* studies may have little relationship to complex living reactions.

Much remains to be done. Earlier epidemiological studies, which have provided most of the impressive evidence about the beneficial effects of high-fibre diets, considered intakes of crude fibre but there is no constant relationship between this entity and dietary fibre. More importantly, standard values for the dietary fibre content of foods are still missing for many foodstuffs and the difficulties of acquiring reliable information on actual food intakes (a problem to be addressed in detail in section 3.2.2) make cross-cultural comparisons far from accurate.

Southgate, noting that he was once accused of taking the simple concept of dietary fibre and making it complex, retorted soundly that 'the real danger is in taking a complex component of the relationship between diet and health and disease and treating it in a naïve fashion beause therein lies the risk that the hypothesis will be rejected because of inadequate evidence.' No statement could be a better introduction to a discussion of another high-profile nutritional story of the 1970s, a controversy with a lively continuation into the 1980s, the big cholesterol debate.

Since the early 1950s, as a result of an international epidemiologic study of heart disease mortality and fat levels in the diet, the main line of reasoning was appealingly linear: heart and blood vessel (cardiovascular) diseases are the major cause of death in affluent Western nations, accounting for over half of all mortality in many countries ('number one killer' in journalese); atherosclerosis—a multifocal process involving the innner layer of arteries with the intracellular deposition of fat, particularly cholesterol ester, and culminating in extracellular cholesterol ester crystals, copious fibrous tissue and thickened plaques—was irrefutably implicated in the great majority of these deaths. Since the diet of affluent nations has been including larger amounts of animal foods which have much higher cholesterol contents than the earlier predominantly plant diets, a reduction of dietary fat should decrease the rates and frequencies of atherosclerotic changes; and, indeed, the epidemiologic surveys confirmed that the nations with the largest cholesterol intakes have the highest rates of cardiovascular mortality.

Dietary treatment, switching from saturated animal fats to polyunsaturated plant oils, substitution of butter by margarine, fewer eggs, less beef, and more chicken and fish, no whole milk—this was the unchallenged, recurrent advice for a generation as heart foundations, scientists, and the media kept spreading the message. And the changes did come: in the United States consumption of butter fell to half of the 1950 value by 1970 while that of plant oils, largely eaten as margarine, rose 70 per cent during the same period; the demand for eggs weakened by about 20 per cent while poultry sales rose nearly two-fold.

These figures indicate deep and rapid dietary change in desirable directions and for those seeing the high-cholesterol foods as the principal cause of heart disease the mortality statistics tell an impressive story. When those death rates are scaled to fit the age distribution of the United States population acording to the 1940 census (an adjustment enabling continuous comparisons in spite of changing age-structure), their steep rise, culminating exactly in 1950, is followed by downward fluctuations in the 1950s, and then a steep decline in the 1960s and even steeper decrease in the

1970s. And, indeed, the great dietary adjustments continued during the 1970s, with poultry up by 25 per cent, eggs down by 12, plant oils up by 16 per cent, butter down by a further 20 per cent. Are any further proofs of the efficacy of dietary prevention and treatment needed?

Not for the proponents of low-fat, low-cholesterol salvation—but their simplistic reasoning has been subjected to a great deal of questioning since the early 1970s and too many facts are obviously incompatible with a view that manipulation of the diet alone holds the answer to the prevention of coronary heart disease (CHD). Epidemiological evidence which provided the original foundation for the CHD-cholesterol link is not, as it is so often mistaken to be, a proof of any cause-and-effect relationship and, even more importantly, atherosclerosis is clearly a multifactorial process. The often cited Framingham study (during which 5209 men and women aged 30–62 years have been followed biannually since 1949) found the higher risks of CHD powerfully correlated not only with total serum cholesterol but also with body weight and significantly increasing with age, higher blood pressure, lack of exercise, smoking, stress, and family history. Consequently, the isolation of a single factor results in irreconcilable discrepancies once the search for relationships is extended and intensified.

Ancel Keys' original correlations, showing the Americans and the Finns with the highest CHD mortalities, and the Japanese with the lowest, were later supported by the much-quoted studies of the migration of Japanese to Hawaii and thence to California: as the migrants' diets became more Americanized, their serum cholesterol levels rose and so did the incidence of CHD. Another much-quoted work on American Seventh-Day Adventists demonstrated the advantages of a low cholesterol diet. But for each of these highly publicized studies one can cite contradictory arguments and surveys pointing in no clear, or in a completely opposite, direction. Since 1950 the Japanese diet has undergone an unprecedented rapid transformation but the shift, to be detailed in section 3.2.2, has been decidedly in an American direction—yet the CHD mortalities continued to fall. The best explanation—earlier diagnosis and much improved treatment of high blood pressure, a traditionally widespread condition in the nation.

The Japanese intake of polyunsaturated/saturated (P/S) fats is thus now well below unity (about 0.6, compared to American 0.4), a ratio recommended as desirable by George McGovern's 1977 Senate Select Committee report on dietary goals—and the Israeli intakes, as calculated from food balance sheets and determined by adipose tissue analyses are far superior. In the mid-1970s Blondheim and colleagues found out that the P/S ratio of Israelis was 0.88 for Ashkenazic Jews and 1.13 for non-Ashkenazim, the highest reported for any population on a free-choice diet—yet Israeli mortality owing to CHD is nearly six times the Japanese mean. Clearly, the beneficial effects expected from the highly polyunsaturated diet are completely eliminated by other factors, dietary or non-dietary.

And in a yet more intriguing sequel a year after the publication of Blondheim's paper, Abu-Rabia's survey demonstrated that desert Bedouins' subcutaneous P/S ratio of 0.76 rose to 0.95 after they settled in Israeli towns—but so did their CHD mortality. On this basis a simplistic correlation argument may be then made for polyunsaturated fats contributing to CHD!

If epidemiological considerations fail to sanction any simple diet–CHD link its defence becomes even more difficult once one considers the possible physiological effects of the 'prudent' diet. The bulk of the cholesterol is always produced by the body itself: typical ranges for the population of North America with its high cholesterol intakes show 19–22 per cent coming from meat, 5 to 6 per cent from dairy products, and some 5 per cent from all other foods—with 67–71 per cent synthesized in the body, above all in the liver. Manipulation of the dietary intake will thus influence no more than a third of the total cholesterol input and differences in absorption will further lower this share.

Once biliary cholesterol passes into the gut it may be absorbed better than dietary cholesterol: intestinal perfusion studies have shown absorption of endogenous cholesterol in the range of 60–80 per cent, while isotopic tracer measurements of dietary cholesterol reveal absorption rates varying from 20 to 80 per cent. Consequently, even a 50 per cent reduction of dietary cholesterol may mean no more than a few per cent reduction in the total amount of cholesterol entering the gut. Ahrens calculated that reducing American fat intake from 44 to 30 per cent of the mean diet, raising the P/S ratio from 0.4 to 1.0, and lowering cholesterol intake from 500 to 300 mg a day (essentially the 'prudent' diet recommendations advocated by the American Heart Association) would, with the perfect 100 per cent compliance of the country's population, lower plasma cholesterol by about 12 per cent—which means that a more likely (but still optimistic) 50 per cent compliance would bring a change of a mere 6 per cent.

In answering the question: is this small decrease worth the enormous effort required to achieve it (just recall the success rate in the United States Government anti-smoking campaign—has it eliminated half of the habit after two decades?) one must also take into account the large, and unpredictable, differences produced by dietary changes in different people and the inapplicability of data from metabolic studies to 'free-living' individuals. For example, in Ahrens' experiments the substitution of lard (P/S 0.43) for corn oil (P/S 8) in thirteen patients' eucaloric diets for two months caused plasma cholesterol to rise by as little as 5 per cent and as much as 39 per cent, and butter changed them by 18–84 per cent. And even more notably, people not confined to metabolic wards and eating ordinary foods on free-choice diets have not been responding to variations in fat and cholesterol intake in the way experimental subjects do. Not surprisingly, mixed diets contain a variety of compounds influencing serum cholesterol in different ways.

In December 1979 Ahrens' paper in the *Lancet* concluded that 'adoption of a "prudent diet" will cause various reactions in different segments of the population. No one can say today what proportion of the general population will experience a decline in plasma cholesterol levels, nor of what degree; how many will increase cholesterol synthesis to match a decrement in cholesterol absorption, nor to what degree. . . . I believe it is anything but a service to the public to postulate *one* dietary solution for hyperlipidemia, no matter how well-meaning one is in advocating it.'

In the same year, Sir John McMichael, writing in the *British Medical Journal*, was even blunter in his conclusions: 'I do not subscribe to the view that atheroma is a nutritional disorder caused by faulty dietary habits. It is a complex process influenced by heredity, infection, and other factors, but basically it is a wear-and-tear disease, in which mechanical stresses and injury at special points of turbulence in the blood stream break the vascular endothelium. . . . The time has to come to reject advice to substitute polyunsaturated fats for animal and dairy fats in the nation's diet.'

And by 1980 the broad scientific consensus echoed these opinions. In spring 1980 the Food and Nutrition Board of the National Academy of Sciences concluded that there is no persuasive evidence to indicate that healthy people should restrict their cholesterol intake, no proven benefit in reducing fat intake except in the control of obesity and weight, and that 'at this time' it does not seem prudent to advise an increase in the proportion of polyunsaturated fats except for individuals in high-risk categories.

In May 1981 the Consumers' Union of the United States published in their widely read *Consumer Reports* a survey of the diet–heart development designed as a guide for the perplexed layman. Its concluding recommendation: 'Since the safety and efficacy of fat-controlled diets have yet to be proved and perhaps never will be, CU's medical consultants cannot recommend such a dietary policy to the entire population. Its implications for women have not been studied, nor is anything known about the long-term effects of such diets begun in childhood or adolescence.'

This was the first time the question of safety of the 'prudent diet' was publicized in such an influential mass-circulation source. What a turn-around—after more than two decades of increasingly confident pitches for changed diets (sustained not by health faddists but by such prestigious organizations as the American Heart Association), after reducing the butter industry to a fraction of its former self, and elevating margarines almost to the status of health drugs guaranteeing cholesterol-free arteries, the people were told that 'prudent' dieting was not only a misplaced but possibly also a dangerous effort.

Low-cholesterol diets could bring undesirable changes in the composition of common diets as energy supplied by eggs, meat, and milk may often be replaced by low protein high-fat food items resulting in the overall increase in intakes of plant lipids. This would lead to even greater consumption of hardened oils whose fatty acids, converted into 'trans' forms (from natural 'cis' double bonds), would yield much more intensely sclerogenic esters than those resulting from liquid lipids. Trans fatty acids might also contribute to altered characteristics of cell membranes, and animal experiments have shown the unsuitability of some oils to heart muscle cells. Consequently, Sir John McMichael stated that margarines (hydrogenated plant oils) may be actually more damaging than natural animal and dairy fats.

Much more shocking has been the identification of a significant association between low serum cholesterol and increased cancer mortality (above all for colon cancer). No less than sixteen serious studies showed this link in populations and patients who have had low cholesterol levels for a number of years, leading Eliot Corday, a former president of the American College of Cardiology to call the findings 'frightening' and warning that we should be careful when advocating cholesterol-free diets because we still do not know why there is an increase in the cancer rate.

The risks may, in fact, be minimal and they may be outweighed by benefits, but the question put by Thomas N. James, a former president of the American Heart Association, cannot be evaded: 'Do we really know enough about the long-range consequences of such dietary advice to be confident of its merit?' This concern is mainly over the possible long-term effects on children whose mothers were eating low-fat diets during pregnancy (British studies show no correlation between birth weights and maternal protein intake, but a significant relationship between consumption of essential fatty acids during pregnancy and newborn weights) and whose growing nutritional needs are to be covered by 'prudent' diets whose effects are unknown—hardly a prudent scientific approach.

The next major threshold in diet–CHD studies came in 1982 with the release of the final results of the Multiple Risk Factor Intervention Trial (MRFIT). This study followed 12 866 men aged 35–57 years who were at increased risk of death from CHD: half of them received 'special intervention' aimed at cessation of cigarette smoking, reduction of blood pressure and lowering of serum cholesterol through initial (since 1973) reduction of saturated fat intake to less than 10 per cent of all energy and the limitation of dietary cholesterol to no more than 300 mg/day, and subsequent (since 1976) further restrictions to 8 per cent of saturated fat in the diet and 250 mg cholesterol a day.

The conclusions of this extensive (250 investigators in 28 North American centres) and costly ($110 million) study were—inconclusive. The special intervention group had CHD mortality 7.1 per cent lower than the control set but its overall cardiovascular disease mortality was just 4.7 per cent lower, and its mortality from all causes was actually 2.1 per cent *higher* than among the 'usual care' participants: 'The overall results do not show a beneficial effect on CHD or total mortality from this multifactor intervention.' Although dietary cholesterol intake was just one of the controlled factors the failure to prove any significant benefit from the combined management effort helped to strengthen the growing doubts about the possibility of successful dietary intervention in CHD.

Moreover, since the late 1970s renewed interest in lipoproteins (cholesterol-carrying macromolecules) further complicated the diet–heart story as it was realized that the relative distribution of very low-, low-, and high-density lipoproteins (the former apparently promoting the atherosclerotic process, the latter actually protecting against the degradation effect) rather than the total serum cholesterol level may be the best predictor of CHD risks. The study of the relationships between diet and lipoprotein fractions is still too little advanced to make significant conclusions but exercise, normal weight, and no smoking have been shown to elevate the amounts of 'good' high-density lipoproteins.

The tentative nature of these findings was best summarized at the 1983

Symposium on High-Density Lipoproteins and Coronary Artery Disease where the reports concluded that in spite of the strength of epidemiologic associations between high-density lipoproteins and CHD risks there is no evidence from experimental studies or clinical trials to establish that low HDL levels are causally important in atherogenesis and that, although HDL levels correlate positively with total energy intake, dietary cholesterol, total and animal fat, and alcohol (a line-up destroying decades of scientific consensus!), the relationship of HDL manipulation to cardio-vascular health is yet to be defined.

That was in 1983. On 20 January 1984 the *Journal of the American Medical Association* published the final results of the Lipid Research Clinics Coronary Primary Prevention Trial (LRC-CPPT), a long-term $150 million study which tested the efficacy of lowering cholesterol to reduce the risk of CHD in 3806 asymptomatic middle-aged men with primary hypocholesterolemia (i.e. only men aged 35–59 years with plasma cholesterol in excess of 265 mg/dL). The treatment group received cholestyramin resin, a drug sequestring cholesterol from the body, for an average of 7.4 years.

The results showed an impressive 24 per cent reduction in CHD deaths and a 19 per cent decrease in non-fatal myocardial infarction for the treated group whose total plasma cholesterol and low-density lipoprotein cholesterol levels fell, respectively, 8 and 12 per cent relative to levels in placebo-treated men. Moreover, the incidence rates for new positive exercise tests, angina, and coronary bypass surgery were reduced by 25, 20 and 21 per cent in the cholestyramine group. Overall conclusion: 'The LRC-CPPT findings show that reducing total cholesterol by lowering LDL-C levels can diminish the incidence of CHD morbidity and mortality in men at high risk for CHD because of raised LDL-C levels. This clinical trial provides strong evidence for a causal role for these lipids in the pathogenesis of CHD.'

The trial was not designed to assess directly whether lower dietary cholesterol prevents CHD (both groups ate a moderate cholesterol-lowering diet) but 'if it is acknowledged that it is unlikely that a conclusive study of dietary-induced cholesterol lowering for the prevention of CHD can be designed or implemented' the study's findings 'support the view that cholesterol lowering by diet also would be beneficial.' These quotes were necessary to establish precisely what the study did and what it concluded.

Nevertheless, two months after its publication *Time* ran a cover story reaching tens of millions with its heading 'Hold the Eggs and Butter' and asserting that 'anybody who takes the results seriously may never be able to look at an egg or a steak the same way again.' But the study was neither about anybody, nor about eggs: it concerned high-risk middle-aged men and it looked at their response to a cholesterol-lowering drug. There is no lesson in this whatsoever for children, women, young men, and all those middle-aged men whose cholesterol levels are not at all or only slightly elevated (after all, about 480 000 age-eligible men were screened to select 3806 *high-risk* participants with primary hypercholesterolemia): no lesson at all for all those eating balanced diets. *Time*'s subtitle 'Cholesterol is proved deadly, and our diet may never be the same' was a deplorable, deliberate journalistic misrepresentation of realities assessed by scientific research.

Middle-aged, hypercholesterolemic men (and the study says nothing about why they may be so) clearly benefited from drug treatment—but that is patently no basis for recommending a drastically changed dietary pattern for the general population. Preference for simple solutions is hard to eradicate, be it among the media-men or scientists. Publicity around LRC-CPPT is just one more attempt to resurrect a simplistic causal diet–CHD link as if nothing else, from generic endowment and normal ageing processes to scores of ingredients of life-style, mattered.

This is not the last place in this book where I will remark on the apparent ineradicability of false news once it enters the public perception as (misrepresented) scientific findings via the media. As soon as the LRC study was out Fleischmann's margarine bought bold advertisements claiming that the trial 'outlined the danger of foods that are high in animal fats such as eggs and butter.' Although the trial did not do this at all, unsuspecting consumers were undoubtedly impressed (after all, a $150 million study should have the weight). And, to bring the development to the very end of that eventful year, in December 1984 a meeting at the National Institute of Health recommended for the first time specific cholesterol guidelines: no more than 200 mg/100 ml of blood for men over 30 years.

And so the fat and cholesterol saga continues, but I have to leave its many intriguing aspects, including studies of the hypocholesterolemic effect of dairy products (after all, Kenyan Masai with their huge milk consumption have low cholesterol levels and extremely low incidence of CHD!), or dramatic LDL and VLDL declines in hyperlipidemic patients eating fish oils (Greenland Eskimos rarely have CHD—but fish oil diet also significantly reduces platelet aggregation times and so one might keep on bleeding!). But we are not yet leaving controversial topics—the next section on vitamins and minerals is yet another example of our ignorance of what is really best for us.

3.1.3 *Vitamins and minerals*

* Malt	80	bushels
* Sour Krout	19 337	
* Salted Cabbage	4 773	
* Portable Broth	3 000	Pounds
* Saloup	70	
* Mustard	400	
* Mermalde of Carrots	30	gallons

* The articles marked thus (*) are antiscorbuticks and are to be issued occasionally

From Captain James Cook's victualling list for HMS *Resolution* for its 1772–5 voyage.

The famous seafarer was not the first one to recognize the need for regular consumption of antiscorbutics on long voyages—that distinction belongs, most notably, to James Lind. Moreover, Cook's rejection of citrus fruit, which he thought excessively expensive for general crew use, led him to load up his ships with tonnes of cabbage, and his reliance on portable broth and saloup proved completely misplaced—but he was undoubtedly the first captain to enforce the regular consumption of preserved vegetables and, during the stop-overs, fresh fruits and greens.

We of course do not know what amounts of these fresh foodstuffs were eaten, but when *Resolution* left Plymouth in July 1772 its anti-scorbutics were good enough to supply, according to my calculations, almost 20 mg of ascorbic acid each day for 112 men during their three year voyage—while modern experiments have shown that 10 mg is enough to prevent scurvy. If fresh sources added another 10 mg, Cook's men had 30 mg of vitamin C a day, exactly the intake recommended by FAO/WHO expert consensus as the minimum for adult moderately active men.

Whatever their actual intakes they had basically enough—not one of them died of scurvy and unlike their mates on *Adventure*, where Captain Furneaux was much more lax in enforcing prevention, they suffered much less from the disease's early manifestations. What I find most notable about this story is that James Cook victualling his ship today could still not be given unequivocal advice about how many barrels of 'sour krout', gallons of lemon juice or, now most likely, artificially synthesized pills of ascorbic acid, to take aboard. During nearly a quarter millenium after publication of James Lind's *Treatise on Scurvy* (1753) we have amassed thousands of studies about the essentiality of vitamin C in human nutrition but as to the optimum daily allowances extreme recommendations continue to differ by more than an order of magnitude.

This controversy, so much in the public focus during the recent past, is a perfect example of major uncertainties surrounding the needs for vitamins and minerals. Rather than just touching on each of the ten vitamins and six minerals for which the United States National Academy of Sciences now makes daily dietary recommendations, I will take a more detailed look at the C debate as ascorbic acid has been certainly by far the most overused vitamin, and after a glance at a fashionable vitamin E controversy I will deal with the most notable mineral deficiency, the lack of iron.

Vitamin C follies are perhaps the best example of the major nutritional trends which swept much of the rich world during the 1970s—the fad of megavitaminosis. In the case of ascorbic acid this vogue of megadoses was not launched by an obsure nutritionist or a quack dietician reaching for fame. The man responsible was, after he devoted a lifetime to the study of biochemistry and to the advocacy of various social causes, one of the world's most famous scientists: Linus Pauling who ushered in the vitamin C megadose controversy in 1970 when his *Vitamin C and the Common Cold* came out.

The book recommended large doses of ascorbic acid as both a prophylactic and therapeutic agent for the common cold (aborting a cold in its very early stages and suppressing symptoms of a fully developed one). Sales of the vitamin pills soared— but Pauling did not contribute to this rise. Although he himself went on a regimen of 4–8 g of ascorbic a day he was buying the compound in bulk ('to save money,' he said) and putting it into his orange juice whose plunging pH he occasionally elevated with some baking soda.

But Pauling went considerably further with his claims and by the early 1980s he and his disciples had the following good things attributable to megadoses of vitamin C: increased feeling of well-being; improvement of immune status; lowering of blood cholesterol; protection against anaemia, intestinal cancer, and the harmful effects of smoking; prevention of bleeding gums, dental caries, and stomach cancer (since it prevents conversion of nitrites in food and saliva to carcinogenic nitrosamines); a treatment for osteogenesis imperfecta, familial polyposis, infectious hepatitis, infectious mononucleosis, syphilis, snakebite, and disorders of collagen synthesis; decreased atherosclerosis; faster wound healing and more rapid resolution of urinary infections; improved cancer survival and reduction of the toxicity during cancer chemotherapy (by inreasing the rate of excretion of harmful compounds); beneficial effects on mental health; improvement in diabetes; and better tolerance of higher altitudes.

From trivial to life-saving—it appears there are few health dangers ascorbic acid would not prevent or set right. Not surprisingly, very large numbers of people in North America and Europe are now supplementing their diets with high doses of vitamin C but, in spite of numerous biochemical, clinical tests and trials (involving a disproportionately large allocation of research resources to this fashionable topic), we still have no decisive resolution regarding the merits and the risks of this practice. Although people take vitamin C megadoses for a variety of reasons, Pauling's initial claim, prevention of colds or their abortion, remains perhaps the single strongest motive and as the common cold is indeed a major health problem—responsible for much of worldwide respiratory morbidity, and billions of hours lost each year from work and school—it would be most desirable to offer good scientific guidance.

As there has been no shortage of studies on the effects of vitamins C on the common cold (the earliest ones dating from the 1930s) we have now several thorough reviews appraising their results and none of these concludes more pungently than Thomas Brunoski's 1983 survey according to which, 'the case, then, seems to be closed, other than for the emotional issues. The health-food advocate, with his beatific vision of less illness through vitamins, does not have a scientific leg on which to stand in his advanced–design running shoes.'

Some rather casual, and many rigorously double-blind studies involving altogether thousands of people followed for weeks or months have failed to uncover any prophylactic benefits and differences in the frequency or severity of colds, or come out with only clinically forgettable benefits. And as far as those many people who are absolutely sure that the megadoses work for them are concerned, Brunoski offers a logical explanation: ascorbic acid does have a mild antihistamine effect so that the course of the cold may seem shorter and more benign, the placebo effect in these self-administered cures is undoubtedly quite strong, and people who eat bulk vitamins do in general take better care of themselves and should be expected to have fewer minor illnesses. Hence his conclusion: 'In other words, taking vitamin C may in fact be associated with fewer and milder colds, but not causally.' Other amazing claims made on behalf of vitamin C rest on similarly shaky bases as can be seen by examination of critical appraisals now running into hundreds.

The wide acceptance of vitamin C megadoses, is due not only to the amazing effects claimed on its behalf, but also to its purported harmlessness which fosters the 'I have nothing to lose' attitude. To Pauling it is 'one of the least toxic substances known'—but it is worth noting that three efficient processes guard the human body against excessive levels of ascorbic acid in the first place. First, the efficiency of vitamin C absorption decreases with larger doses (to the extent that megadose consumers may lose up to 94 per cent in faeces and may be actually getting only little more of the substance than modest users!); second, at high doses vitamin C is not reabsorbed in the kidneys and its urine excretion rises fast; and, finally, high C levels trigger speedier enzymatic destruction of the vitamin.

The inescapable conclusion would seem to be that the body clearly tries to prevent overloading with the substance: if its defences weaken or fail, very large doses have been shown to carry disturbing side-effects. The most publicized of these hazards is the formation of renal stones but there are more troublesome ones: increase in red blood cell lysis, decreased leukocyte count, lowered plasma levels of vitamin B_{12} (destroyed by high doses of ascorbic acid during digestion), false positive reactions on common tests for urinary sugar (hence a possibility of overdosing by insulin), neonatal scurvy in babies of mothers on C megadoses, and rebound scurvy in adults after abrupt cessation of the therapy (both of these deficiencies being caused by highly active catabolic enzymes), and, certainly the most worrisome risk, the possibilities of infertility, abortion (in the first three months), and even teratogenicity.

These hazards would seem to justify much more research into the consequences of vitamin C megadoses: the real question is not whether several g of ascorbic acid a day will make somebody's nose feel less stuffy and his spirits more bouncy—but rather the long-term safety of these self-administered therapies for which there is no unequivocal proof of real health benefit. Vitamin C is obviously a critical micronutrient but, as with allowances for other nutrients, pinpointing single values for its minimal and optimal daily intake is an elusive, and in many ways an impossible, task. Much work has been done toward this goal but more must be done in setting out the maxima beyond which there is little or no benefit and growing odds of risks.

Minima have been clearly spelled out: 10 mg a day experimentally established as the lowest intake to prevent scurvy; 20 mg for children up to 12 years, and 30 mg for adults, recommended by the FAO; 35 mg for infants, 45 mg for children up to 10 years, and 60 mg for men and women according to the 1980 daily dietary allowance of the Food and Nutrition Board of the National Academy of Sciences. The last level should prevent any conceivable deficiency in a normal adult. It will maintain a body reserve of about 1.5 g of ascorbic acid and it can be easily ingested as a part of normal, varied diet-a 200 ml glass of orange juice or a small sweet pepper will alone suffice to fill the need.

But experiments with various population groups show predictably wide differences among the studies and among the individuals. Mean intakes needed to prevent scurvy range from just 6.5 to 45 mg a day, whole blood saturation was achieved with anywhere between 50 and 77 mg and blood plasma saturation with 70-130 mg; urine excretion measurements indicated needs as low as 25 and as high as 125 mg. Unfortunately, as Isabel Irwin and Bobbie Hutchins point out in their exhaustive survey of research on vitamin C requirements, in spite of a huge corpus of studies (they cite 449 of them) most of this evidence comes from detailed work with only small numbers of subjects in any age groups, and re-evaluations are bound to come.

The latest NAS adult allowance of 60 mg a day issued in 1980 represents one such change, a 33 per cent increase from the previous level of 45 mg; at the same time, the new RDA report concluded that there is some evidence that intakes of up to 200 mg a day may be beneficial—though it did not recommend intakes at that level. The upper bounds of sensible, rewarding intake thus remain open but only in the range between sixty and a few hundred mg a day: until a very large mass of experimental evidence has been shown to be uniformly faulty, scientific imperatives demand a sharp rejection of megadose prophylaction with intakes of several grams a day.

But to me the most persuasive arguments against megadoses of any micronutrients have always been the evolutionary ones. First, until the past few decades it has been impossible to consume vitamin C in megadoses: nature does not offer any such concentrated sources and hence even the most determined eater would face a mighty challenge in digesting the requisite doses. Taking Pauling's lower dose of 4 g a day would mean it would be necessary to drink 8 L of orange juice (normal daily total liquid intake is about 1.5-2.0 litres), or eat one kilogram of ripe red chilli peppers; about 5 kg of cauliflower would also do but it is clear that even the most inventive dieticians would be at loss to combine these rich sources of ascorbic acid into balanced meals.

Second, ascorbic acid deficiency has never been a health problem among even poor populations as long as a reasonably balanced diet was eaten; scurvy was certainly widespread in some parts of temperate regions but it disappeared once just 10-30 mg a day became available even during the winters—and for this potatoes or sauerkraut will suffice at the cheapest consumption level (having, respectively, 14 and 20 mg of vitamin C per 100 g). Third, the very fact that humans were able to leave the tropics (with their abundance of antiscorbutics) and to settle successfully in temperate and even arctic regions, where foods containing vitamin C are rare in winter, proves that normal requirements for this micronutrient are relatively low.

Supplies of the order of 10^1, perhaps just into 10^2 mg are easily available in varied temperate diets but intakes of the order of 10^3 mg would be impossible to provide naturally either in France or in Northern China, to name just two places in northern temperate latitudes where human settlements go back many thousands of years. These populations, as Thomas Jukes pointed out, must have been under strong selection pressures for low requirements for ascorbic acid. The fact that today's tropical populations do not have any higher requirements for vitamin C than their more northerly counterparts is best explained by incessant genetic mixing.

Finally, Pauling's argument that the loss of ascorbic acid synthesizing ability in the common ancestor of man and other primates (which he places some 20 million years ago) 'provides strong evidence that the optimum rate of intake by man is about 2 or 3 grams per day or more' is well answered by the studies which show that the distribution of the evolutionary loss of this ability in warm-blooded vertebrates is quite scattered, perhaps sporadic, and that in *Homo sapiens*, as suggested by King and Jukes, it was a nonadaptive, neutral evolutionary change: when consuming a normal varied diet the loss of this ability is functionally neutral.

The simplistic attitude 'if a small amount of vitamin is good for you then a large amount must be even better' has not, of course, stopped at ascorbic acid. Every known vitamin has been touted as a key to perfect health and longevity and most are taken in megadoses by unknown numbers of people (but surely totalling millions)—while critical examinations show not only lack of the claimed effects but identify numerous dangers.

Starting alphabetically, many food faddists who have been taking vitamin A in the range of 100 000 to 500 000 IU a day (5000 units is the American RDA) are undoubtedly flirting with the prospect of future liver diseases (hepatic fibrosis, cirrhosis). Hypervitaminosis-A induces abnormalities in the structure and bio-chemistry of red blood cells and their membranes, and in 1983 came the first evidence about sub-clinical liver damage with doses exceeding 100 000 IU. Another recently (1982) identified side effect of high dosage vitamin A therapy is premature epiphyseal closure, bone pains, and slowed growth in children.

The B group, except for B_{12}, have fortunately low toxicity and rapid excretion rates so that, in Thomas Jukes words, their 'large doses . . . serve primarily to enrich the urine of patients and the pocketbooks of those who merchandise the products'. The latest fashion is a vitamin looking for a deficiency disease: as there is plenty of E (tocopherol) in plant lipids (above all in all ordinary oils) its intake could be insufficient throughout the rich world where large quantities of such oils are consumed regularly only in the very rare instances.

This fact did not in the least prevent the rise of veritable cult around this common substance. *Runner's World*, unhappy with federal restrictions on any public claims regarding vitamin supplementation, touts vitamin E which 'ensures fertility, protects us from heart disease, maintains pulmonary capacity and keeps us feeling young and physically active'. But even the National Cancer Institute is interested and awards more than a million dollars to an MIT researcher to study the effects of tocopherol, together with vitamin C, on the blockage of nitrosation along the gastro-intestinal tract.

Indeed, judged by summaries of the claims made on its behalf, tocopherol vies with vitamin C for the status of the true elixir of life. Pop-science writers see in it the very essence of all vitamins because 'it puts oxygen quality in each blood cell' and assure us that its therapeutic and prophylactic actions are beneficial in ischemic heart disease, occlusive peripheral vascular disorders, thrombophlebitis, leg cramps, fibrocystic breast disease, vascular complications of diabetes, assorted dermatoses, sterility, polyomyositis, and in the slowdown of aging, the last property leading some of them to advocate tocopherol's addition to food in 'project Methuselah'.

No wonder then that some patients and many healthy people will resort to truly gigantic doses. Hyman Roberts, writing in the *Journal of the American Medical Association*, noted that he frequently obtained a history of the daily ingestion of 800–1600 units from his patients! While Pauling's megadoses of vitamin C are 'just' 65–130 times the current American recommendation, Roberts' patients and thousands of other faddists have been eating up to 200 times the daily allowance recommended by the National Academy of Sciences (FAO has not yet set any international minimum for vitamin E).

Yet the list of clinical disorders attributable to large E doses, verified and published in numerous scientific journals, is not any shorter than the array of the supposed benefits: it ranges from the just annoying (chapping of the lips, urticaria, headaches, fatigue) to the worrying (hypertension, intestinal cramps, hypoglycemia, vaginal bleeding) to highly hazardous (pulmonary embolism, thrombophlebitis, apparent aggravation of diabetes). In general, the potentially harmful side-effects of vitamin E megadoses are much greater than those resulting from very large intakes of ascorbic acid: the daily consumption of many hundreds, even thousands, of units of tocopherol in an unsupervised way means taking unnecessary risks for no clearly proven benefits. Besides, physiological and evolutionary imperatives against any large supplementation are in this case even stronger than with vitamin C: scurvy is induced easily with otherwise nutritious C-poor diets but vitamin E deficiency is difficult to demonstrate or to induce even in elderly persons with manifest multiple deficiencies of other vitamins!

In view of the lively megavitamin cults one might expect the existence of megamineral schools: after all, deficiencies of essential minerals are not uncommon and their symptoms can be in many ways no less crippling than shortages of vitamins yet, strangely enough, there is no Linus Pauling pushing trace metals in megadoses. There are, however, large-scale efforts at iron supplementation, consisting mainly of 'fortification' of white flour and breakfast cereals and production and promotion of a wide variety of oral iron preparations to treat iron-deficiency anaemias.

In spite of these efforts, and in spite of the fact that only a mere 1.0–1.5 mg of iron have to be replaced each day (amounts far smaller than even a relatively iron-poor diet will contain), iron deficiency is a common finding even in the rich countries and it is prevalent throughout much of the poor world. The principal reason is the inefficient absorption of much of the food iron. Haeme iron derived from haemoglobin and myoglobin of animal tissues is absorbed easily but bio-availabilities of plant iron, non-haeme iron in meats, and soluble inorganic iron salts are relatively low and depend critically on the overall diet composition with ascorbic acid and meat protein enhancing the absorption and carbonates, oxalates, and fibres inhibiting the process.

For this reason FAO's recommended iron intakes are specified as a range, 5–10 mg a day for children up to 12 years, 9–18 mg for boys 13–15, 12–24 mg for girls of the same age, 5–9 for adult men, and 14–28 for adult women with the lower value applicable when more than 25 per cent of food energy comes from animal foods, the higher one when these foods provide less than one-tenth of all energy intake. In the rich countries, where animal proteins are present in substantial quantities, single values of RDA (recommended daily allowance) will suffice: United States levels are 15 mg for children up to three years, 10 mg between 4 and 10 years, and 18 mg for teenagers up to 18 years; the last value also applies to women up to 50, while adult men's need is put at just 10 mg.

A quick appreciation of the global extent of iron deficiency can be gained from a survey of iron availability in FAO's food balances: of 150 nations 70 have average daily per capita supplies below 15 mg and as most of that iron is in non-haeme form anything less than 20 or even 25 mg a day may mean a deficiency. But iron deficiencies are surprisingly common even in the rich meat-eating nations, especially among children and women.

With infants and small children it may be largely a feeding problem: they eat relatively small quantities of solid food and are not eager to try most of the high-iron meats and vegetables. For this reason the American Academy of Paediatrics recommends iron-fortified infant cereals but, although these have been just as much available for older children, their iron deficiencies are often incredible. The Ten State Nutrition Survey conducted during 1968–70 revealed that among the children of low-income families a mere 2 per cent received their recommended daily allowance while two-thirds consumed less than half the desirable amount.

Incredibly, ten years later the situation had hardly changed. Between 1976 and 1980 the National Health and Nutrition Examination Survey, of which much more will be said in section 3.2.2, revealed that 90 per cent of all 1–2 year old boys and girls in families with incomes below the poverty line received less than the recommended daily intake and that half of them got less than 50 per cent of it. And the deficiency is not eliminated by higher incomes, just attenuated. Figure 3.8 traces the fiftieth percentile of atual iron intakes for all 10 000 women in the 1976–80 surveys versus the RDA: only the youthful eaters of fortified cereals meet the need for a few years and the gap remains large for most of the women's lives.

These are serious shortfalls but even so they cannot be simply translated into comparable iron deficiencies. Iron deficiency progresses in three successive stages: during iron depletion reserve forms of iron in muscle, liver, and reticuloendothelial cells are reduced; in the second stage the level of the haeme precursor rises in the red cells indicating an insufficient supply of iron for the haeme synthesis; and only in the third stage, during iron deficiency anaemia, is the supply of iron to the erythroid marrow so restricted that the numbers of circulating red cells drop. Because haemoglobin levels do not fall until iron efficiency has gone relatively far their measurement is a rather poor, though still the most frequent, means of

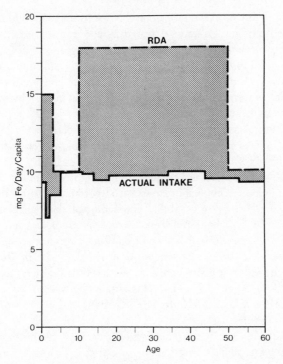

Fig. 3.8 The recommended daily allowance for iron and actual intakes of the metal by American females established by the surveys between 1976 and 1980. Except at the ages of 5-11, the time of eating iron-enriched morning cereals, American females appear to have huge iron deficits throughout their pre-menopausal lives.

detailing iron deficiency status. Moreover, low haemoglobin is not a specific consequence of iron deficiency and normal variation of its levels is considerable.

Consequently, Jean Bowering, Ann Sanchez, and Isabel Irwin concluded that 'the wide range of iron intakes and hematological values that appear to be compatible with good health suggests that adequate nuriture is achieved through a remarkable set of compensating mechanisms in the human body' and these wide ranges also suggest that survey data from various population groups 'are of questionable value in determining the human requirement for iron'. Normal individual variation rather than deficiency calling for treatment may thus account for many low iron intakes and haemoglobin levels lower than uncertain standards.

Iron deficiency, whatever its actual extent, is not only the most frequently observed shortage of mineral element in man-it is also a serious and widespread problem in crop production. Consequently, it would seem that raising the iron content of soils could lead to lower iron deficiency but, counter-intuitive as it may be, such is not the case. Iron deficiency in crops is only rarely caused by absolute shortages of the element in the soil: almost always iron's ion solubility is to blame and this difficulty is not effectively treated by additions of soluble iron (it will rapidly revert to insoluble forms), but rather by changing soil pH to raise acidity and so raise iron solubility (incorporation of sulphur is a common procedure).

Plants grown with more soluble iron in the soil are greener, grow larger, and yield more but their iron concentration per unit weight of dry phytomass may not be any higher than in stunted, chlorotic growth with inadequate Fe. Hence there is little or no direct relationship between common iron deficiencies in crops and in people, but there is at least one important benefit from higher availability of soil iron as it raises the concentration of provitamin A.

Eating foods with a high iron content is only slightly more efficacious advice. Nutritional guidebooks invariably stress the importance of offal meats but this advice goes unheeded by most of the population of the rich countries whose preference is now definitely for red meat. No other food has such high iron content as pork liver (21 mg Fe/100 g) but consumption of this superior source of the mineral has been declining not only in North America but also in Western Europe; in the Orient, however, it is still consumed with relish. Veal liver has 14 mg, beef and chicken livers less than 9, and various kidneys between 6 and 7 mg/100 mg. Virtually all other animal foodstuffs rate less than 5 mg Fe/100 g, and all plant staples have just around 1 mg.

Fortunately, for those unable to face regular platefuls of kidneys and liver there are three excellent plant substitutes: dried apricots (5.5 mg), sunflower seeds (7.1 mg), and sesame seeds (10.5 mg), richer in iron than anything but pork and veal liver. For most people less palatable vegetal alternatives would be a regular consumption of rather substantial quantities of legumes (when cooked mostly between 2-3 mg Fe/100 g), whole-wheat bread (2.3 mg)—or the proverbially efficacious spinach (2.2 mg).

But most of this plant, non haeme, iron will not get absorbed without concurrent consumption of vitamin C and, as this effect does not depend on the overall ascorbic acid status of the individual but is directly related to the amount of the vitamin in the meal, the vegetarian paths to iron sufficiency lead only through combinations of high-Fe and high-C foods.

This advice may be a sensible one even for a relatively poor Western family eating less meat to balance the budget: a sprinkling of seeds in the morning cereals accompanied by a glass of orange or apple juice (now usually fortified with ascorbic acid to about two-thirds of the natural level in oranges) is not a financially crippling proposition. But in poor countries this balancing act would be beyond the means of all those hundreds of millions eating monotonous diets of grain or tuber staples enlivened so very infrequently by consumption of finer foods. Equally unrealistic would be to think of regular intake of iron and vitamin C supplements in such circumstances.

Clearly, iron deficiency will remain a widespread and chronic feature in both rich and poor countries: in the first because people do not eat what is easily available and financially accessible, in the second because they cannot. Extensive and serious iron deficiencies in the rich world are certainly more surprising and together with the surfeit of sugars and lipids discussed in the preceding section they are the hallmarks of deteriorating quality of diets in countries which could easily provide excellent nutrition.

But so far I have concentrated on food needs in terms of total energy and principal macro- and micro-nutrients, introducing actual consumption figures only in several places when dealing with sugars, fats, ascorbic acid, and iron. A systematic appraisal of harvests and meals in the next two sections will show the inequalities both in terms of food production and in actual everyday consumption.

3.2 Unequal bounty

> The problem of regional differentials in production and consumption of food
> ... is not simply one of plenty in one corner and tremendous poverty in
> another. The picture of plenty is as much a problem as that of poverty.
>
> J. Chandrasekhar, *Hungry People and Empty Lands* (1954)

Historical statistics of food production are available in great detail for over a century for many Western nations, but a satisfactorily reliable global look can be offered only since the early 1950s: pre-World War II coverage by the publications of the League of Nations was never exhaustively global and the first few FAO *Production Yearbooks* in the late 1940s still had many gaps. But more than three decades of consistent (although often still disputable) output series offer plenty of fascinating information illuminating many strange realities of global food production.

As noted at the opening of this chapter, if ours were a single uniform global civilization then the record would be definitely satisfactory: Fig. 3.1 charted the steady increases of all five major plant food groups, between the early 1950s and the early 1980s, and in energy terms (multiplying by 3600 kcal/kg for cereals, pulses, and sugar, 900 kcal for tubers, and 8800 kcal for oils) these harvests would translate to 2.75×10^{15} kcal in 1950 and 7.25×10^{15} kcal in 1980, equivalents of about 3000 and 4500 kcal a day per capita.

What is surprising in the real world with its endless inequities is not the existence of disparity between the production of essential plant foods in rich and poor countries—but rather the continuation of disproportionately high staple crop harvests in the rich nations. In per capita terms the situation of the poor countries improved appreciably, with the harvests of the five major food groups rising from about 2500 to 3200 kcal between 1950 and 1980, a nearly 30 per cent increment achieved in spite of the doubling of the poor world's population. But during the past three decades the poor countries have not enlarged at all their share of global food production: in the early 1950s they produced (rounded to the nearest significant fractions) half of all cereals, three-fifths of all tubers, oils, and sugar, and four-fifths of pulses—and these shares have remained surprisingly stable since then.

In the early 1980s the poor world's shares of global food output were between 48 and 49 per cent for cereals, 59 and 65 per cent for tubers, 57 and 60 per cent for sugar, 55 and 58 per cent for oils, and 72 and 79 per cent for pulses, the ranges reflecting natural yield fluctuations but displaying no secular rise. And so the rich world, whose share of global population slipped from 36 to less than 25 per cent betwen 1950 and 1985 continues to produce just over half of all staple cereals and almost exactly half of all food energy from the five leading categories of plant foods. This vigorous upholding of the disproportionately large production share has increased the availability of major plant foods to the rich nations from 4000 kcal per capita in 1950 to 7900 kcal in 1980.

Obviously, people cannot consume so much food directly—and so a steadily increasing portion of these rich harvests has been fed to domestic animals and eaten indirectly as meat, dairy products, and eggs—or exported. The first choice has led to diets where animal foods—supplying less than 10 per cent, and often even less than 5 per cent of all food energy, and protein in traditional farming societies—are now providing up to 35 per cent of all energy and 70 per cent of protein. Moreover, as they are inevitably associated with higher fat content they also raise total lipid consumption to unprecedented heights—for example in the United States to an average of 40 per cent of all energy, and for many middle-aged males up to 55 per cent.

As repeatedly noted in section 3.1.2 the evidence is far too complex to find any simple relationship between this historically anomalous diet and health and longevity, but deriving such large portions of energy from a single nutrient must be clearly seen as unbalanced nutrition, a situation which is especially unnecessary when one considers the variety and price of foods now available in the affluent nations.

Feeding of large parts of crop harvest to livestock has also fundamental energy implications: inescapable losses during the conversion of grains and legumes to meat, milk, and eggs mean that only a small fraction of the original energy will be consumed in animal foods. Only for an approximation of the right order of magnitude it is sensible to apply the venerable '10 per cent law'. Modern domestic animals convert feed nutrients to edible products with efficiencies ranging between 5 and 10 per cent for beef, 10 and 15 per cent for pork, and 15 and 20 per cent for poultry (depending, obviously, on the breed, quality of feed, degree of confinement and health of the animal), while feed-to-milk energy transfers are between 15 and 25 per cent.

Growth and maintenance of sire and dam animals and disease losses must be accounted for by cutting all of the cited values by at least 20 per cent. Consequently, energy losses ranging around 80 per cent are the inevitable price of milk drinking while blood-oozing steaks may represent as much as a 96 per cent loss. Obviously, any shift away from the current meaty Western diets—not a transformation to lacto-ovo-vegetarianism, or to pure vegetarianism, but just a reduction in the overall meat consumption (after all, when an average Western adult male now eats one-and-a-half times to twice his weight in meat there is certainly room for decline)—would have profound implications for the overall global availability of food.

Basic quantification of the United States commitment to animal farming illustrates best the gains possible even with relatively painless (in terms of nutritional composition) adjustments. In recent years the country's domestic animals were fed annually a staggering mass of feeds: aside from pasture grasses about 130 million tonnes of hay (with about two-thirds of it from alfalfa), and 10 million tonnes of milling residues (various brans and hulls) were fed for roughage—and, of course, there is no better use for these commodities than to feed them to animals since bran for fibre enthusiasts constitutes a forgettably small portion of all bran production, and digestion of alfalfa except for its young sprouts is not easy for humans. However, reduced meat consumption would enable part of the land now used for roughage to be planted to food crops and I will make an estimate of this effect presently.

Oilseed cakes are like hays—not very palatable to humans, but soybeans (of which some 20 million tonnes are now fed in high-protein meals) can be switched from feed to food (whole or as oil) easily enough. So could grains, of which some 140 million tonnes are now used as feed annually, although it would often be preferable to plant more desirable kinds or preferred food cultivars on the land released from growing crops; overall yields would remain very similar in the second case (for example durum wheat instead of soft wheat), but as corn (maize) dominates American grain feeding but wheat is much more sought after as a food, this exchange would carry a substantial yield decrement. How big this difference would be is not easy to estimate: current average yields are about 8 t/ha for feed corn and just over 2 t/ha for wheat, but wheat grown on part of the Corn Belt's soil instead of corn, and fertilized at Western European level, should yield twice as much as today's mean (the West European mean is now nearly 5 t/ha).

Just for the sake of an example, I will assume that the current United States meat consumption would be reduced only by one-third, that is to some 65 kg a year per capita, leaving it still at a level more than twice as high as the current Japanese consumption. This shift would release directly some 7 million tonnes of soybeans, and open up nearly 20 million hectares of feed grain crops, and about 8 million hectares of haylands for further food cultivation. Planting of soybeans instead of alfalfa, and wheat (yielding just twice the current average) instead of corn would produce an additional 16 million tonnes of the former (for a total food disposal of nearly 25 million tonnes), and 80 million tonnes of the latter.

How huge these totals are can be shown in two ways: firstly by comparing them to current export levels, then by expressing them in daily food equivalents. Recent American exports of all wheats have been fluctuating between 40 and 50 million tonnes a year, so that having an additional 80 million tonnes would boost them by 60 to 100 per cent; soybean exports averaged lately 20 million tonnes a year so that they could grow by 125 per cent. For nutritional value I will not make a straight energy-content calculation as few people would choose to live on wheat flour and soybean-based foods, but rather I will assume that 25 per cent of these increased exports would be fed to animals which would convert this feed with average conversion efficiency of 15 per cent so that the diet would have a minimum of 5 per cent of food energy coming from animal foods—very much a pattern now prevailing on the Indian subcontinent.

The result of these realistic assumptions is stunning: of 105 million tonnes of wheat and soybeans (total of about 3.8×10^{14} kcal) 9.5×10^{13} kcal would be converted to feed, providing 1.4×10^{13} kcal of animal foods so that the total food availability would be almost exactly 3×10^{14} kcal. Current Indian food availability is only about 2100 kcal a day per capita, almost certainly a shortfall compared to requirements more compatible with a healthy and longer life. A mean of 2300 kcal would most likely cover the average need and dividing 3×10^{14} kcal by this average would give enough food for the whole year to 350 million Indians—nearly half of the world's second most populous nation!

And, it needs repeating again, every American could still eat 65 kg of meat a year. I introduced this detailed example to illustrate what I consider to be perhaps the most surprising reality of the current global food situation: an astonishing 'slack' in the system which would result from the adoption of somewhat less meaty diets in the rich countries. This realization is especially essential in view of the chronic comments by assorted agricultural experts who keep warning us about the insecurity of the present global food situation, inadequate grain stocks, uncomfortable dependence on good weather for the harvesting of decent crops, and the difficulties of keeping up with population increases in the production of staple cereals.

Surely it makes little sense to keep on raising the spectre of 'impending global food shortages', and trying to forecast the dates when the world demand for cereals will outstrip production, when the rich world is now feeding annually over half a billion tonnes of grains to its domestic animals—to 'enjoy' diets which are being increasingly implicated in the etiology of major diseases and whose modest modification would release masses of grain surpassing the total amount of current world grain exports.

Quite obviously, global food supply shortfalls are a matter of unequal distribution arising from large differences in output and from the inability of many poor nations to manage their affairs sensibly and to purchase more food on the free market to fill temporary shortages—but it has nothing whatsoever to do with any global physical shortage. Equally obvious should be the potential for much larger harvests throughout the poor world where only a few crops in less than a score of countries, above all rice in the Far East, are currently receiving fertilizer inputs sufficient to produce yields approaching or even surpassing those in some rich nations.

Largely as a result of these improved cereal yields the average nutritional status of not only the Korean and Malaysian, but also the Philippine and Chinese populations is now considerably better than a generation ago and these encouraging changes have been more recently spreading into India, Pakistan, and Indonesia. Still, this is a slow process, especially in those countries where relatively high rates of population growth have barely started to moderate. The nutritional gap between rich and poor countries thus remains one of the most fundamental inequalities separating the affluent fifth of mankind from the less fortunate populations of Latin America, Africa, and Asia. As such it is worthy of a closer scrutiny.

3.2.1 *Rich and poor diets*

> One-year-old boars we keep for ourselves,
> The three-year-olds we offer to our lord.
>
> *Shi jing* (Seventh Century B C) (I. Y. Lo translation)

If prescribing the optimum quantities of essential nutrients is not at all easy it might seem that at least the task of determining actual food consumption is straight-forward and fairly simple. But it turns out that finding out what and how much people eat is surprisingly complicated and as a result we do not have, even for rich countries with abundant statistical bases, very accurate figures of actual food consumption.

This may come as a surprise to all those who have seen abundant FAO statistics with long printouts of daily per capita food supplies, broken down not only into energy and protein categories but also listing contributions of plant and animal foods and available amounts of all major minerals and vitamins. A quick check of FAO's *Production Yearbook* will show these figures for no less than 150 nations, even with short historical coverage: Americans are now averaging over 3500 kcal a day (just 5 per cent more than around 1970), Egyptians are at nearly 3000 kcal after adding nearly 20 per cent in a decade, Haitians below 1900 kcal, no improvement in the past decade.

Unfortunately, none of these values tells us how much the Americans, Egyptians, and Haitians are actually eating as all of the FAO's meticulously tabulated figures are just computerized calculations done in Rome, the best quantifications prepared on the basis of carefully constructed national food balance sheets. A closer examination reveals not a few problems with these exercises. Where the requisite output and conversion statistics are reliable, and where virtually all of the food travels through commercial channels, carefully interpreted food balance sheets are very good guides to the nutritional status of a nation. In the absence of good farming and food processing data, and in those cases of significant reliance on foods whose consumption is difficult even to estimate (wild meat, invertebrates, forest fruits), it is possible to end up with distorting under-estimates.

Fortunately, in most cases national food balance sheets present a comprehensive and also a fairly reliable, though generalized, pattern of food supply—but not of actual food consumption. Every food balance sheet must account for output, trade, and stock changes in all important vegetal and animal products; it must exclude non-food uses of these commodities (that is the fractions going for seed, feed, and industrial processing), as well losses occurring during storage and transportation. Finally, the sheet must incorporate appropriate extraction rates in all cases where some processing (above all cereal milling and oil pressing, also meat dressing) must be done before the product is ready for consumption.

The resulting totals indicate the availability of individual foodstuffs as they enter the households and commercial food establishments, and do not account for any subsequent, and difficult to trace, preparation losses (which may often be very high), household, and institutional spoilage, and distribution patterns (all family members do not, obviously, share the prepared food equally). Consequently, in most instances the national food balance sheet averages will overstate the actual food consumption, and they may also differ somewhat owing to small disparities between conversion factors, and as a result of inclusion or exclusion of food consumption by the military.

Actual food consumption will be best revealed only by careful and systematic household surveys, but to be representative such studies must include a large number of families of various socio-economic groups, they must last sufficiently long or be repeated periodically to capture seasonal variations (which are still very important in the poor countries), and they must account for food eaten outside the home (now a major portion of total Western consumption: for example, in the United States almost half of people questioned during nutritional surveys now eat at least one meal a day away from home). Consequently, such surveys are expensive and are available repeatedly only for some rich nations—but their comparison with food balance sheet values is a revealing experience.

Even when one is aware of possible discrepancies, comparisons of the best available figures still surprise by their differences as can be impressively demonstrated by recent American figures. An extensive United States food consumption survey—the National Health and Nutrition Examination Survey (NHANES) II which examined over 20 000 persons aged form 6 months to 74 years during 1976 and 1980—found that the daily energy intake for all males averaged 2381 kcal (and for those between 25 and 34 years, 2734 kcal), for females only 1579 kcal (1643 kcal for those 23 to 34 years old).

In contrast, United States Department of Agriculture statistics on per capita food consumption—prepared for the civilian sector in terms of retail weight and including estimates of the produce of home gardens—show values of 3260 kcal in 1975 and 3420 kcal in 1980, while the FAO food balance sheets for the country list for the same years, 3552 and 3652 kcal. FAO's American food balances are thus 7–9 per cent higher than USDA's retail consumption estimates, a reasonably close agreement for calculations of this kind—but both of these figures are far above what people actually claim they eat.

Of course, even USDA's statistics must be further reduced by household losses and wastes, and (a peculiarly important consideration in the United States with its millions of dogs, cats, gerbils, hamsters, etc.) by use for pet food, but these reductions would have to add up to well over 1000 kcal a day per capita to balance the means. Taking a weighted mean of the NHANES II study (there were 10338 females and 9983 males) results in an average of 1973 kcal, or a mere 58 per cent of USDA's 1976–80 average. And USDA's own Nationwide Food Consumption

Survey (NFCS) conducted during 1977–8 showed that females between 23 and 34 years of age consumed 1616 kcal a day and males of the same group merely 2449 kcal, the latter mean (based on 770 observations) being a further 10 per cent below the NHANES II average.

As both the NFCS and NHANES means are based on large national samples selected statistically to represent the country's population, and as the daily intakes were determined not only by a 24-hour dietary recall, but also by the keeping of food records, it would have been impossible to leave out food worth over 1000 kcal—but it is equally impossible to postulate a mean daily household loss of that magnitude. Some American household food use data suggest that a loss in the food marketing system may be equivalent to about 600 kcal/capita per day, or roughly 15 per cent of the amount indicated by USDA's or FAO's calculations of supply availability.

Accepting this still leaves a gap of 1000 kcal, or a loss of over 30 per cent of foodstuffs brought into the kitchen, but studies of waste, appropriately based on analysis of garbage, indicate household losses of around 10 per cent of purchased energy. Losses in commercial and institutional cooking may be often much higher but not so high as to close the gap. Perhaps another quarter, or even a third of this gap can be explained by a failure to record all of an individual's food intake: current food survey techniques are relatively poor but, again, it would take improbably large and, above all, uniform under-reporting to fill the remaining gap of at least 400 kcal/person a day.

Whatever the combination of wastes, losses, and shoddy accounting, the surprising fact remains: the actual average per capita food consumption in the United States is not known with satisfactory accuracy and the discrepancy between unexpectedly low food consumption survey data and food balance sheets remains difficult to reconcile. Judging by this example, FAO's food balances (which are the most often used source of international nutritional statistics) would have to be reduced by as much as 40 per cent to bring them closer to actual food intakes—but other comparisons indicate somewhat lower, though still large reductions.

Japan's National Nutrition Survey, repeated for five days each year and encompassing about 15 000 households, is perhaps the most reliable one in the world as all food consumed by all household members is weighed and recorded and the records are checked every day by a dietician. National per capita averages from these annual surveys were 2188 kcal a day in 1975, and 2084 kcal in 1980—while the corresponding FAO food balances gave, respectively, values of about 2800 and 2900 kcal, differences of 23–28 per cent. Such gaps are much easier to explain in terms of wastes and losses, and reduction of FAO means by 25 per cent would thus appear a sensible, conservative correction—at least for the rich countries.

Generalizing on a correction factor for the poor nations, where the construction of food balance sheets must so often start with only approximate output values (which may overestimate as well as underestimate the actual harvests), and where in many cases collected wild forest plants and wild animal meat may be of essential nutritional importance, is much harder. Further errors can be introduced by inaccurate population estimates: for example, until the Chinese renewed their releases of official statistics in 1978, FAO used the lowest available estimates of the country's population which, together with some exaggerated output figures, resulted in considerable overvaluation of China's per capita food availablity during the late 1960s and most of the 1970s.

The Chinese have been releasing official statistics again since 1978, and as food waste in the world's largest nation is certainly incomparably smaller than in the rich Western nations, the difference between food balance calculations and actual consumption should be much smaller. FAO put the national mean of 1978–80 food supply at 2472 kcal, the Chinese State Statistical Bureau (SSB) quoted 2311 kcal for 1978, my calculations, based on the SSB output data, result in 2130 kcal brought into the kitchen where little is wasted. More recent Chinese figures show gross availability of 2779 kcal/day per capita in 1982—while the SSB's first detailed food consumption survey showed 2371 kcal eaten in the capital, 2512 kcal in its suburbs, in 1983. Consequently, even in China there will be about 15 per cent difference between the widely cited food supply means and actual kitchen use.

Across-the-board reductions of FAO food availability calculations ranging from 15 to 40 per cent are thus necessary to bring them closer to real per capita consumption but otherwise these figures, derived by a uniform method, remain the best comprehensive source of global food consumption data and the best foundation for international comparisons. As already noted, FAO publishes annually its summary, containing data on total per capita food supply and its split into vegetal and animal products (all in kcal/day), protein and lipids (again split into the two categories, in g/day), as well as data on key minerals (calcium and iron) and vitamins (retinol, thiamine, riboflavin, niacin, and ascorbic acid).

About 150 nations have been lately included in this regular summary and the organization also brings out at irregular intervals detailed food balance sheets where all the production and disposal data and conversion assumptions are presented in full detail to facilitate a calculation of average food balances for a period of three years, and as many as 164 countries are included in the latest volumes. Even a quick review of these values shows the obvious distinctions between the poor and the rich countries—but, unlike the case of primary commercial energy consumption (Fig. 2.5), the global distribution is not hyperbolic.

As shown in Fig. 3.9 the set can be split conveniently along 'natural' breaks (that is in the places of low frequency) into three major groups. The best-off, where the average daily per capita food supply surpasses 3000 kcal, contains 33 countries: the whole of Europe (with the exception of Albania), United States, Canada, Australia, New Zealand, Isreal, and Argentina—and recently Libya's oil money and high food imports also put that country into this top category.

The middle group, 37 countries strong, includes all the richer industrializing countries, from Venezuela and Chile to Iran and Syria; Japan is near its top and Brazil just gets in at the bottom. The largest group of poor countries with less than 2500 kcal per capita a day contains all but a few African nations and the poorest

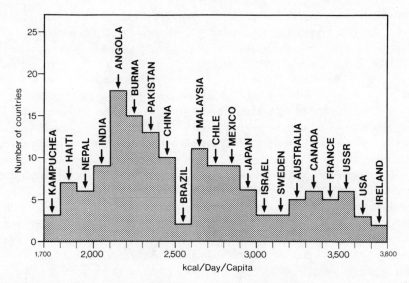

Fig. 3.9 Global distribution of daily per capita food availabilities derived from the food balance sheets for the early 1980s. Distribution is not skewed as uncomfortably as the one for commercial primary energy but the number of nations with food availabilities below adequate growth and activity needs (that is mostly below 2200 kcal) is still very large.

countries of Asia and Latin America; 16 of these 82 nations in this category have a daily food supply below 2000 kcal, with Ethiopia, Kampuchea, and Maldives below 1800 kcal. As the highest supply (in the countries such as the United States, Ireland, Bulgaria and East Germany) surpasses 3600 kcal there is an almost exactly two-fold difference between the topmost and the lowest values.

After discussion of food energy requirements (section 3.1.1) it should be obvious that all nationwide average availabilities in excess of 3000 kcal a day are enormously excessive—unless just about everybody is engaged in heavy manual labour or in everyday strenuous exercise. The previously mentioned and astonishingly high percentages of obesity, now increasing greatly even among children and adolescents, attest that this is, obviously, not the case. And a detailed look at American, East German, or Irish food supply reveals common undesirable dietary habits in very high meat consumption and heavy intake of animal fats (in excess of 100 g per person a day).

On the other hand, daily diets below 2200 kcal are clearly too low to energize heavy field work and to sustain healthy and vigorous life-but it is impossible to estimate with any confidence the share of the world's population which suffers energy-protein malnutrition, or whose total intake and relatively poor quality of food provide a barely sufficient nutrition.

Still, such attempts have been made repeatedly, ranging from Lord Boyd Orr's 1950s estimate that 'a lifetime of malnutrition and actual hunger is the lot of at least two-thirds of mankind', through FAO's *Third World Food Survey* in 1963 according to which '10 to 15 per cent of the world population are undernourished and up to half suffer from hunger or malnutrition', to Jean Mayer's assessment that in the mid-1970s the number of people who were poorly nourished or undernourished was roughly one-eighth of the human population (about half a billion people at that time), and that billion people would have benefited from a more varied diet.

The progression of estimates from two-thirds to one-half to one-eighth of humanity suffering malnutrition, (a reduction of 80 per cent in just 25 years) is perhaps the clearest proof of the unrealistically high notions of nutritional requirements which dominated earlier thinking as well as an impressive warning against liberal tossing around of claims about starving billions. The real share remains unknown but approaches slightly more sophisticated than the simple weighing of food availabilities against theoretical food requirements indicate that the proportion of global population on inadequate diets is definitely closer to the low rather than to the high end of the published estimates. I will mention just one such attempt, Shlomo Reutlinger and Harold Alderman's calculations of the prevalence of energy-deficient diets in 36 countries of Latin America, Asia, and Africa.

They took FAO/WHO recommendations of energy needs as the yardstick, although they were well aware of their possible shortcomings, and used FAO food supply data in constructing indices of energy deficiency which took into account variations in intakes and requirements among individuals by assuming normal distributions of the two variables within separate income groups. This is a sensible approach as numerous food surveys have shown close relationships between income and food intake and the authors refined it further by constructing two indexes, the first being the proportion of the total population belonging to income groups which, on the average, have a caloric deficiency, the second the proportion of the total population having deficits based on a probabilistic appraisal of an individual's intakes in relation to requirements.

Unfortunately, the basic inputs into this exercise are clearly faulty: as noted earlier in this section FAO food supply values are at least 10 per cent higher than actual consumption (although in some cases the food balance sheet may under-estimate the intake!), and the use of requirements based on international reference bodies which are obviously too heavy for the studied populations brings in even more serious bias. However, as these two errors work in opposite directions they may largely eliminate the bias but I think that rather unlikely; more realistically, the net result will be to portray the actual situation still somewhat worse than it really is—but unless one were to treat all nations by calculations similar to those used in section 3.1.1 for China, all estimates of the deficit based on the mechanistic contrast of FAO supplies and needs remain dubious even when their finessing through considerations of income-related consumption and probabilistic intakes brings them, at least conceptually, much closer to the elusive reality.

Detailed surveys of household food consumption in relation to income are an essential tool for understanding the extent of malnutrition. We still do not have enough of this revealing information available at regular intervals to judge the trends, but careful study of the published distributions uncovers some unexpected results. While, for example, the Indonesian rural data show the expected steady increase in consumption of rice with rising income mirrored by a steady decline of cornmeal intakes, and swiftly climbing consumption of sugar and fats (so that families spending monthly three times the minimum per capita food expenditure eat four times more sugar and ten times more vegetal oils!), Brazilian data from Rio de Janeiro reveal a more complex pattern as households with incomes between 1.0–1.5 times minimum salaries eat per capita actually 10 per cent more fat and 15 per cent more sugar than those households earning 5–8 times the minimum.

But even these detailed surveys tell us nothing about one of the generally neglected yet very important inequalities of consumption which results from uneven food distribution within households. In those areas where food supply is not abundant and its quality is relatively poor—be it today's Africa and Southeast Asia or Europe five or more generations ago—children and women almost always fare worst as their diets consist almost solely of monotonous staples while most of the meat, fish, and better vegetables are eaten by men. This traditional and nearly universal bias has, naturally, far-reaching consequences for children's development.

For example, a survey of low-income families in Lagos (Nigeria) showed adult males consuming more than twice as much protein as 7–12 year old children who need more of the nutrient for healthy growth than the adults. In Indian studies (from Uttar Pradesh) men of low and middle-income families ate not only more bread, lentils, and vegetables than women but also consumed all of the small amount of available meat. And several African studies found satisfactory energy and protein intakes for adults of both sexes but considerable shortchanging of small children.

In comparison to the elusive account of energy-protein malnutrition, historical developments in food consumption and important shifts in eating patterns are much easier to chart. Figure 3.10 is based largely on FAO's food balance sheets and it offers an overview of half a century of food energy supply in six nations exemplifying developments in rich and poor nations. Except for wartime reductions, food energy supply in Europe (represented here by France) and North America (United States) has continued to rise as it did, very spectacularly, in postwar Japan. On the other hand, even Brazil, representing here those industrializing nations where abundance of land and relatively low population density have made large-scale food output expansion much easier, has a mixed record of ups and downs and its high population growth has kept per capita food supply virtually stagnant since the mid-1950s.

With densely settled populous poor nations such as India and China food supply fluctuations—caused by natural disasters, wars and, in China of the early 1960s, also by a near collapse of the overstrained economy in the wake of the mad expansion dash during the Great Leap years—have been more pronounced. Fortunately, since the mid-1970s both countries have been doing better and China's farming reforms have brought unprecedented levels of food availability after 1978. For the first time

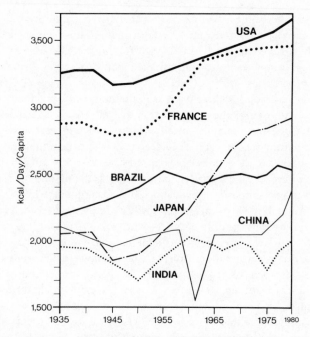

Fig. 3.10 During the two generations charted in this graph average per capita food availabilities rose from already high levels in rich nations (here exemplified by the USA and France), shot up in Japan after 1945, stagnated after an initial growth in Latin America (here represented by Brazil) and have hardly changed in Asia, except for the latest spectacular advance in China.

in their long history the Chinese produce enough food to satisfy the growth and activity requirements of their huge population but the typical composition of their diet remains very different from that of the West and a comparison of the recent detailed Chinese data with American means provides a revealing opportunity to contrast the make-up of food supply in the world's most populous, still over-whelmingly rural, and relatively poor nation with that in the richest industrialized society.

This comparison will be made in terms of average availability at retail level, and the previous comments regarding discounts to actual consumption levels, as well as the cited means of nutritional surveys, should be kept in mind. Figure 3.11 shows at a glance that the American food supply has several major ingredients since there is about as much meat (including poultry and fish) as vegetables, or roughly as much weight in cereals as in fruits; in contrast, the Chinese pattern is dominated by cereals and vegetables, and the other foodstuffs have only a marginal place. This difference becomes, if possible, even more impressive by comparing the shares of energy and protein derived from plant and animal foods: while in the American food supply

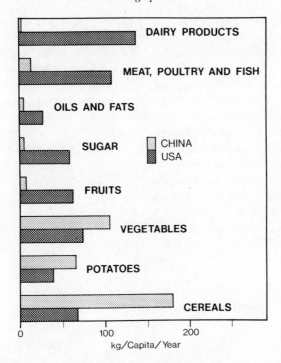

Fig. 3.11 The rich and poor pattern of eating is well contrasted by comparing average annual availabilities of major food groups in China and in the USA. The Chinese diet can be best described as grains and tubers enlivened with vegetables and sprinkled with occasional meat, the American one as a dairy and meat diet well sweetened and accompanied by fruits and vegetables.

animal foods provide nearly 40 per cent of total energy, and almost two-thirds of all protein, in China these shares are, respectively, a little over 10 and 20 per cent.

These are fundamental differences as one essentially herbivorous nation, exemplifying the poor three-quarters of humanity, is enriching its overwhelmingly plant diet by irregular eating of very small quantities of meat—while the other's eating habits are now about equally split between consumption of plant and animal foods with meat and dairy products providing almost half of all energy and surpassing that mark in proteins.

Yet at the beginning of the last century the Chinese pattern was dominant everywhere. Until very recently the Chinese peasants ate meat only during the New Year and *qing ming* festivals and at weddings or funerals, a handful of times a year—and so did French villagers in the last years of Napoleon's empire. The great dietary shift paralleled the rapid rise of industrial civilization in nineteenth century Europe and historical data, although not as longstanding and as complete as one would wish, are sufficiently accurate to trace the transformation (the inevitable imprecision of some values matters here much less than the unfolding of obvious trends).

The earliest British data are available only with great temporal gaps but, imprecise as this historical reconstruction for cereals, potatoes, meat, and sugar may be in many details, it reveals very clearly the general trends for the whole of the past century and, a point of great interest, it illustrates those sharp temporary reversals during World War II and the re-establishment of the pre-war course since the late 1940s (Fig. 3.12). American food supply figures have been available annually since 1909, constituting the finest unbroken and consistent record of its kind. Interestingly enough, there is no perfect trend for the whole period of 70 years shown in Fig. 3.13 as even the deep declines of cereals and potatoes were, though only very slightly, reversed during the 1970s, and other foodstuffs had their ups and downs during earlier decades.

But there is certainly no record more fascinating, and at the same time so very accurate, than that demonstrating the unprecedented rapidity with which Japan

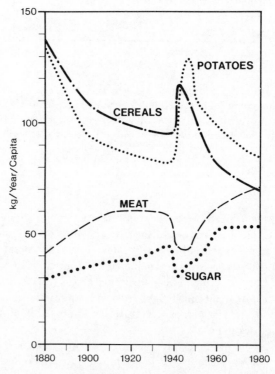

Fig. 3.12 A century of nutritional transition in England: the predictable pattern of falling consumption of cereals and tubers and rising consumption of meat and sugar was temporarily reversed during World War II and the long-term trends were not resumed until the late 1950s.

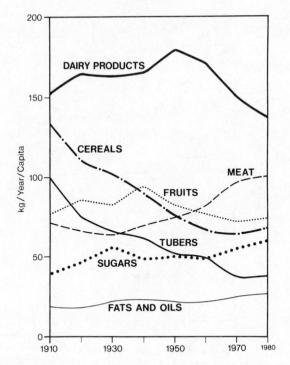

Fig. 3.13 American dietary changes since 1910 are a mixture of expected and surprising shifts. Consumption of cereals and tubers are down but so are dairy products after the early 1950s and fruits after the early 1940s. Sugars and meats still continue to rise much faster than fats and oils.

altered its traditional food ways after World War II as it shifted so decisively toward the Western diet. Excellent National Nourishment Surveys, already praised in section 3.1.1, enable an exhaustive tracing of this stunning transformation accomplished in less than two generations.

With the consumption of some foodstuffs going up by an order of magnitude in just three decades, these changes have been so large and so rapid that I had to use a semi-logarithmic graph with three cycles (!) to show the shifts clearly (Fig. 3.14). Milk drinking has been the most notable instance of this growth, from less than two litres per person in 1950 to over 40 litres in 1980, and consumption growth rates for fruit and meat (excluding fish) almost match milk's rapid rise. As might be expected, rice and potatoes are down and consumption of oils and fats, though still very low compared to Western levels, is up more than seven-fold.

Altogether these changes represent the most dramatic shift of eating habits in modern history, a change also well reflected in the steep rise of per capita food energy availability (Fig. 3.10) and in the more vigorous growth of the younger

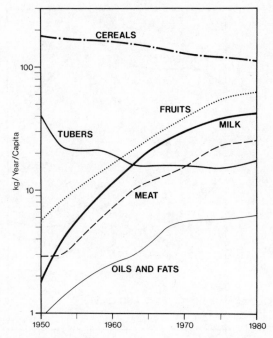

Fig. 3.14 To chart the Japanese dietary changes after 1945 I had to use a semi-logarithmic graph as the consumption of meat, milk, and fruits rose by an order of magnitude in less than two generations. Cereals and tubers declined in importance as expected but fats, although about six times above the pre-World War II level, are still consumed in moderate amounts compared to the rich West.

generation. In the early 1980s, an average 60-year old man measured 163.3 cm, an average woman 152.0 cm, but their 30-year old counterparts were, respectively, 4.0 and 4.1 cm taller (see also Fig. 1.5 for growth of young boys). These changes appear to be a sufficient testimony to the nutritional adequacy of the new Japanese diet, but a closer look uncovers some shortfalls in comparison with recommended daily nutrient intakes.

Recent Japanese consumption of over 1700 International Units of vitamin A is 2.2 times higher than the FAO/WHO allowance and still nearly twice the American consumption but riboflavin (B_2) intakes are about 40–50 per cent below the best accepted recommendations, and iron and calcium are also insufficient. Similarly, NHANES II diet results show that Americans are consuming much more vitamin A than is needed (on the average about five times the recommended amount), whilst calcium levels are barely adequate, and iron, although just about sufficient as a nationwide average, is highly deficient in females and children (only 50–60 per cent of requirements covered). However, according to the NFCS sample, vitamin A needs were covered in only half of the cases and the shares were even lower for iron (43 per cent), magnesium (25 per cent), and vitamin B_6 (20 per cent). Clearly, there are major, widespread nutritional deficiencies even in the world's most prolific producer of foods.

On the other hand, there is the previously discussed surfeit of fats and sugars and the general trending of rich diets in the direction opposite to the emerging consensus about nutrition and minimizing the risks of premature deaths owing to cardiovascular diseases and cancer. Clearly, the major challenge for the rich nations is not an enlargement of their food supply and a boosting of their overall food consumption—as only relatively small increments will be needed to cover the additional needs of very slowly-growing populations and as the current food available is over-abundant by any reasonable measure—but rather a qualitative, compositional reform of their typical diets. But the former goal is very much needed for virtually all of the world's still overwhelmingly rural and very poor nations (mainly in Africa), as well as for the rapidly industrializing but still far from well-off countries of Latin America, Africa, and Asia.

The remaining sections of this chapter will assess the options and the chances of such an expansion which must come primarily from the nations most in need of a larger and more varied food supply a great deal of which however will still inevitably originate in the rich countries, through both expertise and new approaches as well as higher output and bigger exports.

3.3 Expanding the harvests

> Agriculture is a science common to them all in general, both men and women, wherein they are all expert and cunning.
>
> Thomas Moore, *The Utopia* (R. Robinson translation)

On that new isle of Moore's imagination worries about the next year's harvest would hardly occur: that whole 'expert and cunning' population, steeped in the science of agriculture and the realm's other enlightened ways, would have been solid guarantees of abundant harvest. Ours is a more precarious state. In poor countries there are still too many subsistence peasants whose only expertise consists of repeating their ancestors' routines (some good, some bad), while in the rich nations lawyers and civil servants vastly outnumber those with expertise in the science of food production as this critical process is left to a small fraction of the population occupied in farming, and agricultural research and management.

But these shrinking castes (relative numbers of peasants and farmers have been declining all around the world and, while the numbers of agricultural researchers and extension workers are now the largest ever, they, too, have been slipping in relation to industrial, management and military research, and development expertise) have done well, obviously with much help from producers of assorted inputs making up new industrialized farming, to bring in ever larger harvests and to supply more food per capita in every successive generation of this century, in spite of the unprecedented rates of population growth.

But these impressive gains were achieved, not solely but to an overwhelming degree, by large-scale simplification, and by concentrating on the genetic and cultural improvement of a handful of cereal crops. The total number of crops cultivated around the world has never surpassed 300 to 400 plant species, that is something of the order of 0.1 per cent of all higher plants, but this century, and especially the post-World War II period, has seen the emergence of an ever greater dependence on a very small number of staple grains.

Global food production statistics collected by the League of Nations show that in the early 1930s the three top grains—wheat, rice, and corn—accounted for 70 per cent of the total cereal harvest, while in the early 1980s their share surpassed 80 per cent. Signs of this simplification are everywhere. Before World War II the Japanese harvested each year nearly 2 million tonnes of barley; two generations later they grow less than 0.5 million tonnes. Similarly, Russian harvests of rye declined from 25 million tonnes a year during the late 1930s to about 8 million tonnes in the early 1980s.

Moreover, the rising importance of the three major grain varieties has been accompanied by a relative decline of the other two staple food groups, tubers and pulses. During the half century between the early 1930s and the early 1980s, cereal production rose about 2.9 times but harvests of pulses and tubers went up only about 1.9 times each. Nearly 95 per cent of food and feed energy currently harvested worldwide in cereals, tubers and pulses comes from grains, an increase of about 5 per cent in just two generations. And the shift toward greater dominance of just a few grain crops has been even more pronounced in the rich countries. For example, the combined share of wheat and rice in the United States and Canadian cereal harvest rose from 73 to 83 per cent during just one generation, since the early 1960s.

This unfortunate trend, so obviously increasing the vulnerability of our agro-ecosystems to disease and pest attack, and so unnecessarily narrowing our nutritional options, will almost certainly continue and so any informed discussion of larger future harvests must start and be largely devoted to the possibilities and limitations of expanding grain production.

3.3.1 *Cereals: potentials and limitations*

> A poor woman with a young child at her side
> Follow behind, to glean the unwanted grain.
> Her tattered basket strung upon her left shoulder,
> She tells her tale and the hearer is filled with grief.
> Her farm was lost entirely to taxes and her family starves.
> And these bits of wheat will be all they live on . . .
>
> Bai Chuyi, *Watching the reapers* (807) (L. S. Robinson translation)

In the second year of the Yuan He reign, in AD 807, Bai Chuyi, a poet and a newly promoted military officer assigned to Shaanxi province, watched the poor peasants picking up wheat grains that fell unnoticed to the ground remembering, ashamed, that on his good salary 'every year finds me with more food than I can eat'. At that time the grain harvests were low in the rainfed fields of North China, as low as they were for millenia before—and for another eleven centuries after.

We have enough information about ancient farming practices to calculate or to estimate fairly reliably that early yields of grain crops in China were no more than 400–750 kg/ha, identical to those gathered on dry lands of the Roman Empire, or on the fields of medieval England. True, there are some records of much higher yields for irrigated crops in the Middle East but for millenia most were scarcely higher than yields obtainable from collection of the wild cereal progenitors in Galilee.

Average yields started to rise only slowly with the spread of better farming practices—abundant recycling of organic wastes, planting of green manure crops, manuring, following careful seed selection—so that by the end of the last century grain yields in the most intensively farmed areas were double or triple their long-lasting historic means. Fertilization and mechanization, and later the use of pesticides, contributed greatly to these improvements but it was the introduction of new cultivars which made it possible to increase average harvests so fast that in some countries gains during one decade far surpassed the improvements during the preceding century. Instead of skimming the basics concerning all major cereals I will illustrate this transformation as well as other relevant changes, with a closer look at wheat.

That wheat should get special attention is easy to explain. Although corn is now the grain crop with the widest global distribution (about 140 countries grow it, compared to some 115 for rice and 100 for wheat), and although the worldwide corn harvests are now virtually at the same level as wheat harvests (while at the beginning of the 1970s wheat output was around 340 and corn production about 300 million tonnes, by 1981 the two totals were, respectively, 458 and 452 million tonnes), wheat's lower yields make it still the world's number one crop in terms of area (240 million hectares compared with 150 million hectares for rice, and 135 million hectares for corn) and its higher protein content (on the average 11 per cent compared to corn's 9 and rice's 7 per cent) means that no other crop provides more protein (directly or indirectly) to human diets.

The crop's most notable agronomic characteristic is that its winter varieties seeded in the fall can germinate before the winter, survive even very low temperatures (under snow cover up to −30° C), and resume their growth and mature rapidly before the peak heats of summer. This hardiness stands in a marked contrast to corn and rice cultivars intolerant of even light frosts. Wheat is thus assured of its position as the world's principal food grain but its dominant current cultivars are very unlike their traditional ancestors.

Traditional wheat plants were tall (usually 125–50 cm) and their long stalks contained much more biomass than the grains. Harvest index (the ratio of the weight of the grain to total dry above-ground phytomass) of common wheat varieties was no higher than 25–32 per cent, and the yields were not much more than 500–700 kg/ha for extensively cultivated wheat in poor countries and around 1 t/ha in European nations: French national averages for the years 1820–1940 and American yields before 1940 exemplify very well this long period of marginal changes (Fig. 3.15).

French yields in 1940 were not even 10 per cent higher than a century earlier (!) and lower than the records of the 1890s; similarly, American yields in 1940 were only about 10–20 per cent above those of the 1870s. Virtually all the pre-1960 advances, such as the Dutch take-off after 1890, were due to better agronomic management rather than to breeding and the greatest frustration rested in the fact that higher nitrogen applications made the tall plants even taller, further aggravating the lodging troubles and causing greater tillering and thicker stands which favoured diseases.

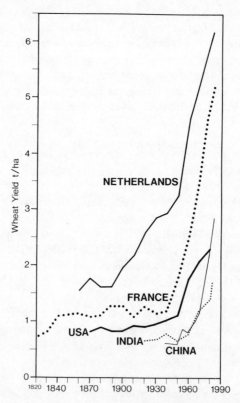

Fig. 3.15 Wheat yields show very similar secular trends: long periods of fluctuations at low levels came to an end only with the advent of better seeds and greater use of fertilizers between the 1940s and 1960s. Dutch yields are the only early exception to this trend, a consequence of much greater use of fertilizers, and intensive cultivation already well established some 100 years ago.

The Japanese were dwarfing their wheats for generations, selecting for short stalks able to bear heavy heads (resulting from copious application of organic manures) in the windy climate of the islands. Between 1917 and 1935 they first crossed one of the local dwarfs (Daruma) with an old Mediterranean variety (Fultz) imported from the United States, and then they re-crossed Daruma-Fultz with Turkey, another old variety brought by the Mennonites to the United States from southern Russia, to produce Norin 10, a 52–5 cm tall, unusually high-yielding plant. Released in 1935 the new variety was sent to the United States only in 1946 where, after years of extensive crossing, first emerged semi-dwarf winter varieties (Gaines, released in 1961) and, almost concurrently, further disease resistance was also bred into Norin 10-based spring semi-dwarf wheats developed in Mexico (Pitic 62 and Penjamo 62).

These Mexican semi-dwarfs were just the first in what is now a long series of successful cultivars, 30 of them between 1960 and 1980, which revolutionized wheat farming in Latin American, Africa, and Asia where these Norin 10-based varieties now make up nearly two-thirds of the spring-wheat crops. Many of these Mexican spring semi-dwarfs found an enthusiastic acceptance also in spring-wheat areas of the rich world, where new winter-wheat varieties bred in the United States, Europe, USSR, and China contributed to the stunning take-off of national yields charted in Fig. 3.15.

These huge gains can be attributed to short stalks (50–100 cm), higher harvest index (commonly almost 50 per cent), sturdier straw and stronger crown roots, more fertile flowers (120–50 grains per head are possible compared to 50–65 in old varieties), more profuse tillering, faster maturity, vigorous fertilizer response (easily up to 20 kg of additional grain for each kg of nitrogen up to the first 50–70 kg, as opposed to just 8–12 kg in traditional cultivars), insensitivity to day-length (hence ability to grow in various latitudes), adaptability to different environments, and resistance to some common diseases.

However, these advantages of new varieties do not emerge automatically. Careful management is a must to bring out the high yield potential: timely sowing, irrigation or good soil moisture retention practices, adequate fertilizing, and necessary weed and disease controls will all contribute to a richer harvest. And precisely because some of these ingredients are suboptimal wheat yields in many poor countries still have a long way to go to match the advanced performers. The Chinese have recently surpassed the Americans but the Indians still average only about two-thirds of Chinese performance and the African mean is merely 1000–1100 kg/ha. But to look at the prospects of future yield increases requires a sweep beyond agronomic practices, and intensified inputs, as there are both possibilities of much more fundamental changes (achievable only over the long term but with enormous pay-offs) and limitations which we currently cannot foresee how to overcome.

Until evolution develops an entirely new way of fixing atmospheric carbon nothing at all can be done about the overall efficiency of the process (gross photosynthesis), but recent advances in biochemistry and plant genetics make it possible to contemplate improvements in net photosynthetic performance by reducing the respiratory loss of CO_2 (occurring through photorespiration and dark respiration). This notion is especially appealing because photorespiration limits photosynthesis in all C_3 plants (labelled so because their primary photo-synthetic products are three-carbon compounds)—and both wheat and rice (as well as other major food crops such as barley, potato, cassava, soybean, peanut, sugar beet, or banana) have C_3 metabolism and net rates of their production are usually only about half of those in C_4 plants (forming first four-carbon acids) which are represented among the leading food crops by corn, sorghum, millet, and sugarcane.

Experimentation in this potentially highly rewarding area (much of it done with tobacco cells) has uncovered many promising leads and attempts to control photorespiration including selections of mutagen-treated cell lines or cell lines with higher intracellular concentrations of metabolites expected to inhibit photorespiration, and regulation of respiratory pathways in plant mitochondria, but none of this research has as yet approached practical transfer to the breeding plot. Glenn Johnson and Sylvan Wittwer, in their recent appraisal of agricultural technology until the year 2030, conclude that although at that time we will have a better understanding of the processes involved we will still not be able to regulate them. Whatever the actual time of practical breakthroughs, reduction of photo-respiration is clearly only a very long-term hope.

Even more difficult would be another often talked-about fundamental genetic manipulation—the transfer of nitrogen-fixing genes into cells of grain crops so they could supply a large portion of their own nitrogen requirements. This amazing feat would be of little use if the new cultivars would not also have at the same time excellent yielding ability and resistance to pest and environmental stresses. This leads directly to suggestions of fusing protoplasts of different species to make an almost unlimited genetic variability available to crop breeders but this would be an achievement rivalling evolutionary selection and hence one not to be considered a part of even any long-term (up to fifty years) outlook.

From the implausible to the realistic: what are the improvements and changes which will readily contribute to higher grain productivity? Nothing can be done with the length of growing season, total insolation and basic soil types (although proper soil management may actually improve the tilth and soil structure over a long period of time), but much can be gained from the controls of weeds and pests and by boosting the tolerance to commonly encountered environmental stresses.

Field losses caused by weeds and by heterotrophs—viral, bacterial and fungal diseases, nematodes, insects, rodents—reduce the current global harvest of major cereal crops by at least 35 per cent. The best available, but necessarily uncertain, estimates assign losses of about 10 per cent to weed infestation of rice, wheat, and corn and another 10 or so per cent to various diseases; losses from insects cut the corn yield by more than 10 per cent, wheat harvests by an average of only 5 per cent, but rice yields by more than 25 per cent.

Returning again to wheat as a detailed example, there is, above all, a continuous need to introduce new varieties with greater resistance to major fungal diseases. No matter where wheat is grown, and regardless of assiduous agronomic practices, leaf and ear diseases can still cut the yields by a large margin, commonly by 10–30 per cent in endemic areas, and cause a complete crop loss during periods of rapid diffusion. Globally by far the most destructive of fungal wheat diseases are rusts: their rapid mutation rates and easy airborne diffusion make them especially difficult to control.

Stem rust (*Puccinia graminis*) is most severe in Southeastern Europe (the lower Danube basin), the Kenyan and Ethiopian highlands, south India and Brazilian Paraná; leaf rust (*Puccinia recondita*) is relatively rare in Europe but common in Mexico, the Indian subcontinent, and China, while stripe rust (*Puccinia striiformis*) strikes hard in northwestern and central Europe, the South American highlands, the Middle East, and India. Testing of numerous varieties to find those with broad genetic resistance and then crossing them to get even higher concentrations of rust-resistant genes has been the standard defence but staying ahead of the constantly mutating pathogens (as new biotypes are continually formed through hybridization and recombination) has not been easy.

New approaches include multilines, varietal mixtures and special placement of resistant varieties. Multilines are simply mechanical mixtures of similar wheat varieties (a dozen or so) which have different genes for rust resistance; when a new rust strain attacks a multiline only one or two of its components may be susceptible and the infestation may not progress to a destructive scale before the harvest. Full commercialization of multilines will soon show their effectiveness. Varietal mixtures (mechanically mixing two to five cultivars with similar maturity) are even more diverse than a true multiline but also pose more problems for the grower and the miller. During special placements, known resistant varieties are planted as a barrier to check the emergence of the disease long enough to escape the worst damage.

Criticality of the broad genetic base is evident: without numerous resistant strains crossing, the preparation of multilines and mixtures and planting of barriers would be impossible, and preservation of genetic diversity is thus a task whose importance cannot be over-emphasized. Of course, the very success and rapid diffusion of semi-dwarf cultivars have been responsible for the drastic reduction of genetic variation of the world wheat crop as new varieties, each being just a single genotype, replaced the traditional land races (regional species) with their great genotypic variation. Moreover, most of the genetic variability of the cultivated wheats has been already used during the reent decades of intensive breeding while attempts to increase the variability of new cultivars by artificially induced mutations (either by ionizing radiation or by chemical treatment) have not been encouraging.

Fortunately, it will be possible to counteract these limitations and to enrich the gene pool of the cultivated wheats by crossing them with the plant's wild relatives belonging to the *Triticeae* tribe of the grass family. As not only many species of *Triticum* but species of other genera in this tribe can be crossed with common and durum wheats to produce viable hybrids, the genetic variation of the entire tribe appears to be potentially exploitable for the improvement of the crop.

And this potential is very large indeed because the wild relatives of wheat grow in climates ranging from the very dry to the quite humid, and on various soils, and even their partial preliminary screening has shown that they could be used to improve the quality and amount of wheat protein, to impart greater resistance to diseases, drought, salt water, and lodging, and even to confer directly higher productivity.

If in the future the available spectrum of wheat cultivars does not possess resistance sufficient to overcome attacks by new virulent rust races, large-scale chemical treatment (by oxycarbonin) may then be necessary to save the crops from obliteration. Nevertheless, resistant cultivars are the best defence and it is encouraging to note that stem rust infestations (which caused huge losses in the United States in 1916, 1935, and 1952-4), and leaf rust (which cut the harvests in 1938, 1961, 1949, and 1954) have not returned for three decades—and that continued improvements in broad-spectrum resistance are virtually certain.

Insects do generally much less damage to wheat than they do to the world's rice and corn crops, but there are still no fewer than 200 species belonging to seven orders which can occasionally cause serious local or regional damage (for example *Eurygaster* or sunn pest in the southern part of Russia and parts of the Middle East, or the Hessian fly in Europe, the USSR, and United States). As with fungal diseases, a combination of traditional controls, ranging from selection of resistant cultivars to crop rotation and uses of insecticides, will continue.

There are no wheat-specific weeds and there are large regional differences in the presence of unwanted growth. In the United States the more important weed species encountered in wheat fields include some 30 annuals (from barnyard grass to yellow rocket) and about a dozen perennials (from buttercup to wild onion), but globally there are only about a dozen species of overall importance. Annual wild oats (*Avena fatua* and *A. sterillis*), troublesome in summer wheat, are specially important owing to their worldwide distribution.

Controls are not easy and can be successful only as a package starting with planting of weed-free seed (which can eliminate such typical seed weeds as cheatgrass, corn cockle or darnel ryegrass), good seedbed preparation (tillage timed to prevent seed formation or to delay weed emergence), and proper time and rate of seeding and adequate fertilization. This last practice is to enable the crop to overgrow and shade out the weeds, but intensive fertilization has also changed weed composition and practically all weeds in well-fertilized regions are now nitrophilous and benefit from fertilization at least as much as the crop; moreover, many of these weeds can wind up the wheat plants so as to overgrow them. Combination of all these measures is usually insufficient to control the weeds and herbicides are now considered essential for securing maximum yields, and a wide variety of selective as well as broad-spectrum chemicals are now available for both pre-emergence and post-emergence application.

None of the common environmental stresses reduces the yields of cereals more than insufficient water and, as with photorespiration losses, C_3 crops are at special disadvantage. Plants cannot absorb CO_2 without losing water and the rate of this inevitable exchange puts high demands on water supply: even under the theoretically most favourable conditions no fewer than 109 moles of H_2O would have to be lost for each mole of CO_2 used in photosynthesis. In reality the exchange is even more lopsided. Species with C_4 metabolism transpire 450–600 moles of H_2O for each mole of fixed CO_2 but C_3 plants usually lose 800–1000, and up to 4000 moles of water for each mole of CO_2.

Availability of water thus puts a fundamental limit on the photosynthetic performance and only by varying internal CO_2 concentrations has a plant any control over the water losses it has to bear in acquiring CO_2: C_4 species reduce the concentrations of CO_2 inside the leaf air spaces to nearly 10 ppm from 340 ppm in the atmosphere, and the greater difference of the two partial pressures cuts water losses to an average of about 40 per cent of those incurred by C_3 plants.

Depending on climate (which determines the maximum evapotranspiration rate of the crop) the top yields of maize (6 to 9 t/ha of grain), when soil moisture is not limited, require 500–800 mm of water which means that anywhere between 0.8 and 1.6 kg of grain are produced for each m^3 of evapotranspired water. In contrast, the water needs of wheat and rice for top performance unlimited by moisture are 450–700 mm, but their lower commercial yields mean that water utilization efficiencies are no more than 0.8–1.0 kg/m^3 for wheat, and 0.7–1.1 kg/m^3 for rice.

Relationships between relative yield decrease and relative evapotranspiration deficit vary quite considerably with different growth periods. Looking again at wheat's specific needs, a slight water deficit during winter wheat's vegetative period may have little effect on subsequent crop development, water deficits during the yield formation period cut into the harvests much more seriously, and water shortages during the time of flowering are by far the most crippling (Fig. 3.16). Heavy water stress at that time seriously affects pollen formation and fertilization, reduces root growth, number of heads per plant, their length and grain density—all losses which cannot be recovered by adequate water supply during later growth stages.

Increased efficiency of water use in cereal farming would thus be a most welcome development, but each of the five possible options to achieve this goal appears to have fundamental limitations. Altering the biochemical processes of C_3 species will not, as already noted, be easy; influencing stomatal behaviour to get mid-day closure may, even if successful, lengthen the growing period and actually increase yield losses owing to pests and environmental stresses; breeding for hardiness in cooler weather may have similarly negative yield consequence; further substantial improvements of harvest index are unlikely; and increasing the proportion of transpired water by greater rooting depths would most likely lower the harvest index, and may also lead to more rapid exhaustion of soil water.

Further extension of irrigation and breeding of drought resistant crops are thus the two most realistic options for overcoming water stress. Irrigation is usually expensive—without proper erosion control the lifetime of costly reservoirs and

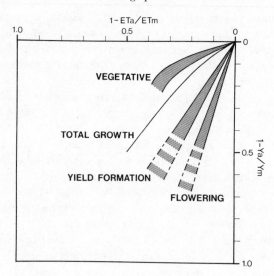

Fig. 3.16 The relationship between relative yield decrease (1−Ya/Ym) and relative evapotranspiration deficit (1−ETa/ETm) in winter wheat. Moisture stress during the vegetative period has a vastly smaller effect than drought at the yield formation and especially at the flowering stage ('a' are actual, and 'm' are maximum harvests).

canals may be just a fraction of the anticipated span, and many countries would be hard pressed to extend significantly their irrigated cropland. On the other hand, investment in irrigation may be the most cost-effective step toward higher yields (in poor countries average yields of irrigated wheat are double those of the rainfed crop), and the planting of trees (beneficial for many other reasons) and grasses to control erosion is a highly effective protection against silting. The irrigation potential of many nations is still very large (India is perhaps the best example).

In arid and semiarid regions irrigation may lead to soil salinization and the traditional expensive engineering solutions of this spreading problem in the near future may be greatly aided by breeding of salt-tolerant crops. Barley is the most salt-tolerant of all cereals: in California the best selections grown experimentally under irrigation with undiluted seawater applied to dune sand (supplemented with nitrogen and phosphorus) averaged almost 1000 kg/ha compared to the crop's global mean yield of about 2000 kg/ha. Wheat is far less salt-tolerant, but screening of just 5000 cultivars from the USDA collection uncovered 34 lines of spring wheat capable of producing grain at a salinity of 50 per cent that of seawater and further improvements seem likely.

The second most important inadequacy of the natural environment is shortage of nutrients and higher global yields will be impossible without further increases in the application of nitrogen, phosphorus, potassium, and critical micronutrients. More will be said about this later (in section 5.3) but the disparity between poor and rich countries here comes once again to the fore: while only a small fraction of African cereal crops receives nitrogen, phosphorus, or potassium applications, virtually all grainfields in Europe and Japan, as well as in the American Corn Belt are fertilized (however, only about one half to two-thirds of wheatfields on the American Plains and Canadian Prairies receive regular fertilizer applications). Both the global extent and the average intensity of fertilization will have to increase to bring higher yields.

Better cultivars, more irrigation and fertilization, and more effective pesticides will still leave a number of anthropogenic factors which will work against the higher yields. These are, above all: losses of topsoil caused by advancing soil erosion and losses of some of the best productive farmland to non-farming uses (see section 4.2.1 for discussion of both), decline of soil organic matter and worsening soil structure caused by reduced recycling of organic matter and compaction by heavy field machinery, and environmental pollution ranging from accumulation of pesticide residues in soils to the effects of anthropogenic gases and particulates absorbed by and deposited on crops and soils.

The redistribution of photosynthate, that is further shortening of straw and additional improvement of harvest index, is unlikely, especially with wheats in drier regions where higher levels of plant residues are essential to control erosion; moreover, extreme shortness is detrimental to harvesting and difficulties with the emergence and development of shorter cultivars would not be easy to overcome.

And, naturally, climatic variations will continue to have an indisputable effect although in retrospect it has been extremely difficult to separate this impact accurately from the contributions of production input combinations. Still, there are interesting indications of greater yield stability even in some environments notorious for their climatic extremes. For example, a study of winter wheat yields grown under fallow systems during three decades at three Central Great Plains agricultural experimental stations (in Colorado, Nebraska, and Kansas) looked at water-use efficiency, that is wheat yield divided by two years' precipitation (fallow and crop years), and discovered that the index, which is independent of variations in precipitation, has been increasing steadily at the same rate as the crop yields.

During 1926–35, a period including the worst years of the great 1930s drought, average precipitation at the three locations was 432 mm a year, wheat yields were 1225 kg/ha, and water efficiency averaged 14 kg/ha · cm; 40 years later, during 1966–75, 427 mm of rain and 2730 kg/ha of wheat produced a water-use efficiency index of 32 kg/ha · cm, more than double the level of the 1930s and an impressive 40 per cent up compared with the years 1956–65 when the average annual precipitation was 12 per cent higher! The authors could not credit this advance to any single phase of production but felt that improved fallow systems and better crop varieties, followed by improved planting and harvesting equipment, explained most of the gain.

To identify the factors which will push the yields up, and those which will constrain the growth, or make it irregular is easy; to offer quantitative conclusions about the relative contributions of individual segments is very difficult; and to put down forecasts of specific yields is impossible as concatenations of technical, social, and environmental circumstances can produce results much above or much below the trend expectations. During the discussion of energy prices I tried to make clear that no confident forecasts of these essential variables are possible and, as they will critically determine the costs of key energy-intensive farming inputs, it is impossible to say to what extent and how long different countries will be able to shoulder the costs of productivity gains. Japan or France have obviously greater latitude in meeting this need than South Korea and Malaysia which are, in turn, a class above India and Nigeria.

As with energy forecasts, there has been no shortage of erroneous prognoses of cereal production—which is naturally taken as the best indicator of global food supply. During the mid-1960s two years of drought in the Indian subcontinent and the resulting huge grain deficits in India and Pakistan raised a prospect of deepening Asian food shortage which the first successful wave of the green revolution replaced, less than five years later, by unrealistic expectations of lasting food security. The Sahelian drought and other crop failures of the early 1970s promptly reintroduced and potentiated the gloomy outlooks, but before the end of the decade production rebounded, stocks were rebuilt, and grain prices fell, in real terms, to pre-1973 levels. Finally, the early 1980s brought, alongside the ever higher outputs in rich countries, an impressive second wave of the green revolution in China, India, and Mexico but a continuing dismal record (more man-made rather than weather-caused) in Africa.

One way to gauge the potential is to compare the current productivities in various environments, and under various levels of management, and then try to estimate the rapidity and extent of the climb to the higher rungs of this productivity ladder. For the last time I will use wheat as an example. Traditional dryland cultivars grown unfertilized and with little weed control by subsistence farmers yield just around 500 kg/ha: unfortunately, too many African wheatfields are still in this poorest category, and the continent's mean yield in the early 1980s was barely above 1100 kg/ha.

The recent mean for all poor countries is around 1700 kg and the global average is about 2000 kg/ha. Among the rich countries the two large North American producers average 2300 kg, but Europe harvests, on average, no less than 4000 kg/ha, and two of its top performers, with large areas in the crop, the United Kingdom and West Germany, go over 5000 kg/ha, whilst the Netherlands, the world's leader, has topped 7000 kg/ha. The highest yields in reliably rain-fed fields in Western Europe and in irrigated fields in the subtropics have been recently between 9000 and 11 000 kg/ha, and the world record yield was harvested in 1965 in Washington State where Gaines winter wheat variety on an irrigated 2 ha plot brought, after 330 days, 14 100 kg/ha. This performance is not so far away from the 20 000 kg/ha calculated to be the theoretical maximum.

The last two values are not sensible yardsticks—even the record performance was an unrepeated oddity as the yield was about twice as high as in the neighbouring fields planted with the same variety—and the real span for large-scale improvement thus lies between the 1000 kg/ha of the African average and the 4000 kg/ha of European performance. Instead of engaging in speculations as to how much will the poor performers close the gap separating them from the best producers, and how far the latter will advance during the next generation, I will just cite what I think is the fairest semi-quantitative assessment offered by the world's top three wheat geneticists, Norman Borlaug, Haldore Hanson, and Glenn Anderson.

Writing about the prospects of poor countries they concluded that new agricultural technologies will make it possible to meet the countries' food needs 'until the end of the century largely from domestic production, even if population rises 50 per cent. We believe the world has only begun to exploit the potential for higher agricultural productivity.... The odds are against dramatic breakthroughs of the type scored by semi-dwarf wheats, however. Steady increments rather than quantum jumps are to be expected.'

Increases in the global production of cereals between 1980 and 2000 may not then match the 1960–80 rise when wheat output rose nearly 1.9 times and the total grain harvest went up almost 1.6 times, but if the gains in the future generation do fall far short of this past record it will not be because we have approached the limit of agro-ecosystemic productivity, or nearly exhausted our breeding options and cultural improvements, or could not turn out enough essential inputs to sustain the yield improvements.

Climatic fluctuations, national and regional environmental and economic disparities guarantee an uneven progress but the early years of the new century should be able to see an impressive global advance in the harvests of the world's most important foodstuffs, an advance which should, as in the preceding generation, stay well ahead of population increase—and which could be further strengthened by reasserting some valuable old ways, and introducing some new options into the world food supply.

3.3.2 *Old ways and new possibilities*

'What did they live on?' said Alice, who always took a great interest in questions of eating and drinking.
'They lived on treacle,' said the Dormouse, after thinking a minute or two.
'They couldn't have done that, you know,' Alice quietly remarked; 'they'd have been ill.'

Lewis Carroll, *Alice's Adventures in Wonderland* (1865)

Alice's remark confirms her nutritional acumen: indeed, they—the three little sisters Elsie, Lacie and Tillie—would have been ill on such a strange diet, but modern nutritionists have come out with diets on which the girls would not do any better. Perhaps the most notable of these scientifically deep but practically offensive ideas have been the persistent attempts to prove that microbial cells make an excellent food for man. After all, are not the amino acid scores of algae, yeasts, and bacteria much better than those of staple grains, are not their crude protein contents averaging 35–60 per cent (only soybeans can compete at the lower level of this range), and cannot these micro-organisms be cultivated with impressive rapidity on a variety of cheap substrates?

All true, but mass cultivation of microbial cells (a task with no small problems in itself) to provide a major protein source for man works so much better in sci-fi novels than in the real world. Some algae species have strong and disagreeable flavours (when unextracted), and their deep green colour is difficult to mask for them to be used as food ingredients, but the real obstacles to large-scale consumption of microbial cells are their high content of nucleic acids and their numerous gastro-intestinal effects. Eating just 30 g of yeast, or algae, a day produces serum and urinary uric acid concentrations much like those found during gout, while as little as 20 g a day of unextracted green algae cause gastric distress owing to the irritation caused by fibrous cell walls.

Severe gastro-intestinal disturbances, headaches, and general weakness follow the eating of 25–50 g of washed, sterilized bacteria, while eating larger quantities of dried yeast brings nausea, diarrhoea, and symptoms similar to food poisoning just after one to four weeks of eating a mere 20 g a day. Clearly, the extensive and costly processing of all these microbial protein sources would be imperative before they could be seen as a major human protein food—but the preparation of isolates free of nucleic acids and wall cells would increase so much the overall cost of microbial protein that soybeans would easily retain their current cost-benefit edge.

Microbial protein is thus a fine example of a rather rich class of new food possibilities whose impracticality and excessive costs make them perennial features of writings on future food options. I will not spend pages on these revelations but I must at least note that these options embrace not only unorthodox terrestrial processes, such as gathering tree leaves and separating their voluminous indigestible wall fibres to get edible protein, but also grand marine schemes. Perhaps the best examples of the latter class are Howard Wilcox's proposals for kelp farming of 'arable surface waters' (estimated to cover up to 260 million km^2) whereby each square kilometre of cultivated ocean would contain 100 000 giant kelp plants whose canopies would be harvested by special vessels every three to six months.

Annual yields of 65 000–90 000 wet t/km^2 would be processed to provide enough food for 1000–2000 people as well as provide feeds, fertilizers, methane, ethanol, light oils, lubricants, waxes, fibres, paper, plastics, industrial gums, and chemicals. With about the same chances of future success one could also contemplate the mass domestication of manatees, or the genetically engineered synthesis of amino acids, but I will do neither! Risking the opprobrium of taking insufficiently inspired looks into the future I shall touch on a few more sensible novel choices—but not before looking at some outstanding old stand-bys.

None of these is more fundamental than breastfeeding. Vigorous revival of this so unfortunately slipping practice would go a long way to deal with what is now undoubtedly the most widespread and potentially the most crippling nutritional deficiency in too many places in the poor world—infant energy-protein malnutrition. The advantages of this practice are numerous and no commonly available alternatives can better them. Breastfeeding, the sole source of infant nourishment throughout human evolution, provides not only virtually perfect nutrition during the most vulnerable period of life but it also offers well-documented physical, psychological, neural, and immunological advantages.

The extent of these benefits is most impressive, ranging from the near absence of gastro-intestinal infections (owing to humoral and cellular anti-infectious factors) to the very low incidence of obesity (thanks to a small risk of overfeeding and the low carbohydrate content of the milk); from the creation of the otherwise irreproducible bond between mother and child to protection against sudden infant deaths. What is perhaps most notable is that maternal nutrition has a surprisingly small effect on the quality of the milk and often even the quantity is not much affected as even plain, inexpensive, and unhygienic foodstuffs and water are transformed in women's bodies with admirable efficiency into the nutritious fluid which is superior to any substitute put together by the most advanced food technologies.

Yet the development of milk formulas trying to simulate human milk and their aggressive commercial promotion, first in industrialized countries and then also throughout the poor world, led (together with changed women's lifestyles and disdainful attitudes to breastfeeding) to a rapid worldwide decline of this most sensible nutritional practice. This most unfortunate trend has been arrested and even reversed in some Western nations since the mid-1970s but it is still very strong in poor nations, especially among new migrants to big cities.

This is the worst imaginable case because illnesses or harmful effects come rapidly where artificial feeds are not prepared with all due care. Almroth and Greiner put it best by writing about 'poverty characterized not only by limited money but also by constraints on the time, energy and patience of mothers, who are likely to overdilute powdered milk to make it go further. This can lead to severe under-nutrition and eventually marasmus. Poor mothers are unlikely to understand the need for sterilizing bottles, and seldom have the necessary facilities, fuel or time to clean them well. It follows that the more limited a mother's time and resources are, the greater the value of breastfeeding will be to her.'

An interesting approach to the evaluation of breastfeeding then is to take human milk as just another food commodity and compare the expenses of breastfeeding with the cost of an alternative source of infant nutrition. The cost of breastfeeding may be expressed simply as the cost of all additional food needed to produce the milk and the value of the time spent breastfeeding, while the costs of artificial feeding encompass the expenses for the food, feeding bottles, fuel and utensils necessary to clean and to sterilize the bottles and, again, the value of the time spent on these tasks.

Such analyses based on field observations, and appropriate estimates to fill the inevitable data gaps, have repeatedly shown the very considerable financial benefits of breastfeeding. For example, the total costs of two years of artificial feeding in the southern regions of the Ivory Coast were calculated at between $(1976) 910–90, while the total cost of breastfeeding during the two years was no more than $260–310. Savings of $600–730 are impressive indeed, as the average annual wage of Ivorian plantation workers was then only about $300 a year—and it must be remembered that only time and goods entered these calculations which do not include the obvious, but impossible to quantify, mental and health-producing effects of breastfeeding, and the disease-inducing consequences of artificial feeding.

If breastfeeding deserves all imaginable support and encouragement as the superior start to a lifetime of good nutrition, then appraisals of maintenance and diffusion of milk-drinking after weaning cannot be far behind. Milk is an inexpensive source of ideal protein, in terms of energy by far the most efficient animal product and there is a wide choice of milking animals suiting various environments and playing other convenient roles in traditional cultures.

Although the absolute differences are not large, relative ranges of the major constituents of whole, good quality milk of various milch animal species are surprisingly wide (even if one leaves aside reindeer whose milk is really different!). While typical Friesian cow milk has 3.5 per cent of fat, and Guernsey milk up to 4.7 per cent, mare's milk has only 1.6, and buffalo and ewe's milk 7.5 per cent; nonfat solids range much less, just between 8.5 and 10.9 per cent (8.7–9.1 for cow's milk); protein values fit between 2.1 and 3.9 per cent (3.2–3.7 in cow's milk); calcium concentrations are between 0.08 and 0.2 g per litre (0.11–0.13 for cows); and energy content, mainly a function of fat presence, ranges from as low as 450 kcal per litre in mare's milk to 1000 kcal in buffaloes and ewes, with 600–750 kcal for cows.

Moreover, milk composition may differ rather widely within any one species, and even within its breeds or races. The individual animal's metabolism, quality of nutrition, stage of lactation, age, and season of the year will determine the rate and composition of milk production so that the representative values quoted in the preceding paragraph are just good general guides and their use in calculations of potential milk production is accurate just for heuristic approximations.

Differences in protein, calcium, and vitamin content cannot make any particular animal milk a special source of nutrients; higher fat content can be objectionable, but it is rather easy to adjust it to desirable levels even outside modern dairies. Consequently, acceptability is mainly a matter of cultural conditioning, acquired taste (even when they contain the same amount of fat, milks can still taste different)—and of ability to digest lactose, milk's disaccharide (a combination of glucose and galactose) which must be broken down in the gut into simple sugars before it can be absorbed.

Cow's milk contains about 50 g of lactose per litre while human milk has almost exactly 70 g. Lactase, an enzyme present in all suckling young, is the specific compound to do the unravelling and unless there is a congenital disorder children of all countries and cultures have enough of the enzyme to perform the conversion. But in the years immediately after weaning we begin to differ: some of us continue to drink plenty of milk and our liking of that smooth, sweetish taste lasts a lifetime; others, although they had high lactase activity as infants, and have not suffered any intestinal disorder, disease, or surgery since then, are unable any more to hydrolyze lactose in quantity and hence cannot consume appreciable amounts of milk and other lactose-rich dairy foods without intestinal problems which may be relatively mild but still unpleasant (flatulence, intestinal cramps), or rather debilitating (diarrhoea, nausea, vomiting).

This primary adult lactose malabsorption has a striking geographic distribution with virtually all of the relatively unmixed ethnic groups of the world falling into two major categories, those with low (0–30 per cent) incidence of lactose malabsorption, and those where the prevalence of the syndrome is very high, at least 60 and commonly near 100 per cent. Frederick Simoons undertook a thorough review of all available lactose malabsorption studies which shows clearly the split with the highest prevalence rates (85–100 per cent) in hunting and gathering people (Inuit, American Indians, Bushmen), in farming populations of non-milking cultures (concentrated in sub-Saharan Africa, and Southeast and East Asia), and the lowest incidence among peoples who have consumed milk and lactose-rich dairy foods for a long historical period (above all Europeans and their overseas descendants, people of the Indian subcontinent, and some Near Eastern and African pastoralists).

The obvious question is then: can milk drinking be promoted among all those hundreds of millions of people who would benefit from this excellent nutrition but who are lactose malabsorbers? The positive answer may seem to be in doubt when one considers that, of all hypotheses advanced to explain the distribution of lactose malabsorption, evidence for the genetic one is overwhelming while the induction theory can no longer be supported as all efforts to induce lactose activity through milk feeding after weaning have been unsuccessful. Moreover, some studies have clearly shown that many adult malabsorbers are well aware that they would have gastro-intestinal symptoms after drinking larger amounts of milk and hence they restrict the intake, and even the malabsorbing black schoolchildren in American tests consumed significantly less free milk provided during lunches than their white schoolmates.

These deliberate restrictions of milk consumption are only one part of the problem: even when no overt gastro-intestinal effects are present, lactose malabsorption may cause significant energy losses, as well as lower utilization of calcium, so that even malabsorbers with subclinical ill-effects will not be able to take full advantage of this excellent nutrition. And yet the prospects for widespread promotion of dairy products are surprisingly good.

The high prevalence of lactose malabsorption has not prevented such intakes in the past. The simplest adjustment is to limit the consumption of lactose-rich products to quantities causing no or few distress symptoms: a small glass of milk a day is well tolerated by most adult malabsorbers and even a mere 100 ml can still provide about ten per cent of average daily protein needs! A much better way is, of course, to consume fermented dairy products. Compared to 50 g of lactose in a litre of milk, buttermilk has about 45 g, and yoghurt less than 35 g, but it is the fermentation of cheeses which reduces lactose content to less than 2 g/kg in ricotta, to just 0.5 g in cottage cheese, and to virtually undetectable amounts in fully ripe cheeses, be they soft (such as Roquefort or Camembert) or hard (Cheddar, Edam, Emmenthaler).

Consequently, lactose malabsorbers can eat more yoghurt and unlimited quantities of cheese, thereby obtaining the full benefits of dairy products without symptoms of illness. The diffusion of milk production, yoghurt- an cheese-making beyond their traditional realms is thus most desirable and while one should not expect that the new adopters will ever match their tolerant counterparts in overall consumption, rapid and relatively large gains are possible as best illustrated by the post-World War II Japanese experience with annual per capita consumption moving from less than two litres per capita in 1950 to about 40 litres in the early 1980s (Fig. 3.14), that is about a 20-fold expansion!

This experience is a most encouraging model for the current Chinese efforts to boost milk drinking. Shortages of grazing land and grain feeds, cultural aversion (milk being associated with the inferior pastoral invaders), and high rates of lactose malabsorption have combined to make the Chinese a virtually milk-free society, the only such case among the world's major civilizations. By the late 1970s annual per capita milk consumption averaged a mere 950 ml, and hundreds of millions of Chinese had never drunk milk. Since then the promotion of dairy production has nearly doubled the country's annual milk output but this leaves the rather meaningless average per capita daily consumption still at a mere 5 g a day, a nutritional contribution of utter insignificance compared to daily mean intakes of over 100 g even in such a non-dairy country as Japan, and to almost 300 g in the United States.

While turning the Chinese into even modest consumers of dairy products will take a long time, selling them more soybeans should be a simple task: after all, the world's most populous nation has always relied heavily on this plant to supply good vegetal protein. The many ways of eating soybeans—green as a vegetable, dry seeds as a crunchy snack, sprouted, processed into flour, 'milk' and oil, milled and then coagulated in bean curd (which can be eaten fresh, pickled or dried), and fermented to yield soy and other sauces and pastes—are the best testimony to their traditional popularity.

Yet even the Chinese have shown that they prefer more sugar, oil, and meat rather than more soybeans: the country's per capita production is now less than half of the harvests of two generations ago, and this can only mean that in many other cultures, where legumes never found such a high approval as soybeans did in China, the trends must be still less encouraging. Indeed, the global output of all food legumes (or pulses) has remained at about the same level for the past quarter century and continental disaggregations show large absolute declines in pulse areas and outputs in North America and Europe, stagnation in Asia and Latin America, and only Africa's production growing at a moderate pace.

Yet legumes have so many advantages as nutritious and easy-to-grow foods. The first is their great variety, embracing such nearly universal species as *Phaseolus vulgaris* (commonly called kidney, french, or haricot beans), *Pisum sativum* (peas, or garden peas) and *Lens esculenta* (lentils) as well as many lesser-known plants whose cultivation is largely restricted to the tropics, or to portions of one or two continents (for example, bambara groundnuts in sub-Saharan Africa, rice beans in Southeast Asia, or pigeon peas in India or East Africa).

Second, when compared with cereal crops most of the legume cultivars require less production management: their symbiotic *Rhizobium* nitrogen fixation obviates costly fertilization (for more on this see section 5.3), they rarely require irrigation, and are fairly pest-resistant. Their average yields are not as high as those of cereals but there is a great potential for improvement in their performance should they receive only a fraction of the breeding attention spent on grains during the post-World War II period and should their largely subsistence cultivation throughout the poor world be better managed. Even with the current low yields, legume cultivation produces larger quantities of protein, the third great advantage. While cereal protein contents range between 7 and 11 per cent, those of legumes are commonly between 20 and 25 per cent (soybeans, of course, top all scores with 38 per cent). Compared to cereals, common legumes also have much higher amounts of riboflavin and thiamine, more calcium and iron (groundnuts and soybeans also have much more lipids), and about the same overall energy content.

And, lastly, legumes can be eaten in an appealing variety of ways as immature pods (Chinese snow peas are perhaps the best known instance), cooked dry seeds (ranging from well spiced Indian *dal* to Boston baked beans), fermented foods (the already mentioned Oriental bean curds, also Indonesian *tempeh*), as well as roasted and milled and used as excellent sources of oils (peanuts, soybeans). All of this adds up to an impressive justification of the further substantial expansion of such crops as a key part of long-range strategies to assure global nutritional needs—yet worldwide legume outputs, except for soybeans which go overwhelmingly for animal feeding, are stagnant.

What is then wrong with pulses as food? Biochemical analyses show that their abundant proteins are deficient in sulphur amino acids (methionine and cystine) but as these amino acids are well supplied in grains, whose lysine and threonine deficiency is, in turn, abundantly complemented by legumes, this consideration hardly detracts from their appeal (Fig. 3.17). In fact the amino acid complementarity of *tortillas* and beans, rice and *doufu*, and *chapatis* and *dal* has been the key to the nutritious vegetarian diets of the Americas, the Orient, and the Indian subcontinent: such combinations provide amino acid patterns surprisingly close to that of milk!

Much more significant is the presence of metabolic inhibitors, above all antitrypsins and haemagglutinins, as well as substances causing flatulence. Trypsin inhibitors have been identified also in common cereals and tubers, but only in leguminous seeds can their presence cause serious nutritional deficiencies. Experimental animals fed raw legume flours use proteins inefficiently, fail to thrive, and may even die rather rapidly. But these enzyme inhibitors are thermolabile, or disappear in various fermentation processes, or during soaking so that eating of properly cooked or fermented legumes should not cause any trypsin inhibition. Similarly, the growth inhibition effects of haemagglutinins are removed by cooking

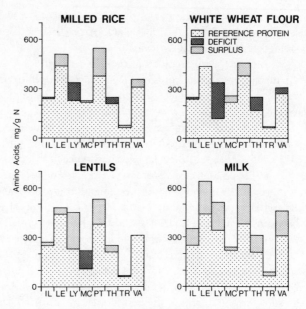

Fig. 3.17 Amino acid composition of four proteins compared to the reference score. Milk satisfies and surpasses the ideal pattern for all amino acids; cereals are always short in lysine and somewhat less so in threonine; while legumes are deficient in methionine and cystine. Co-digestion of cereals and legumes offers a balanced vegetarian diet but the digestibility and availability of proteins is still lower without any animal foods in the everyday diet. Abbreviations are, in the same order as they appear in the graphs, isoleucine, leucine, lysine, methionine and cystine, phenylalanine and tyrosine, threonine, tryptophan and valine: the ten essential amino acids.

(or even just by heating). In healthy people digestion and nutrient absorption of properly prepared legumes should thus be as effective as in the assimilation of cereals.

Flatulence associated with legume consumption has been for long recognized as the main restriction on their consumption, above all by young children, and oligosaccharides (stachyose, raffinose, verbascose) and peptids in the pulses, acting together with *Clostridium perfringens* bacteria residing in the gut have been identified as the causative factors. Unfortunately, heating does not destroy these substances and the only practical way to control the effect, at least until we have succeeded in breeding cultivars with low content of the unwanted compounds, is to limit the intakes. Important as they are, anti-nutritional properties and digestive complications associated with legume consumption cannot preclude recommendations of regular intake of pulses in low or moderate quantities.

Declining consumption of these nutritious foods must be ascribed to more than just these largely avoidable problems. The flavour of many cooked pulses, and even their smell, are often disagreeable to many would-be consumers, the cooking complications (overnight soaking, long simmering times) put off modern urbanites as well as those villagers whose improved economic status enables them to get more rice or wheat-flour (moreover, soybeans, because of their high fat and low carbohydrate content, never cook as soft as other pulses).

Cultural traditions and social acceptance have very important effects. The Chinese may like bean curd but never look to lentils, soybeans have never been accepted as food in the Western world (although the United States is now by far their largest producer), and legume consumption in all cultures declines with higher family incomes. The image of traditional legumes as the fare of the poor and the deep aversions of people to the adoption of new foods, especially when their tastes are far from universally appealing and their preparation may be cumbersome, are perhaps the most potent obstacles to the regular consumption and further diffusion of pulses.

Similar perceptions and cultural considerations will be of no less importance in all attempts to expand the harvest of food from the ocean. While the total population of marine fish appears to be several times (two to four times) greater than the current global catch, the already clearly demonstrable overfishing of many conventional species in several major productive areas around the world (from yellow croakers in the East China Sea to North America's salmon), and economic constraints on this relatively expensive way of food production, will necessitate a shift toward unconventional fish species and other marine organisms if the food output from the ocean is to grow for decades to come.

For example, actual global catches of salmon have been around 90 per cent of the potential harvest during the past decade and the share has been almost as large for crustaceans (crabs and lobster), and it surpassed 70 per cent for flounders, cod, haddock, and large tuna, all widely preferred species. On the other hand, no more than about 2 to 4 per cent of the potential sustainable catch of cephalopods are taken and harvesting of krill removes only a fraction of one per cent—yet it is very safe to predict that in the early 2000s squid and octopus are not going to adorn the weekday tables of millions of new consumers (although their potential harvest is two orders of magnitude above that of salmon!), and that frozen krill will not displace a large chunk of beef consumption.

Mariculture, the most controlled high-yield route to an expanded ocean harvest, would also be unable to fulfill even a part of its promise without a stress on the cultivation of plants. However, seaweeds, the principal autotrophs suited for mariculture, appear to be much like soybeans: nutritious, and deserving of extensive diffusion, but not about to conquer the world beyond their limited areas of traditional consumption. Japanese like their *nori* (*Porphyra sp.*), and *wakame* (*Undaria*), and the Chinese have been expanding their cultivation of *Laminaria* kelp, and hydrocolloids present in many red and brown algae are finding wider markets as stabilizers and emulsifiers in food and cosmetic industries—but unless the affluent West acquires the less than alluring habit of munching on often strongly scented tissues literally crackling with sodium, potassium, calcium, and phosphorus there is little chance that seaweed consumption will expand greatly.

Indeed, the importance of the involvement of rich countries in promoting new foodstuffs cannot be overestimated not only because much of the needed research and development funding, and a good part of the scientific manpower, must originate in the affluent world but, even more essentially, examples set by the wide adoption of non-traditional foods, or by elevation of formerly marginal foodstuffs to notable sources of fine nutrition, would have a profound influence on global diffusion.

One has only to recall the global consequences of such vigorously promoted American innovations as hamburgers and soft drinks to appreciate this allure: in the first instance, tasteless, fatty frozen beef microwaved amidst a squeezable bun, which a detached degustator could not mistake for anything but emergency food of the last resort; in the other case sticky, oversweetened, fizzy liquids with queer after-taste which actually *cause* the thirst—and yet the world lies prostrate. Recently even the Chinese started to write about the desirability of having convenient fast foods such as hamburgers!

And so without a newly-found American passion for foodstuffs such as squids and bean curd it would be a tall task to push the desirable nutritional innovations which would expand the availability of good food in a sustainable way. Lack of deep commitment on the part of rich nations will similarly lower the probabilities of successful widespread commercialization of scores of currently underexploited subtropical and tropical plants with promising economic value.

The Board on Science and Technology for International Development, a part of the United States National Research Council, has since the mid-1970s repeatedly examined the possibilities of such commercializations and its panels have gathered large amount of fascinating information but I doubt that we shall see any major new adoptions of bambara groundnut (*Voandzeia subterranea*), a commonly cultivated but otherwise neglected 'poor man's crop' of black Africa, where the cultivation of major established African commercial leguminous crops have been declining, with the continent's recent peanut harvests 10–20 per cent down below those of the early 1970s (and Nigerian production, formerly the largest, down by 65 per cent). Unfortunate as it is, only those minor established or newly introduced foods which elicit extensive commercial interest and eager consumer response (and the latter may be simply the straight result of the former sold by mass advertising campaigns) in the rich countries have a good chance to make a future difference on a global scale.

I will close this section by offering just one unorthodox example which may eventually fit these requirements: the eating of cultivated mushrooms. Although there are more than 2000 known edible species of mushrooms (filamentous fungi is the proper term) only some 25 are widely accepted as food, only 10 are cultivated, and a mere 4 are truly important. White mushrooms (*Agaricus bisporus*, best known as the French *champignon*), Black Forest mushrooms (*Lentinus edodes*, Japanese *shiitake* now available in many Western specialized food stores), and the straw and oyster mushrooms (*Volvariella* and *Pleurotus*) confined almost solely to the Orient.

Champignon is well ahead of the rest and now grown in controlled environments in some 80 countries (the United States and France, not surprisingly, lead) mostly on the substrate of pretreated (composted) wheat, straw, and horse or chicken manure. Experienced growers can achieve substrate conversion efficiencies of 60–75 per cent, five to six times higher than with other cultivated mushrooms. *Volvariella* grows just on wetted straw in the open, while *Lentinus* and *Pleurotus* are cultivated on tree logs.

That mushrooms taste good is abundantly attested by their high culinary status in cultures Oriental and Occidental—but their high nutrition status comes almost always as a surprise: after all are not fresh mushrooms rather like vegetables, mostly water and hardly any protein, without the benefit of vitamin C? True, mushrooms like vegetables, are low in food energy but *Agaricus* is rich in vitamin C, *Lentinus* in D and, most importantly, because they are heterotrophs (unlike vegetables) they contain all essential amino acids and are especially rich in lysine, an amino acid present in inadequate amounts in staple grains, and only milk, eggs, and meat have better amino acid indices than mushrooms.

Perhaps even more surprising is the difference which the greatly expanded cultivation of mushrooms could make to global nutrition. To begin with, there is no shortage of substrates for cultivation, from sawdust to banana leaves, from cereal straws to animal manures. A few years ago I calculated that using one-fifth of the world's production of cereal crop residues for mushroom composting (a fraction which would leave plenty of straw for the protection of soils against erosion, and for feeding and bedding), with conversion efficiency of just 30 per cent, would yield annually about 180 million tonnes of cultivated mushrooms, or a little over 100 g a day per capita.

Each of these portions would contain nearly 4 g of excellent protein, compared to the average daily per capita availability of 24 g of animal proteins in global terms, and a mere 12 g in the poor world. Consequently, cultivated mushrooms would extend high-quality protein supplies by about 15 per cent on a global basis and by a third in the poor world. Higher conversion efficiency, and the use of a larger share of crop residues and other organic wastes for cultivation, would further increase this already impressive share. Most importantly, this nutritious and tasteful contribution would become available by tapping the existing phytomass wastes, by an efficient and relatively simple delay in nutrient flows.

But formidable limitations, largely of an environmental and logistic nature, assert themselves in this case no less than with all previously discussed innovations. Harvesting, transporting, and treating cereal straws, and cultivating and harvesting the mushrooms are fairly labour-intensive activities, and in temperate latitudes, where most of the straws and most of the potential consumers are concentrated, efficient mushroom growing must be done indoors in a controlled environment. The natural ingredients of the process may be cheap but production requirements make the costs rather high. Moreover, fresh mushrooms cannot be stored for long, while the canned and frozen ones lose nearly all their taste—and even though the cultivation has been rising steadily and the mushrooms sell very well throughout the rich world it is doubtful if they would be ever accepted as an everyday food by most consumers.

Clearly, both the resurrection of old food ways and the introduction of new ones are far from being as promising as the arguments of nutritional benefits, and appraisals of outstanding production potential, would indicate. Of one fundamental conclusion I feel confident: we would do much better to try to re-introduce or to extend many old ways rather than to pretend that exotic novelties have more promise. And for novelties the common-sense evaluation should be the same as in the case of dubious alternative energy technologies: why bother when there are so many, and still better, traditional options?

Finally, the most fundamental consideration of all: if we continue to do well in staple grain production and reduce somewhat our high consumption of animal foods there will be no need for any shocking food novelties for generations to come. And this conditional optimism will also be found in the next chapter, where I will examine above all the changing perceptions of the state of the global environment.

4

Maintaining a Habitable Environment

Pliny's descripton of the world we live in is perhaps the most elegant definition of
the environment ever conceived. His adjectives—sacred, eternal, immeasurable—
may be seen as scientifically incorrect only by those whose arrogance blinds them to
the still so obvious limits of our understanding, and whose lack of the true
knowledge makes them unaware that we stand just at the doorstep of compre-
hending the environment *totus in toto*, wholly within the whole.

Fortunately, we know enough to appreciate that so many of the everyday actions
that we have taken to house and to feed ourselves, to heat our homes and to produce
goods, essential and frivolous, can change the environment in ways that will often
diminish the initial benefits, and that in many drastic cases will bring overwhelming losses which may be sometimes beyond our powers to remedy. Some of
these realizations, resting just on careful observations, are ancient—Plato's description of Attica as 'a country of skin and bones' resulting from soil erosion, or Meng
Ke's (Mencius) laments about the disappearance of Chinese mountain forests.

Making the deductions in many other cases called for sophisticated science; there
is, for example, no obvious, simple connection between the mercurial fungicide
used to coat wheat seeds and the trembling arms and the gradual loss of vision
experienced by heavy consumers of fish caught in a lake far downstream and
receiving the runoff from the wheat fields.

Pre-industrial civilizations caused many spectacular and lasting transformations
of their environment through deforestation, overgrazing and poor farming
practices, but vastly greater and much more concentrated flows of energy which
became available to industrial civilizations from the recovery of fossil fuels
accelerated and broadened the processes of environmental degradation and
pollution to an unprecedented degree. Energy industries—the extraction, distribution, and conversion of fossil fuels and the generation of electricity—became
themselves one of the largest causes of undesirable environmental transformation.

That exploratory, extractive, transportation, conversion, transmission, storage,
and disposal activities forming the energy supply systems of modern societies will

have numerous environmental implications is self-evident. All will occupy space, more often than not good, flat, accessible land which used to be in crops or in pasture, some by just sitting on it (power plants, transforming stations, refineries), some by leaving it largely intact but pre-empting other uses (pipelines, transmission lines), some by degrading it and changing it even beyond restoration (surface mines, also deep mines).

Some will use large volumes of water and discharge it either much warmer (thermal power plants), or considerably polluted (coal cleaning plants, refineries). Air pollution is a ubiquitous problem associated with the burning of any fossil or biomass fuels, be it in huge thermal stations or simple household stoves. Although land degradation, water demands, and water and air pollution are the most common environmental complications of modern energy production there are other considerations, ranging from a variety of risks such as explosions at liquified natural gas terminals, and accidental releases of radioactive contaminants from nuclear power plants, to noise and aesthetic objections.

And, of course, some energy projects will cause only one kind of environmental problem—constructing pipelines in harsh mountain or Arctic conditions may bring plenty of disruption but once installed these links are by far the safest bulk carriers available—while others will have multiple impacts. The most ubiquitous example of the latter category is certainly a large modern thermal electricity generating station and a detailed examination of resource requirements for a typical modern thermal power plant illustrates best the environmental demands and degradations caused by this widespread energy conversion.

The model plant will be coal-fired, as most of its real-world counterparts, and Fig. 4.1 delineates its basic processes: the burning of coal in a boiler in whose long tubing feedwater is converted to pressurized and hence superheated steam which passes through a turbine driving a generator; and the condensing of exhaust steam back to water which is recycled to the boiler. Cooling towers are now commonly used to lower the temperature of water used for condensing exhaust steam while electrostatic precipitators and flue gas desulphurization (FGD) systems treat the flue gases before their release to the atmosphere through tall stacks.

All calculations will be done for what is now a very typical North American plant with 1000 MW of installed capacity in two 500 MW units (integrated boiler-turbogenerator assemblies). Average load factor (percentage of time actually generating) will be 70 per cent, or about 6100 hours annually, and the average net heat rate (gross heat rate plus penalty for stack gas reheat and allowance for auxiliaries) 10.8 MJ/kWh, equivalent to a conversion efficiency of 33 per cent.

Bituminous coal burned in the plant will average 24 MJ/kg after partial clean-up, so that annual consumption will total 2.767 million tonnes. If this coal comes (as does most of this continent's steam coal) from a surface mine with a seam 2 m thick, 138 ha would have to be stripped annually. When built right next to the mine the power plant would need only minimum coal storage and its structures, service and clearance areas may occupy as little as 20 ha.

However, when serviced by a unit train, the railway specifically built for that

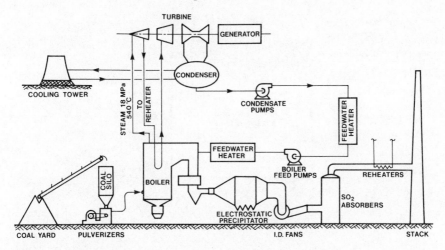

Fig. 4.1 Simplified flow diagram of a typical coal-fired thermal power plant's operations.

purpose would, with a right-of-way width of 30 m and a distance of just 100 km, pre-empt the use of 300 ha of land; the coal-receiving area and coal yard with storage sufficient for sixty days of normal operation would take over another 50 ha, the plot for the flue gas desulphurization system would add about 50 ha, and the area which has to be reserved nearby for the disposal fly ash and desulphurization sludge during the plant's life (at least thirty years) could add up to 300 ha.

Consequently, the fixed occupied land taken by all power plant facilities would add up to about 400 ha, and a unit train link of just 100 km would boost it to 700 ha. Coal mining claims about 150 ha a year but this loss can be reclaimed after a few years, or a decade in properly run mining operations. Another considerable claim on the land arises from the necessity to transmit the electricity to the load centres. A 500-kV line requires at least 4 ha of right-of-way per km so that 100 km link alone would need 400 ha of land to be set aside for a transmission corridor.

Figure 4.2 summarizes these fixed land requirements of the whole transportation-generation-waste disposal-transmission system. Even when considering just the 400 ha of the power plant site noted above, the area is large when seen in terms of food production potential: if it was previously only an average quality of farmland it could support—at the plentiful North American nutritional levels—about 1000 people (and in China it could feed no less than 4000 people). Most of the land expropriated for transmission rights-of-way could be farmed once the construction ends, but the inclusion of railroad rights-of-way would nearly double the calculated potential crop production loss.

As for water, the power plant's condensers would require nearly 42 m³ each minute, an annual demand of over 15 million m³. If not sited near a major river, or on the seacoast, an artificial lake must be created or large cooling towers build (the

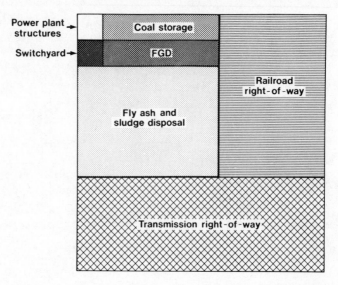

Fig. 4.2 The thermal power plant itself occupies only a small area compared to the space needed for coal storage, desulphurization, and fly ash and sludge disposal, and even these requirements are dwarfed by rights-of-way for railroad and transmission links.

former occupying considerable additional land). And if these huge volumes of water are not cooled sufficiently before being returned into the river or bay changes of aquatic ecosystems would soon follow.

Problematic as this heat disposal might be its management is a relatively straightforward engineering problem. As opposed to once-through systems, cooling towers will minimize the extraction of unwanted heat and although they themselves may cause environmental problems these impacts are strictly local and mostly just seasonal. Similarly, air pollution arising from the storage and handling of coal, and the noise and unsightliness of huge plants can be either minimized by proper management and design or confined to localized impact.

In contrast, the release of water and air pollutants poses much more intractable control problems. Figure 4.3 illustrates all the important processes and flows within a coal-fired power plant which result in direct discharges of liquid or gaseous wastes. As can be seen, there are many potential waste water streams but only five of them-condenser cooling, ash handling, boiler blowdown, air pollutant control, and boiler make-up water treatment-generate wastes continuously.

The volume of these waste streams varies not only with operating levels and maintenance procedures, but also with weather and some of these releases will contain only negligible amounts of hazardous pollutants. The two waste streams of the greatest concern are those from coal ash handling and flue gas desulphurization. Wet handling of fly ash may cause important direct discharges but usually the greatest problem is associated with ponding.

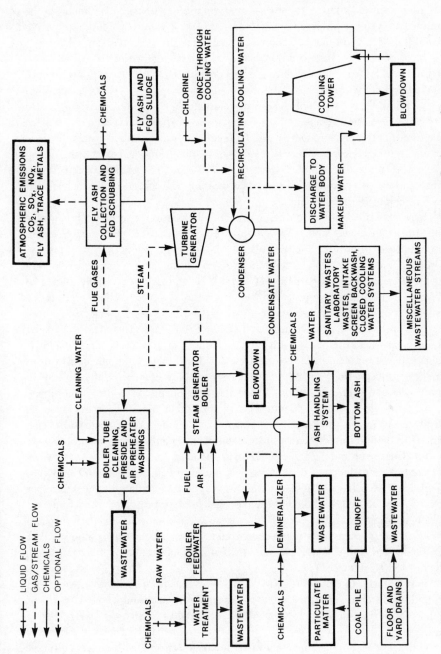

Fig. 4.3 All important liquid, gaseous, and chemical flows are traced in this diagram of a coal-fired power plant's operations.

Fly ash cannot be piled up high, or easily transported to some convenient dumping sites, owing to its obvious air pollution potential so ponding in the immediate vicinity of large power plants is the most common solution requiring considerable areas of flat land and posing potential water contamination hazards.

Most of the fly ash is composed of SiO_2 and a large variety of metallic and alkaline oxides, but virtually all coal ashes contain relatively high concentrations of toxic trace elements, most commonly arsenic (8–45 ppm on dry basis), vanadium (8–67 ppm), lead (8–14 ppm), selenium (about 2 ppm), and mercury (up to 0.5 ppm). On the average at least 75 per cent of these trace elements will be captured in the precipitated fly ash. Consequently, the combined mass of toxic elements in ponded fly ash may surpass 50 t a year in the model power plant, posing serious groundwater pollution risks. Expensive lining of ponds with impervious materials to limit seepage is the only effective solution—but one with which there is still little long-term experience.

Inevitably, any combustion, being just a rapid oxidation of carbon, releases large quantities of CO_2 and some CO. The latter is more abundant where the conversion is less efficient but in large power plants and industrial boilers CO flux is low (an order of magnitude below nitrogen oxides) and as the gas is tolerated by biota in concentrations much higher than SO_2 and NO_x and as it gets rapidly diluted to levels much below any possible health effects, no controls of its emission are needed now or contemplated for the future.

No controls exist for CO_2 either and none would be ever needed for the sake of human health as the gas is completely innocuous at any levels which could be normally encountered in the biosphere; as for plants they could only benefit from higher atmospheric CO_2 levels. The concern lies elsewhere: the absorptive properties of CO_2 in the infrared bands may be fundamental controllers of long-term trends in surface temperatures, and the single most important factor of inadvertent human modification of climate. Our model power plant will release annually about 6 million tonnes of CO_2 compared to only about 1000 t of CO.

In terms of volume the other two most abundant pollutants after CO_2 are oxides of sulphur, virtually all of them being SO_2, and nitrogen, largely as NO and NO_2. Typical emissions carried away in raw flue gas are several hundred ppm of nitrogen oxides and hundreds of thousands ppm of SO_2. In mass terms large modern power plants burning bituminous or sub-bituminous coal in tangentially-fired furnaces (now the most common arrangements) would emit as little as 400–800 ng SO_2/J of crushed, sized and separated low sulphur coal but high sulphur content raises this flux to 1300–2800 ng SO_2/J.

In mass terms the model 1000 MW plant burning steam coal with 3 per cent of sulphur will release at least 150 000 t SO_2 each year (assuming about 10 per cent retention in bottom ash) and of the order of 25 000 t of nitrogen oxides. While nitrogen oxide emissions can be substantially reduced by not very expensive modification of combustion processes, removing SO_2 from the flue gas is a difficult and costly proposition.

In our 1000 MW plant a flue gas desulphurization system able to remove 85 per cent of SO₂ flux will have to use each year about 150 000 t of limestone in reaction which will produce about 400 000 t of dewatered sludge (50 per cent solids), will require large dumping ponds, and will increase the power plant's capital requirement by 20 per cent and total levelled electricity generation costs (including disposal) by as much as 40 per cent.

Finally, our model plant could release each year up to 220 000 t of fly ash (the remaining 20 per cent of uncombustible oxides, metals, and other elements in coal end up as bottom ash) but these potential emissions can be very well controlled: the plant will have a precipitator with a design collection efficiency of 99.75 per cent whose power requirement will be about 1.6 MW—that is less than 2 per cent of the plant's installed generating capacity. Consequently, fly ash ceases to be a visible problem as only a fraction of the smallest particles (diameter below one μm) escapes into the atmosphere—but this success in air pollution control has been translated into the problems described above of disposing of large quantities of potentially hazardous wastes. And even with excellent controls 25–35 per cent of trace toxic metals, and all chlorine and fluorine present in the coal, will be released to the atmosphere. Annual emissions of toxic metals from the plant may thus amount to several tens of tonnes, and those of Cl and F to more than 200 t.

Comparison of the model power plant's fuel consumption with the worldwide burning of coals, crude oils and natural gases, shows that in mass terms the global combustion of fossil fuels is roughly 2200 times larger than the station's needs. However, one cannot simply multiply the plant's uncontrolled emissions listed in Table 4.1 2200 times to get the global totals since hydrocarbons contain more carbon than coals but often less sulphur (actually, most of the coals do not average 3 per cent S either), and almost no ash. Making these necessary adjustments, and multiplying by values listed in the second column of Table 4.1, results in rough but acceptable actual global fluxes of atmospheric pollutants generated in fossil fuel combustion listed in the Table's last column.

All CO₂, and virtually all SO₂ and nitrogen oxides, are released uncontrolled while only about 10 per cent of all fly ash is emitted. These 20 or so million tonnes are only a fraction of total global man-made particulate matter (most likely no more than one-sixth). However, in North America, fuel combustion is the single largest source of particulate emissions, with stone, sand, and gravel industries a rather close second and field farming a distant third in terms of relative contributions and, even on a global scale, the combustion of coals and crude oils mobilizes larger quantities of trace elements than any other human activity.

In fact, some trace elements are now put into the atmosphere at rates of the same order of magnitude as natural fluxes of those elements through water during weathering. Mercury, vanadium, gallium, lanthanum, and cesium are in this category, while for lead, strontium, rhenium, nickel, copper, zinc, and many others the natural weathering flux is at least an order of magnitude higher.

Still, this anthropogenic mobilization of particulate matter is only a small fraction of natural emissions; dust, sea salt, forest fires, and volcanic explosions put nearly 2 billion tonnes of primary particulates in the atmosphere each year and

Maintaining a habitable environment

Table 4.1 Derivation of approximate global emissions of major air pollutants from fossil fuel combustion

Atmospheric pollutants	Uncontrolled emissions from a 1,000-MW model coal-fired power plant (tonnes)	Multiplication factor to calculate global emissions	Approximate average degree of controls (per cent)	Approximate atmospheric global fossil fuel combustion (10^3 tonnes)
Carbon dioxide	7,000,000	2,400	0	16,800,000
Sulphur dioxide	150,000	1,200	5	170,000
Nitrogen oxides	25,000	2,600	5	62,000
Chlorine	1,000	1,000	0	1,000
Fly ash	220,000	1,000	90	22,000
Zinc	60	1,000	75	15
Boron	30	500	75	4
Vanadium	15	2,500	75	9
Arsenic	5	1,000	75	1

secondary aerosols—sulphates, nitrates, ammonium, and hydrocarbons—nearly rival that amount so that the annual natural loading exceeds 3 billion tonnes.

A single volcanic eruption can put into the atmosphere more particulate matter than that annually generated in hundreds of large thermal power plants and, moreover, the mode of injection—a rapid transfer of fine particles all the way to the lower layers of the stratosphere—will ensure global or at least hemispheric distribution of these ejecta during the months following the explosion. But, as far as the gaseous releases are concerned, human additions to natural flows have grown to notable shares and discussions of the possible consequences of CO_2, SO_2 and NO_x releases from fossil combustion will be key concerns of the next chapter.

Environmental consequences of food production are, in contrast to the global and continental reach of atmospheric pollutants from combustion, overwhelmingly limited to local and regional effects, but combined productivity losses resulting from soil degradation and farmland disappearance may have worldwide repercussions if allowed to continue for a few more generations. Of no less concern is the simplification of ecosystems by the large-scale planting of a limited number of staple food cultivars and the consequential general decline of genetic diversity. All of these topics will be discussed later in this chapter while human interference in the nitrogen cycle, definitely the most significant anthropogenic intervention in biospheric flows owing to food production, will be a part of the more detailed treatment of biogeochemical cycles in the next chapter.

But first I will look at the genesis of our deepening interest in environmental affairs. Until the 1950s the environmental concerns of Western industrial civilization were largely restricted to big cities and only since then, although still intensifying in many fast-growing urban areas and conurbations, they have spilled rapidly into the countryside as well as to scores of still poor but rapidly industrializing countries in Latin America, Africa, and Asia.

By the 1960s environmental pollution and degradation reached a new, worrisome stage—just at the time when there was a widening success in cleaning up the most visible, objectionable forms of pollution which were present during the many decades of industrial expansion in northern cities. Perhaps most notably, coal substitution for household use by liquid and gaseous fuels in the urban areas of rich countries, and the virtually universal adoption of efficient electrostatic precipitators at large industrial boilers and coal-fired power plants, lowered dramatically the amounts of dustfall and concentrations of suspended particulate matter, a most welcome trend bringing more sunshine, more distant horizons, and lower health risks. Similarly, more and better municipal water treatment facilities reduced the waste burden of many rivers and coastal seas.

But these improvements, important and desirable as they were, were predominantly effective only on a limited, local, or small regional, scale—while a new class of environmental impacts was increasingly apparent, embracing very large regions, major portions of continents, even the total biosphere. Persistent pesticides were found in biota thousands of km away from the application areas; the long-distance atmospheric transport of gaseous pollutants from the tall stacks of large power plants was identified as the main cause of acidification in lakes hundreds of km away; methyl-mercury in water was accumulating in the bodies of fish in large catchment basins; nitrogenous fertilizers were thought to cause unacceptable groundwater pollution under millions of hectares of intensively cultivated farmland; and the continuing rapid growth of fossil fuel consumption revived concerns about global climatic change.

These and other concerns (presented by increasingly interdisciplinary environmental studies) combined with spreading public disgust at common pollution—as well as with the desire to conserve many splendid natural environments, and with the growing realization that public goods taken so long for granted must be much better managed if they are to provide sustainable inputs—and thus produced an unprecedented wave of interest in environmental affairs, a wave which caught the interest of the media, academe, politicians, government bureaucracies, business, and international organizations. I will look, necessarily just sketchily, at this explosive rise of environmental awareness, then at some of the most notable fallacies this outburst has produced.

4.1 The raising of consciousness

> The question nowadays is not how we shall live, but indeed if we are going to live at all for very much longer.
>
> T. R. Harney and R. Disch in the introduction to *The Dying Generations: Perspectives on the Environmental Crisis* (1971)

The genesis of that extremely intensive period of environmental concerns in the late 1960s and the early 1970s is a multi-stranded affair a brief survey of which will not be done here for the sake of historical interest—although the story is certainly important on that account alone as it does not happen every decade that the public as well as decision-making elites come to recognize and embrace a new concern vital for human survival—but rather in order to identify the major forces behind the environmental wave because their particularistic points of view have not been yet amalgamated and thus continue to feed several persistent fallacies which will be dealt with in the coming section.

Certainly the oldest, and a very powerful, impulse came not from detailed studies and purposeful science but from impressions of beauty, feelings of awe, efforts to see man as a part of the whole wondrous creation, not apart from it. These emotions were expressed in diverse ways by different civilizations, from tribal worship of trees and skies to eloquent treatises on pantheism, but artists' identification with landscapes and creatures, and their portrayals—be it by pen or by brush, as a realistic record, as an impressionistic capture, or as an abstract transmutation—have been always in the forefront of these appealing expressions.

From Li Bo's lines extolling the mountains as the poet's most faithful companions—'the birds have flown to their roost in the tree/The last cloud has just floated lazily by/But we never tire of each other, not we/As we sit there together—the mountains and I'—to Henry Rousseau's forests full of mysterious flowers, unknown trees, exotic animals, and birds; from monumental descriptions of heavenly phenomena in Lucretius's *De rerum natura* —'now clouds will gather when, as on they fly/Through these high realms of heaven, in numbers vast/Bodies of rougher mould will meet'—to Piet Mondrian's austere geometric landscapes created of coloured order. Art of every civilization overflows with admiration, awe, and reverence of nature.

These attitudes are, undubitably, the mightiest spring of environmental conservation, and sustain the efforts to preserve natural wonders untouched, pristine. First turned into organized form during the last century, the conservation of landscapes, waters, and wild species became a passionately argued goal of numerous organizations ranging from localized and narrowly focused ones (e.g. Save the Redwoods) to the broadly-based and universally oriented (The Sierra Club, The World Wildlife Fund, Greenpeace Foundation, to name just a few).

Science, strangely enough, was a latecomer in comparison with the conservation movement itself. Competent, revealing studies had been produced for decades: to give just a few examples, the first systematic work on acid rain, a problem 'discovered' in the late 1960s and the early 1970s, was published in London by R. A. Smith in *Air and Rain* in 1872; the beginnings of studies of CO_2 influences on climate also go back to the latter decades of the last century; and the writings of Andrew Herbertson or Lewis Mumford were nothing else but brilliant exercises in human ecology. But integrated efforts to present the 'big picture', and willingness to engage in popularization and political lobbying were too few to make much difference.

Professional ecologists, the people who should have been in the forefront of the environmental push were very weakly represented and, anyway, most of them were unprepared to tackle the challenge. As Bruce Welch reminded academe in 1972

the profession of ecology has not attained the ideal of producing versatile generalists to guide our efforts to solve environmental problems. . . . Most ecology programs are housed in departments of botany or zoology, and many departments in each of these disciplines actively discourage students from becoming deeply involved in the other, let alone in the social, physical, or applied sciences.

(How little has changed since then!)

But, emboldened by the public and political interest, some ecologists stepped out to claim their share of media limelight: Barry Commoner turned from being a student of ecosystems to being a political activist (including being a quixotic candidate for the President of the United States), and Paul Ehrlich moved from observations of butterflies to prognoses of environmental horrors soon to descend on our planet. Students had no small part in the academic adoption of environmental concerns. In the United States of the 1960s environmental awareness was raised considerably by the activities of various anti-establishment groups or, perhaps even more correctly, one should say it was aided by the whole generation revolt which germinated from a strange mixture of affluence, fear, and lost bearings on the country's campuses, and for which the identification with anti-pollution, conservation, reduced consumption, and lowered growth sentiments was an almost reflexive way to further strengthen the expression of displeasure with the *status quo*.

Not surprisingly, much of this support, propagandizing, and sloganeering was very shallow but as it was, at least for a few years, fairly persistent and, above all, quite high-pitched, it got considerable media attention and hence an influence much beyond its rather unimpressive factual foundations. And, in curious displays of flexibility seen by the outraged environmentalists as nothing less than acts of sacrilege, large industrial companies hesitated little in joining the environmental movement in a very impressive way—through well prepared advertising campaigns on television and in the press. An uninitiated stranger could be forgiven for believing that these companies were in existence in order to combat pollution. Texaco ('working to keep your trust') surrounded a little boy by a veritable Noah's ark of wild animals and announced 'some good news for all living things on this earth.' Reynold's Aluminium used inch-high letters to say that their can-recycling initiative was working, while DuPont helped 'the alligator save his skin' with its substitute poromeric material. Union Carbide claimed (double-page spread) that 'we're doing what we can to keep pollution off our highways', while Bethlehem Steel came out with more than one big line, they had a complete short story. A retired miner pulling bluegills and large-mouth bass from a new man-made lake told it:

... when they began surface-mining in the area here, I thought, 'Boy, these old hills have had it! But I was wrong. Before they removed the coal, the land just sort of laid there ... pretty, but not much use to anyone. But now that these hills have been mined and reclaimed, they're just as pretty as ever. And we've got something we never had before: a lake for fishing and swimming ... and acres of restored land for picnicking. Now I'm glad they mined this land.'

And some companies used the new wave to float to higher profits. Makers of Arm & Hammer, a venerable brand of washing soda, first listed all major detergents with high phosphate content and if the one you used was on it (nine out of ten times it would be) they implored you to switch to old-fashioned laundry soap and use it with their soda to help 'save our nation's waters because phosphates promote algae pollution—killing fish, stagnating water, turning lakes into swamps' and, at the same time, get 'a cleaner wash even in the hardest water.'

And the politicians were equally flexible. For the Democratic Party the environment was just another item to put on their always lengthy list of government-nursed social programmes but it was a Republican president who acted very quickly to put in place unprecedented institutions and bureaucracies. First, the President's Council on Environmental Quality was set up and when it transmitted its first report to the Congress in August 1970 Richard Nixon's message stressed that it was 'the first time in the history of nations that a people has paused, consciously and systematically, to take comprehensive stock of the quality of its surroundings.'

Then came the much more difficult task of fashioning the new Environmental Protection Agency from fifteen major entities previously belonging to five government departments and independent agencies. In retrospect, this appears to me to have been an even greater change and challenge than I thought it to be at the time, a move—EPA's recurrent mishaps, infighting, and malaise notwithstanding— of truly revolutionary significance for the protection of the environment on the highest, systematic level.

Internationally, the ponderous bureaucracy of the United Nations eventually organized a big Conference on Human Environment in Stockholm in June 1972 but, as with all similar UN efforts, the meeting was a mixture of excellent, factual contributions (perhaps most notably the Swedes unveiled the first detailed national review of acid deposition) and pious ideological posturing (with China in particular playing a sanctimonious role less than a decade before its new policies led to shocking admissions of environmental neglect and harm).

Fifteen months after the Stockholm meeting came the Yom Kippur war, the Arab oil embargo, and the first manipulation of crude oil prices. Energy supplanted environment as the big, intractable bad news, and falsely predicted dire fuel shortages and OPEC's ever-growing might offered scenarios no less catastrophic than those spun out shortly before by environmental doomsayers. But environmental concerns were also in many ways strengthened by preoccupation with energy 'crisis', and the soon-to-come renewed doubts about the world's capacity to feed itself (following in the wake of the Sahelian drought and poor Indian harvests) lent further support to the continuing high-level focus on a multitude of environmental concerns.

Energy stole the limelight for the rest of the 1970s, but attention to environmental degradation and pollution became a permanent ingredient of informed decision-making and responsible management. Asked about the implications of the new environmental ferment, English ecologist Max Nicholson went in 1971 as far as saying that

we are now seeing something that historically, in the long perspective, is a comparable movement to a Renaissance, for it will lead people to look at everything with new eyes; it will lead to a new view of man, in some ways a heightened view, which may make all the materialist values of the Victorian heritage seem irrelevant.

We are still too close to the onset of this possible new era to look back and either to confirm or reject Nicholson's exalted view. But I am certain that the rapid diffusion of environmental awareness and swift bureaucratization of this new, and at least initially enormously appealing, preoccupation have already made us forget that a truly revolutionary shift of perceptions has taken place.

New perceptions do not, could not, always mean new desirable actions even where the scientific, technological, and financial bases are propitious for such moves, but I think that even the most cautious assessment would have to conclude that in Western societies the tide of environmental degradation, advancing with industrialization since the last century has been impressively slowed down; my best appraisal would be that it has turned and a perceptible ebb can be seen already.

In the poor nations, unfortunately, the situation is a great deal less satisfactory but even there the recognition of the necessity to adopt environmentally more benign ways of modernization has gained much ground: the lastest Chinese words supported, at last, by some impressive deeds, Brazil's rising appreciation of limits on the Amazonian development, Indonesia's new World Bank-sponsored environmental review, are perhaps the most significant examples.

Another telling sign of great advances achieved during the first decade of the new environmental era is how many fallacies that were influential during the formative years of the late 1960s and the early 1970s can now be discarded, and how drastically the current perceptions of critical global environment problems differ from those formed just 10–15 years ago. The next two sections will examine these lapses and shifts.

4.1.1 *Fallacies, fads, and frustrations*

Science is the topography of ignorance.

O. W. Holmes, *Medical Essays* (1883)

That such an outburst of interest and worries would produce many dubious explanations, conclusions, and precepts is not at all surprising: all fashionable concerns attract swarms of instant experts who peddle all-embracing, casual, shallow observations as redeeming truths. I shall not waste space on refuting the calls to action summarized by such rallying slogans of the more militant branches of the environmental movement as: 'To clear the air, smash capitalism!' If those sloganeers had bothered to inquire into the state of the air around Soviet smelters or in the Chinese cities, they could have considerably extended the ideological embrace of their intended destruction!

Nor will I try to belabour the obvious elitist bias of Western conservation movements. A single example will suffice. During the high-tide of environmental consciousness the President of the Sierra Club assured a dinner audience that conservationists 'have written their bible' and that its first and most important principle was 'that conservation is mainly concerned with the quality of life measured in quieter and essentially non-commercial and non-materialistic terms'. A statement appealing to an affluent, educated San Francisco dinner crowd which returned at the evening's end to its multi-bedroommed houses already stuffed with all sorts of gadgets—but hardly applicable to Sichuanese peasants who at that time ate meat once a year, owned two sets of pants and shirts, and whose children still had to carry their tiny wooden stools every morning to a one-room school—or else sit on the floor.

Instead, I will concentrate on the much more dangerous fallacies and fads promoted by scientists whose training, appointments, and research record means that their writing will be taken seriously, and too often diffused uncritically, both by most of their fellow reserachers and by the media popularizing scientific developments. Like so many topics in this book, this is a vast territory and so I will have to concentrate on what I consider to be the single most annoying fad of the environmental movement: persistent, unalloyed catastrophism. But before doing so I will deal with two influential but quite erroneous fallacies: explaining the environmental crisis in terms of its historical roots, and the anthropomorphizing of attitudes toward nature. The section will close with musings on the single most frustrating obstacle in studying the environment—its totality.

But now, the two influential fallacies. The first erroneous thesis to be discussed here was offered by Lynn White just a few years before the Western preoccupation with the environment reached its peak. Writing in 1967, White built the following explanatory structure (the simplification is mine). First, he found the victory of Christianity over paganism to be 'the greatest psychic revolution in the history of our culture'. Second, he believed our daily habits of action are still rooted in Judaeo-Christian theology, and in spite of discarding much of Christian thinking and language 'we continue today to live ... very largely in a context of Christian axioms'. Third, 'Christianity is the most anthropocentric religion the world has seen' and with its story of creation it gives the world explicitly into man's dominance: nothing exists save for serving man's purpose. And fourth, Christianity's destruction of pagan animism 'made it possible to exploit nature in a mood of indifference to the feelings of natural objects'.

And White was no less adamant about the solution: 'Hence we shall continue to have a worsening ecologic crisis until we reject the Christian axiom that nature has no reason for existence save to serve man'. Here is an explanation so sweeping, apparently so historically well-founded, so appealing: the Judaeo-Christian tradition made us into callous destroyers of our environment and, unless we part with it, doom awaits. And yet this whole model is ludicrously wrong. Rustum Roy's dismissal is the sharpest one I have come across: 'It is sheer ignorance of history compounded by unbecoming brazenness to equate *contemporary* American habits with the Judaeo-Christian tradition. . . . Surely a much better statistical case could be made to correlate environmental degradation with *lessening* of the hold of the Judaeo-Christian tradition at the *personal* level in America . . .'.

All one has to do to see the untenability of White's argument is to back up a bit into pre-Christian time and read the preserved Greek laments about the ruination of soils and forests. But these examples could be dismissed as inappropriate by pointing out that Judaeo-Christian beliefs owe so much to ancient Greece that, in a way, this comparison remains in the same broad cultural category.

Agreeing or disagreeing with this (not quite persuasive) argument is irrelevant as it is easy to find convincing proofs of the falsity of White's scapegoat theory by looking at the treatment the environment received and receives in two great cultures whose axioms are so unlike those of Christianity: China and Japan. If one limits this inquiry to aesthetic and religious attitudes and ideals, the two great Oriental cultures do, indeed, provide numerous proofs and examples of an enchantingly quiescent, purposefully adaptive approach toward their environment: man as a part of nature, rather than a separate superior being.

In dynastic China both Taoist and Buddhist traditions were the carriers of this non-interventionist spirit and many lines of classical poetry and countless brush strokes attest to the admiration of unspoiled nature. Long is the list of plants and animals venerated by verse and pictures: pines clinging at impossible angles to rocky slopes, plum blossoms and bamboo leaves rendered with masterful touches requiring years of quiet observation and dedicated study, attention lavished on potted and painted magnolias, chrysanthemums, or peonies, myriads of life-like studies of frogs, cicadas, carps, monkeys, ducks, magpies, horses and, of course, the grandest images of all, the traditional Chinese *shanshui*—landscapes full of mountains, clouds, mists, trees, waterfalls, streams and lakes where human works—graceful pavilions, arched bridges, thatched cottages—are overwhelmed by the grandeur of nature.

And, no less significantly, there was even official interest in conservation—decrees against the felling of trees, enforced by ancient institutions of mountain and forest inspectors. But all of this was just one, albeit a very important, strand of Chinese history. The country's environment was drastically transformed by human actions on scales easily surpassing pre-industrial European changes: widespread and ruthless deforestation did not leave behind a single tree over vast areas of hilly and mountainous land, opening the way for erosion and desertification. And much of this cut wood was used to build sprawling rectilinear cities where the only signs of nature were tiny courtyard gardens.

Dynastic China is thus an outstanding example of a deep discrepancy between venerated, idealistic, environmental attitudes and destructive, everyday behaviour, and a perfect refutation of a thesis that the road to environmental degradation leads above all through Christianity. As for modern China, the post-1949 Communist state ignored environmental protection for so long (until the late 1970s) that some degradative trends, above all deforestation and soil erosion, have reached relative rates and absolute levels rarely surpassed elsewhere around the world. (These phenomena, and many other worrisome aspects of China's environmental treatment, from heavy urban air pollution to wildlife extinction, are discussed in detail in *The Bad Earth*, a depressing book I was able to write on the basis of new Chinese, post-1978, candour about the country's many difficulties.)

And the other great Oriental culture with a traditionally reverential attitude to the environment? Japan is another perfect example of stunning contradictions between the environmental ideals and realities. There society has been traditionally preoccupied in so many ways with the harmonious arrangement of space, with its personalized perception, with ubiquitous touches of beauty. *Chanoyu* (the tea ceremony), *ikebana* (the flower arrangements), *hakoniwa* (miniature gardens), *shodo* (caligraphy), *kimono* (the woman's dress), *tanka* (thirty-one syllable poetry), *origami* (paper folding), *noh* theatre, *zen* temples, all captivating examples of refined aesthetic appreciation, all testimonies to a search for beauty requiring a participant whose subjective feelings are aroused by more than just viewing: an active quest for beauty indeed!

The relative absence of monumental scale in classical Japanese architecture is yet another expression of this personalized perception of beauty subsumed in the untranslatable term *shibui*, something being so captivating that the participant is irresistibly drawn into perceiving the perfection of shapes, colours, sounds, textures, arrangements. And yet the people who look with quiet admiration at random patterns of fallen gingko leaves, who carefully place serving bowls (having the more elaborate lacquer *under* the lid!), minutely carved vegetables and artistically sliced *sashimi* (raw fish) on exquisite trays so that an otherwise plain meal would evoke an experience of beauty, and whose morning crowds flowing out of subways and trains are cleaner and more immaculately dressed than any other in the rich world—these very same people have a surprisingly undistinguished modern record in caring for their environment.

In fact, perhaps the most infamous case of lethal industrial pollution preceding the 1984 Bhopal episode was the Minamata disease, chronic methyl-mercury poisoning of thousands of fishermen and their families in and near a small town on the west coast of Kyushu. Starting in 1953, releases of mercury from chemical plants of the Chisso Corporation affected at least 3500 people with nearly 50 deaths by 1970—but, in spite of the clear recognition of the problem's cause in a 1959 study, it took another nine years before the government supported the findings, and fourteen more years before Chisso finally admitted its guilt and started to pay compensations.

Another case of heavy metal poisoning which affected large numbers of Japanese industrial workers and villagers involved cadmium-contaminated water, and the resulting painful diseases of muscle and bone abnormalities became known worldwide under its Japanese name *itai-itai* (ouch-ouch). Air pollution in large Japanese cities vied until very recently with the worst cases anywhere, the habits of supposedly reverential crowds ascending Mount Fuji are embarrassingly pointed out in Japanese publications (garbage heaps along the sacred paths), the absence of careful urban as well as small-town planning and zoning is obvious, and the country's chronic neglect of its sewers became in 1984 a part of the leadership fight within the long-ruling Liberal Democratic Party.

In a bid to oust the sitting leader and Prime Minister Yasuhiro Nakasone, his rival Kiichi Miyazawa promoted a plan to double Japan's neglected infrastructural assets in a decade. His question: why should a Japanese woman wearing the latest Paris fashions have to step over an open sewer ditch to get into her cramped house? Lack of availability of space and a multitude of people ordains that Japan cannot ever aim at five-bedroom bungalows, but it is certainly an incredible sign of infrastructural backwardness and environmental neglect that by the mid-1980s two-thirds of Japanese households still were not connected to main underground sewers!

Enough of cultural comparisons. Modes and rates of environmental degradation—the addiction of some scholars to construe fanciful theses notwithstanding—have very little to do with religious or cultural traditions (or ruling political ideologies). The simple, ubiquitous quest for food and shelter, for a better life (including, paradoxically, better health), more leisure and, after a certain level of economic development is reached, for more luxuries, is sufficient to explain the universal trend of environmental degradation which gets slowed down or reversed only after a society learns some costly lessons from its previous neglect, reaches a sufficient level of understanding about irreplaceable environmental services, and musters enough political will, capital, and managerial expertise to act.

Dispelling the historical roots of environmental degradation appears to be much easier than doing away with a diffuse class of feelings which were formed in Europe largely during the eighteenth and the early part of nineteenth century and which has persisted in the Western intellect to this day. The fallacy of anthropomorphism is my preferred term, the romantic spirit or nature's benevolent character might also do. Whatever the elusive concise labels, the thesis can be stated fairly easily: in contrast to Man's aggressive, exploitative, malicious, greedy demeanour there is peaceful, benevolent, generous, harmonious Nature.

As always with human attitudes, turning to *belle lettres* and classical treatises illustrates the point better than any modern description can. Rather than quoting predictable Jean-Jacques Rousseau (always ready to give 'himself up to the ecstasies this harmony arouses') I shall offer a few perfect English quotes overflowing with the sentiment I seek to illustrate. Shaftesbury's nature-worship comes out in the most exalted manner in *The Moralists*: 'O Glorious Nature! Supremely Fair, and sovereignly good! All-loving and All-lovely, All-divine! . . .'. For Wordsworth Nature 'never did betray/The heart that loved her' and 'Love and Truth compose her train'.

Rather well-off *literati*, gentlemen enjoying the advances of improving urban life, taking leisurely walks through the rich, green woods and meadows of England and France (places of equable climate and few crippling environmental extremes), and sublimating their dissatisfaction with society into exalted admiration of nature were an understandable source of this kind of nature worship, and continuing urbanization and industrialization made later generations receptive to their sentiments (so, for example, Thoreau could see in the wilderness 'universal innocence').

Darwin did some blunt correcting of these feelings—writing about the 'war of nature' and about 'famine and death' from which the higher species derive their existence—and more than a century of unprecedented scientific effort since then, has left little doubt that nature is neither supremely fair nor all-lovely—the truism always realized by those who have had to face flooding rivers, eroded fields, uncontrollable forest fires, crippling droughts, swarming pests, sweeping hurricanes, exploding volcanoes, and other destructive forces so ubiquitous in our environment.

Yet the legacy left by the romantic worshippers of benevolent and harmonious nature is so persistent that its manifestations keep surfacing in various ways. An excellent example of how ingrained are these attitudes is the fact that even one of the world's most respected ecologists, appalled by man's relentless destruction of the environment, makes a similar charge against nature.

Nor can we hold Nature, without such drastic intervention by Man, as being by any means blameless of ecosystem desecration and serious vegetational change, if we recall such alterations of landscape and more as have been wrought over wide areas by, for example, Dutch Elm Disease (*Ceratocystis ulmi*) in recent years and Sweet Chestnut Blight (*Endothia parasitica*) in North America some decades ago,

wrote Nicholas Polunin in 1979.

What an astonishing anthropomorphization and imposition of cultural preferences: surely the pests and parasites, unsightly or destructive as their actions may be, are as normal, organic parts of the environment as the plants they attack, and are patently unable to desecrate anything (that being a sole province of cogitating men)—but just bring about changes, merely to perpetuate those never finished evolutionary processes. Only a particular human perception of beauty, order, and immaculateness would exempt elms and chestnuts from the perils of change, and snuff out the life of many a heterotroph whose daily metabolism produces results offending our egos.

Growing up in Europe I admired every spring the huge white candelabra of chestnut blooms; then, a few months later, I enjoyed prying from their prickly receptacles unblemished shiny chestnuts for building assorted creatures and animals, a highlight of a young boy's Fall. And now, driving through North American cities along long rows of endangered or already withering elms, I would not mind seeing the *Ceratocystis* picking on dandelions instead of on these majestic trees. Yet such emotions have nothing to do with incessant biospheric change where alterations of landscape caused by a few heterotrophs are insignificant episodes in comparison with rearrangements of continents and oceans by moving plates, with temporary burials of tens of millions of square kilometres under giant glaciers, with dramatic climatic changes wiping out established ecosystems and shifting vegetation boundaries thousands of kilometres pole- or equator-ward.

When the unusually rapid movement of a rather small triangular-shaped plate started to scrape away the bottom of Tethys Ocean some 65 million years ago, then closed its V-shaped depths and piled up its rich sediments against Eurasia's flank to create the Himalayas, or when the Pleistocene glaciers covering most of North America started to recede 10 000–15 000 years ago to leave behind vast areas of smoothed barren rock and large lakes, we do not see such profound environmental transformations as 'desecration' of the Tethys ocean or 'deep wounds' to pre-glacial forests. But incomparably smaller changes which we can see unfolding rapidly as a result of our pollution, construction, or farming become a source of concern which sometimes reaches bizarre proportions.

Much of conservation aiming to preserve particular species of plants or animals rests on these inappropriate worries which combine with aesthetic preferences to produce frequent and powerful public campaigns. Although they may seem to be identical preoccupations, I must stress here that these particularistic concerns are entirely different from growing worries about the reduction of species diversity resulting from large-scale elimination of whole ecosystems (they will be the subject of a separate section later in this chapter).

While it is undeniable that the disappearance of most of this planet's tropical rain forests before we have gained detailed understanding of their composition and functioning would deprive us of inestimable benefits, such a case cannot be made on behalf of either the snail darter, a few cm long fish whose endemic occurrence in the Little Tennessee River stopped the construction of Tellico Dam, or the California condor, whose reproductive requirements have become a source of bitter disagreements among the biologists involved in the increasingly expensive rescue mission.

Ian McMillan noted that 'the real importance of saving such things as condors is not so much that we need condors as that we need to save them. We need to exercise and develop the human attributes required in saving condors'. And Paul Fleischman's perceptive essay illuminates the whole concern in definitive terms: 'There is no biological justification for conservation. Nature will not miss whooping cranes or condors or redwoods, any more than it misses the millions of other vanished species. Conservation is based on human value systems. Its validation lies in the human situation and the human heart.'

Similarly, there is no biological justification for the catastrophic faddism which became a recurring hallmark of numerous environmental writings. Soon after Barry Commoner started to study leakage of nitrogenous fertilizers from the farmlands of Illinois, he hastened to conclude that the whole nitrogen cycle in the country was out of joint and that unprecedented action was needed to reverse the degradative trend (for much more see section 5.3). Paul Ehrlich took an assumption about the inhibiting effects of DDT on phytoplankton photosynthesis (refuted within two years after its inception) and was carried away with it all the way to the end of life in the ocean by the summer of 1979.

But this was too weak for Ehrlich, he needed some corpses—and decided to throw them in by hundreds of thousands and with vivid descriptions. 'But suddenly our citizens were faced with nearly 200 000 corpses. . . . The population was terrorized as TV screens became filled with scenes of horrors. . . . Especially vivid was NBC's coverage of hundreds of unattended people choking out their lives outside of New York's hospitals.'

And the end of this phantasmagoria?

By September, 1979, all important animal life in the sea was extinct . . . windrows of dead fish created a monumental stench. But stench was the least of man's problems. Japan and China were faced with almost instant starvation from a total loss of the seafood. . . . On 13 October Chinese armies attacked Russia on a broad front. . . . Most of the people who are going to die in the greatest cataclysm in the history of man have already been born.

Years after his predictions of the dead ocean (in the next section I will have an opportunity to note that the ocean's health actually appears better today than in the late 1960s!), Paul Ehrlich continues to prosper in California, his stature as an expert undiminished by his former hits so wildly off the mark. In fact, it was his earlier catastrophism which made him a widely recognized environmentalist because the worse the news, the greater the coverage of Western media which are always on the lookout for new bad news to adorn front pages and to claim the first few minutes of radio and TV news broadcasts.

Michael Bowen once remarked wryly that should somebody care 'to predict that sonic booms will cause tidal waves to engulf Atlantic City or San Diego, then the front pages of daily newspapers and a prominent place on evening TV news programs will surely be his'. Science is badly served by this dubious process but, curiously enough, disseminators of false dooms are usually let off easily: in the early 1970s there were few comments about crying wolf needlessly, or too often, but the practice has continued—and it has even been strengthened by group doomsaying based on computer models.

The first of these exercises to receive notoriety was the Club of Rome commission to an MIT team led by Dennis Meadows to explore the global future in *The Limits to Growth*. The report featured figure after figure of computer printouts in which black dots charting pollution were zooming up in steep exponential curves precipitating industrial collapse and deaths in billions. The grotesquely unrealistic nature of this model was best illustrated by the fact that 'pollution' was a single variable dependent on 'pollution generation rate' and 'pollution absorption time'—as if one could add up, or average, such disparate variables as releases and effects of sulphur dioxide from coal-fired power plants on chronic respiratory diseases, and mercury conversion to methyl-mercury in lakes, its accumulation in fishes, and its effects on the human nervous system (to name just two of thousands of possibilities!).

The generation of computerized bad news continued with the World Integrated Model also sponsored by the Club of Rome (coming out with massive famines in Asia before the end of this century—while both China and India of the 1980s have had an unprecedented period of record harvests), the United Nations World Model, the Model of International Relations in Agriculture, and it reached its bureaucratic apogee with the Global 2000 Study in which eleven federal agencies and hundreds of researchers participated to construct a world future according to President Carter's administration.

This report's gloomy conclusions—serious worldwide deterioration of agricultural soils, dramatic increase of plant and animal extinctions, severe regional water shortages, massive disappearance of world forests, life more precarious for most people on earth in the year 2000 than in 1980, the world more vulnerable to disruption—received, as might have been expected, publicity much wider than a later effort assembled to refute the depressing forecast, *The Resourceful Earth* edited by Julian Simon and Herman Kahn and published in 1984.

This response contains much useful information correcting many dubious assumptions used in preparing the Global 2000 Study—but it itself offers some embarrassing generalizations and attempts to bend the outcome the other way. I will return to this clash of pessimistic and optimistic outlooks in the closing chapter of this book (section 6.1) but here is the place to note that even with the most scrupulous and skilful use of the best available observations and statistics it is extremely difficult to model environmental developments. The main reason is the inherent complexity of biospheric relationships, and the forbidding challenge of studying our environment in its totality if it is to be understood correctly enough to offer sensible forecasts.

Many people, scientists and artists, have grappled with this uncommon challenge: I quoted Pliny's words in introducing this chapter; nearly two millenia later Paul Klee recorded in his diary: 'I search for a remote, initial point of the original formation in which I forebode a formula for man, animal, plant, water, fire, air and for all the whirling forces as well.'

What Andrew John Herbertson wrote in 1913 in *Scientia* cannot be put better today—and his phrasing is worth an extended quotation.

Environment is not merely the physical circumstances among which we live, important though these are. It is found to be more complex and more subtle the more we examine it. There is a mental and spiritual environment as well as a material one. It is almost impossible to group precisely the ideas of a community into those which are the outcome of environmental content, and those which are due to social inheritance.

It is no doubt difficult for us, accustomed to these dissections, to understand that the living whole, while made up of parts with different structures and functions is no longer the living whole when it is so dissected, but something dead and incomplete. The separation of the whole into man and his environment is such a murderous act. There are no men apart from their environment. There is a whole for which we have no name, unless it is a country, of which men are a part.

And, in a fascinating meeting of minds which travelled to the same conclusion from totally different mental bases, Tolstoy wrote not long before his death (that is shortly before Herbertson's lines were published) that 'the highest wisdom has but one science—the science explaining the whole creation and man's place in it'. How far we have advanced in this quest since then, during the decades of explosive growth of scientific understanding, the emergence and diffusion of general systems studies, the rise of ecology and interdisciplinary research on the structure and function of ecosystems, and the acquisition of computing capabilities which have swiftly reached levels enabling modelling and simulation of very complex realities?

Far—yet far from far enough to forecast with confidence. With inanimate systems governed by orderly and precisely quantifiable variables, advances in computer science and simulation gave us the Moon landings, cruise missiles, superships, laser cashiers talking in synthetic voices, CAD-CAM technologies, and myriads of other large and small innovations: altogether fabulous achievements in just a few decades. In the case of large-scale complex inanimate systems in constant flux—above all the atmosphere and the oceans—our understanding has been increasing rapidly but, as will be abundantly seen, especially in discussions of the carbon cycle (section 5.2), we are still unable to set out satisfactorily reliable forecasts concerning the course of such basic variables as atmospheric composition, temperature, or precipitation.

And with regard to living systems, our successes on subcellular, cellular, tissue, and organ levels (ranging from genetic engineering to organ transplantation) contrast so sharply with the continuing puzzles about the functioning of ecosystems, fluxes of major biogeochemical cycles, and responses of species and communities to environmental changes. Later discussions of the three major cycles will repeatedly show how uncertain we are about the values of many essential fluxes: how much sulphur is transferred to the atmosphere from organic decay in wetlands and coastal waters; how much carbon dioxide originates from global grassland burning; how much nitrogen is lost from the tops of plants or added by free-living fixers. Available estimates differ commonly by up to an order of magnitude. Sometimes we do not even know if a particular reservoir is a net source or sink (most notably, this is the case with global vegetation and CO_2).

And doing better will require much higher research costs and many new non-invasive techniques. The establishment of energy and material fluxes even in small, clearly delimited communities (a cropfield, a tiny river-basin) calls for quantification of scores of variables because extrapolation of generalized values may yield highly misleading results and, in turn, all the impeccable information derived from extensive on-site research may be almost completely inapplicable to a neighbouring field or a nearby river basin.

Perhaps most fundamentally, photosynthesizing plants, the critical energy and nutrient transformers in any environment, are nearly as elusive to study as they have ever been. Clifford Evans expressed this frustration perfectly:

In the broadest terms, the main difficulties lie in the inaccessibility of the plant growing in its natural surroundings—physical inaccessibility because the great majority of methods of investigation involve gross interference with the plant, or with its environment, or both . . . and intellectual inaccessibility. The human mind has no intuitive understanding of higher plants . . . .

Incredible as it may seem in an era of almost fairy-tale sophistication of countless measurement, monitoring and control gadgets whose abilities and reliabilities are so convincingly demonstrated by flawless space shuttle missions, or by prompt and crisp intercontinental communications, we still have no effective way to study even a single tree's responses to environmental changes in their entirety! And even straightforward monitoring of leading indicators of environmental quality, a task incomparably easier than encompassing the interdependent fluxes in plants and ecosystems, remains unsatisfactory.

Consequently, *State of the Environment 1982*, a book prepared by the Conservation Foundation, stated correctly that, even in the rich United States, 'We have no monitoring data sufficient to describe accurately the extent . . . of any environmental problem'. This situation has led, inevitably, to repeated unsubstantiated hypothesizing and to erroneous perceptions and mistaken conclusions about the extent and gravity of many environmental 'crises'—and to subsequent reappraisals and the emergence of new, often diametrically opposite and almost always much more relaxed, consensus. Key global examples of these environmental crises in perspective are worth a detailed recounting.

4.1.2 *Environmental crises in perspective*

> We should be careful not to cry 'wolf' needlessly or too often. . . . Scientific credibility can easily be lost by exaggerated claims and extravagant statements. We need to provide voices of reason, not just of alarm.
>
> S. F. Singer (1970)

During the discussion of long-range energy forecasts (section 2.5) I stressed the great influence which the prevailing, short-term moods are having on the extended outlook. The study of environmental crises as perceived by the consensus of the scientific establishment offers even better examples of this understandable but clearly quite undesirable phenomenon. As this problem is a lasting one—in the 1980s we are certainly as likely to commit similar errors, and fail to come up with enduring perceptions as we were during the 1970s—I will offer a fairly detailed survey and analysis of the past misconceptions: careful study of these misjudgements is invaluable in trying to minimize the future occurrence of the most embarrassing lapses.

The multitude of environmental pollution and degradation processes, so disparate in their duration, spatial extent, and harmful impact, makes it obvious that rigid rankings are completely unrealistic. On the other hand, it is no less obvious that a localized release of even a dangerous chemical cannot be put into the same category of worry as gradual but global alterations of climate caused by human intervention. Clearly, a consensus on critical environmental problems can be sought and found—and it is fortuitous that such assessments, made by interdisciplinary groups of independent scientists, are available both for the 1970s and the 1980s.

The first one is the Study of Critical Environmental Problems (SCEP) sponsored by the Massachusetts Institute of Technology and conducted by forty full-time and thirty part-time participants in June 1970 after extensive preparation. SCEP concentrated on the global effects of pollutants in the atmosphere–land–ocean system and, although there was no attempt to rank-order the problems, the sequence in which they appeared in the chapter summarizing the findings and recommendations of the study clearly indicates their relative importance: carbon dioxide from fossil fuels; particles in the atmosphere; cirrus cloud from jet aircraft; supersonic transports in the stratosphere; thermal pollution; DDT and related persistent pesticides; mercury and other toxic heavy metals; oil on the ocean; and nutrients in coastal waters.

A dozen years later, in November 1982 in Rättvik, when the Royal Swedish Academy of Sciences decided to bring together thirty-five scientists from fifteeen nations to discuss and select what was called 'a strategic set of environmental research and management priorities for the 1980s', only a single item on the ten-subject research priorities list—carbon dioxide build-up and climate change—was shared with the SCEP array. Management priorities included among their ten items dealing with hazardous chemicals and protection of marine environment, two very broad categories implicitly subsuming SCEP's concerns of pesticides, mercury, oil, and nutrients—but not featuring these concerns specifically.

Why did a dozen years make such a difference? Only one person participated in both assessments but this lack of overlap explains little as both of the meetings were carefully prepared and based on large amounts of background evaluation. Even the somewhat different purposes of the two efforts cannot explain a near-total disappearance of SCEP's criticalities from Rättvik's priorities. Obviously, perceptions have shifted, consensus has changed and it is fascinating to trace how and why this has happened.

Curiously, the most hotly debated environmental threat in the United States of the early 1970s, commercial supersonic air transport, turned out to be no danger at all. With the Franco-British Concorde under development and with the Soviets preparing their supersonic Tupolev, Boeing did not want to stay behind in what was shaping as a key plane market of the 1970s, but it ran into extremely active environmental opposition which was instrumental in defeating the project's federal funding (and thus effectively doing away with the plane) largely on the basis of the destructive effects it was predicted the fleet of supersonic transports would have on the stratospheric ozone.

As expert witnesses testified in front of the Congress, and environmental activists flooded the media with vivid stories, a broad consensus emerged that trading-off a serious weakening of the stratospheric ozone's radiation shield for faster transcontinental or trans-Atlantic flights would be a most dubious action. Boeing never took off and the environmental lobby celebrated one of its first big victories. Looking back one must write that a good decision was taken for a wrong reason. The grounding of the American SST was good because after 1973 high oil prices made supersonic planes even more uneconomical than they would have been anyway, and Concordes never flew 'in black'. But if very high fuel consumption and excessive noise on take-off and landing were indisputably correct charges against the plane, the main objection—its purportedly greatly damaging effects on the atmosphere—rested on dubious assumptions.

By 1976 better evaluations made it clear that, even with a huge fleet of 5000 supersonic transports flying daily, the long-term climatic effect of aerosols released by this traffic would be most likely limited to nothing more than a 0.1 °C decline in the mean surface temperature, a decrease well concealed within the natural climatic fluctuations. The main concern, however, was about the depletion of the ozone layer and the most incredible part of this emotional story is that a small group of scientists either suppressed or belittled the observational evidence which was undermining the assumptions and conclusions of computer models prepared to show the declining ozone levels.

Hugh Ellsaesser of the Lawrence Livermore National Laboratory in California wrote a truly shocking account of these developments when not one of twenty-five respondents whose opinions were entered into the Congressional Record during the controversial hearings to cut the funding of Boeing's SST prototype objected to the claim that increased stratospheric loading of water vapour would bring decreased ozone concentration—although among these experts were several scientists whose analysis of the long-term observational record did not lead to such a conclusion.

Perhaps even more incredible was Elsaesser's personal experience when he tried to translate theoretical predictions of increased frequency of skin cancer following ozone reduction into readily understandable risks. The numbers commonly mentioned, 5000–10 000 new skin cancers every year, looked frightening enough but the individual risk, were the predictions of 7 per cent ozone depletion to come true, appeared suddenly in a very different light once Ellsaesser equated it with latitudinal displacement, an easy calculation to do as skin cancer incidence doubles every 1000 km when moving equatorwards.

Anyone can directly relate to such a comparison of risks in a nation where the population of a densely settled Northeast has been transferring itself at a rapid rate hundreds and thousands of km southwards and southwestwards to sunnier climates. Yet, reminisces Ellsaesser, 'each time I tried to insert such equivalences into the appraisals presented to the public I encountered strong opposition from almost everyone, including my own colleagues.' The best example was when he pointed out, at the International Conference on the Stratosphere and Related Problems in Utah in 1976, that a 7 per cent ozone depletion predicted by the National Academy of Sciences was equivalent, in terms of raising skin cancer risks, to moving 135 km south. Russell Peterson, then Chairman of the President's Council on Environmental Quality, replied bluntly: 'That's a silly argument'.

Of course, such a revelation would not fit into the pattern of biased appraisals, even cover-ups, which characterize the SST controversy. Ellsaesser's blunt but inescapable conclusion is that the scientists involved in the case 'took it upon themselves to act as a priesthood by suppressing information which the laity could be expected to interpret for itself to arrive at conclusions different from those espoused by the priests'.

All of this has a no less interesting second instalment: in the mid-1970s came a sharply increased interest in the possibilities of ozone destruction owing to the use of chlorinated fluorocarbons, above all $CFCl_3$ (CFC-11) and F_2CL_2(CFC-12), which are commonly used as propellants in aerosol cans and as cooling liquids in refrigerators. As these compounds have a minimum atmospheric lifetime of several decades, computer models prepared by the National Research Council in the late 1970s indicated that, if the releases were to continue at the 1977 rate, the steady-state reduction in total global ozone would be, in the absence of other perturbations, as much as 18 per cent.

This was a total reduction that was much more threatening than the one originally forecast for a fleet of 500 SSTs, but the threat did not last too long. Not because the releases of halocarbons stopped or slowed down significantly: the latest available long-term trends calculated from measurements done as a part of the Geophysical Monitoring for Climatic Change, show that, between 1977 and 1982, CFC-11 had been going up by 11–13 pptr a year, and CFC-12 by 9–16 pptr annually. What has changed is the scientific consensus. In Rudy Baum's words 'the need for quick answers led to hurried science', and projections based on weak observational foundations had to be discarded very soon.

In 1982 the National Research Council came out with a new appraisal forecasting the halocarbon-caused steady-state reduction in global ozone at no more than 9 but possibly as little as 5 per cent. And in February 1985 yet another NRC assessment put the most likely reduction at only 2 to 4 per cent, a big step down from the 18 per cent offered just a few years earlier! And what has been the atmospheric ozone actually doing since the 1960s? James Angell's analyses of global total ozone data showed a worldwide increase of several per cent in stratospheric ozone during the 1960s, and a decrease of about 0.5 per cent between 1971 and 1980, but the cause of this decline is unknown: considering the 1960s' rise, natural fluctuation is the best explanation but if chemical 'erosion' is involved, a much longer observational period will be needed to prove it.

If the inclusion of the possible stratospheric effects of supersonic transport proved to be more a crisis of information rather than a genuine environmental threat, the addition of cirrus clouds generated by ordinary jet aircraft as an issue of worldwide concern was an even more unnecessary miscalculation: after all, at any time half of this planet is covered by clouds and a bit more vapour in the upper troposphere cannot make a worrisome difference. Since then nothing has been heard about condensation trails as environmental threats and so this 'critical' problem disappeared completely almost as soon as it was elevated among the big items.

In contrast to SSTs and contrails, concerns about man-made particulate matter make sense. Industrial civilization has been releasing growing quantities of particulates from various extractive and processing industries (quarries, surface coal mines, cement works, iron and steel mills) and, above all, from the combustion of solid fuels. Farming, with much new land brought into production in drier regions (see section 4.2.1), has been also an extensive and seasonally very large source of airborne dust.

Global estimates of primary man-made particles can be made with reasonable reliability for the centralized, continuous industrial sources by applying average emission factors to the known amounts of consumed fuels or produced materials, but not for the diffused, intermittent processes such as the burning of forests for shifting cultivation, or the burning of crop residues and solid wastes. Even less reliable are the estimates of secondary man-made aerosols, mainly sulphates, nitrates, ammonium, and hydrocarbons formed by complex atmospheric reactions of gaseous compounds, and comparisons of published global values show two or three-fold differences (see also sections 5.3 and 5.4).

If fixing the man-made total is difficult, coming up with representative natural particulate generation aggregates is incomparably more elusive. Looking just at the major items, available estimates of eroded dust range over two orders of magnitude, as do the values for volcanic particulates and forest fires; disagreements about secondary aerosols are relatively smaller but, as with their man-made counterparts, two or three-fold differences have been common. Obviously, when both the natural and man-made generation rates are so uncertain it is audacious to offer a single estimate of the man-made contribution to the total mass of airborne particulates.

While SCEP put this share at 20 per cent, other studies ended up with values as low as 5 or 6 and as high as 45 per cent. This uncertainty makes any sensible quantitative evaluations of a historical trend of man-made particulates plainly impossible. The best appraisal is that human activities are responsible for only a small fraction of the global particulate burden—something around 15 per cent looks most plausible—and that an increasing trend of aerosol generation corresponding to the growing combustion of fossil fuels and extension of industrial processing was evident earlier in this century, but that it has been at least halted or even reversed after World War II.

The best quantitative proofs of these improvements are improved visibilities, more hours of sunshine, lowered dust fall, and decreased suspended aerosol mass in just about all major urban areas noted earlier for their polluted skies. The conversion of household heating and cooking from coal to liquid and gaseous fuels, the disappearance of steam locomotives, the greater use of coal in large power plants and big industrial boilers where efficient electrostatic precipitators can strip the flue gases of more than 99 per cent of all generated particles, have been the main cures.

Even if there were a gentle upward trend to man-made aerosol contributions—and some estimates suggest a secular increase as high as 0.4 per cent a year (implying a doubling of atmospheric aerosol loading in 175 years)—it would be hidden in great natural fluctuations and its climatic effects would be much less dramatic than those following major volcanic eruptions which can introduce almost instantly 10^4–10^6 t of dust into the stratosphere.

Turning from the atmosphere to land and water, SCEP was much concerned with the ecologial effects of DDT, citing above all its specific effects on the reproductive potential of many birds (owing to the thinning of egg-shells) and its build-up in aquatic organisms; not surprisingly, the report recommended 'a drastic reduction' in the use of the pesticide. Actually, DDT use was already restricted in the United States by 1969 and in 1972 it was banned, except for emergency applications against serious pest outbreaks. Even so, the concerns continued because of the postulated persistence of the pesticide in the environment. With half-life estimates in excess of twenty years it was feared that, even by the end of the century, the most susceptible biota may be still burdened by harmful DDT concentrations.

Yet, in a *dénouement* resembling so much the disappearance of the SST-ozone threat, this danger faded away rapidly. Just one year after SCEP's highlighting of DDT persistence, George Woodwell, Paul Craig, and Morton Johnson published the results of their search for DDT and their observational conclusions did not fit the theoretical expectations at all. Instead of being selectively deposited in lipid-rich biota, most of the lipid-soluble DDT appeared either to degrade to innocuous levels or to be sequestered in places from where it was not free to spread further.

The best quantitative estimate the authors could make was that in the early 1970s living organisms worldwide contained less than one thirtieth of a year's output of DDT in the mid-1960s, an unexpectedly small fraction. Fears of the environmentalists that Antarctic penguins would be turned into walking DDT repositories for generations to come could be forgotten. In later years DDT's rapid disappearanced was confirmed even in heavily polluted areas. Perhaps most notably in 1982 Victor Bierman and Wayland Swain found, on the basis of trends in DDT levels in coregonid fishes in Lakes Michigan and Superior in whose drainage-basins large amounts of the chemical were being dumped for over two decades and where its accumulation in fish tissues caused great public concern, that after the ban on DDT use in 1969 the loss rates 'were more rapid than expected on the basis of hydraulic detention times and the degradation rate for DDT in the environment'.

A simple mechanism insufficiently appreciated in the earlier years of the fears of DDT persisting in biota accounts for most of the loss: the settling of particles from the water column and their effective burial in bottom sediments. DDT residues, without any doubt, were a serious environmental toxin causing major local and regional declines of certain sensitive species but in hindsight the compound can be in no way seen as a dangerous global pollutant threatening survival of any lipid-rich biota.

Although DDT could also be found in increasing concentrations in human adipose tissues, absolute values were very small and the concern remained limited to its effects on wild species. Not so with mercury whose attack on the central nervous system—progressive weakening of the muscles, loss of vision, paralysis, all caused by structural injury to the brain—had been well documented ever since the early 1950s when fishing villages around Minamata Bay experienced what were originally mysterious neurological disorders and scores of deaths. During the 1960s the Swedes started to measure very high mercury levels in their fresh water fish, and in 1970 the concern flared up in Canada and the United States.

The closing of many lakes for commercial fishing in the spring and summer of 1970 was followed by the removal of one million cans of tuna from American food stores in December and the scare reached truly mass proportions. Here was a dangerous pollutant whose concentration had reached such high levels in the global environment that the eating of even deep-sea fish was unsafe! And, once again, there was a surprisingly swift ebbing of fears.

Undeniably, mercury, like all heavy metals, is a dangerous substance and, as it became used in substantial amounts in chemical industries, paint, and paper plants, and as an agricultural fungicide, more of it entered the environment where its conversion into the most injurious organic methyl mercury compounds (by bacteria in detritus and sediments) and its concentration in aquatic food webs led to some very high levels in large fish and large fish-eating birds.

This natural accumulation was found to be the main reason for higher mercury levels in ocean fish: their position at the top of the food chain and their longevity were to blame, not any increased deep-ocean concentrations of the metal. Mercury releases by industries and farming were clearly responsible for local and regional elevation of the concentrations in waters and aquatic species, but even at their peak they were too low to make the metal a global pollutant and a contaminator of marine food webs.

While man-made mercury releases into the atmosphere are at most about 15 000 t a year (but possibly only one-tenth of this total) natural dust contributes ten times as much; similarly, mercury dumped on the land totals some 5000 t compared to global amounts from rain and fall-out of 35 000 t. And continental shelves and coastal waters contain about 4 billion t of the element, making human contributions quite insignificant in global terms. The suffering caused to many children and adults in localities and regions with high mercury levels in fish was avoidable as the risks are limited to only those people who eat fish from heavily contaminated waters as a main part of their diet. More fundamentally, limiting the use of mercury to essential applications and stringent recycling can control the risks quite well.

Mercury in waters was never a critical global problem, just a tragic, localized and avoidable mistake. In contrast, oil on the ocean looked undeniably to be a large scale trouble. As the long-distance shipments of crude oil expanded rapidly during the 1960s and as the average size of tankers more than doubled (to over 40 000 t), as the capacity of the biggest ships approached 500 000 t, and as the first big spill occurred on the coast of southwestern England (*Torrey Canyon* in 1967, releasing over 100 000 t of crude oil), it was natural to worry about the future when much greater volumes of oil were to move across the oceans in even bigger ships. News items describing oil slicks far away from polluted shores became common; Thor Heyrdahl's reports on oil blobs blackening his raft during his trans-Atlantic crossing received nearly as much publicity as the navigator's theories of long-distance historic sailings and, as already quoted, Paul Ehrlich foresaw the end of life in the ocean in September 1979.

Summer 1979 passed and in 1982 a report, *Health of the Oceans*, prepared by the United Nations Environment Programme, found that the world's oceans were much healthier than they were thought to be in the early 1970s. Yet there were many major oil spills in the 1970s, including the world's largest tanker mishap so far when *Amoco Cadiz* went aground on 16 March 1978 off the northwest coast of France spilling about 250 000 t of crude oil, and the blow-out of the offshore Mexican Ixtoc I well on 3 June 1979 which was not capped until nine months later and released at least half a million tonnes, possibly as much as 1.4 million tonnes of oil.

Detailed studies of these two and other major oil spills showed that even where huge areas have been involved the effects in the water column have been few and the oil remained there for only limited periods. Or, as the UNEP report puts it, 'oil spill effects on deep sea communities are rarely drastic; recovery is usually a question of weeks or months'. Several natural processes—evaporation, the formation of emulsions, sinking, auto-oxidation and, most effectively, oxidation by micro-organisms—are keeping the surface of the open ocean and its deep water column surprisingly clean in spite of tens of million tonnes of oil spilled into the seas since the beginning of large scale tanker shipments.

Micro-organisms which attack hydrocarbons are widely distributed, represent many microbial genera, are generally more abundant in chronically polluted areas, and can act over a wide range of temperatures as long as oxygen and essential nutrients are present in the water. Undoubtedly, microbial biodegradation has always been responsible for the removal of hydrocarbons from natural oil and gas seeps.

Oil pollution remains of much greater concern in coastal areas. Even rocky headlands which are quickly cleansed take a few years for a full recovery after a major spill and in soft sediments in shallow protected waters, invariably the habitats worst affected by oil, pre-spill productivity may be re-established quite rapidly, but the complete return of the original flora and fauna may take many years. But the problem is one of local or regional dimensions—not a global one, a part of the wider challenge of managing coastal waters along heavily populated and industrialized shores.

SCEP's inclusion of nutrients in coastal waters among the critical concerns seems to be a sensible one as the coastal zone is used for sewage and solid waste disposal worldwide, and the dumping has been increasing, threatening such rich ecosystems as salt marshes, mangrove swamps, coral reefs, and kelp beds. But, once again, with hindsight of more than a decade it can be seen that the regional label fits better than the global one: semi-enclosed seas—the Mediterranean above all, the North Sea, the Baltic Sea, the Gulf of Mexico—are those worst affected, together with many smaller, localized coastal zones near large cities elsewhere.

And even in these cases no sweeping judgements are possible. To quote once more UNEP's *Health of the Oceans*, 'if the disposal of sewage to the sea is adequately controlled and if the sites are properly selected . . . then the fertilizing nature of the sewage may be regarded as more significant than its potential toxicity'. And in the case of one of the best known examples of seawater quality degradation—the Baltic Sea's falling oxygen content—the primary cause may be an increase in salinity brought about by lower precipitation and runoff in the region rather than simply by increased dumping of organic wastes from cities and industries (above all paper plants). In any case, these regional degradations—reversible and manageable by widespread application of existing control methods, do not portend any demise of the living global ocean.

Finally, the last item of the SCEP list which did not survive the decade, thermal pollution. As with SSTs, or mercury, or oil on the ocean, the news headlines of that time and perceptions of the day had much to do with its selection. Energy consumption was generally forecast to continue its exponential rise for decades to come and to satisfy these needs energy planners were writing about gigantic nuclear parks and contemplating the advantages and drawbacks of thermal power-plant sites where more than 10 GW of generating capacity would be installed on a single site. Obviously, such facilities could generate huge waste heat fluxes and, even if the problems with cooling water could be taken care of, there would still be concerns about possible climatic modification, and this possibility also appeared significant with higher projected power consumption densities over larger urban areas.

Today these worries seem unwarranted. The most intensive 'point' sources of waste heat—colling towers, and power plant and industrial stacks—release energy with power densities between 10^4-10^5 W/m^2 and, although they sometimes generate substantial amounts of local cloudiness, they rarely cause any anomalies of precipitation. Research has shown that in the industrialized temperate latitudes, whose atmosphere is more stable than in the tropics, heat releases causing anomalous rainfall can occur when power densities exceed 1000 W/m^2 over areas larger than 3 km^2 or when the area of smaller facilities (expressed in square metres) multiplied by the power they produce (in watts) exceeds 10^{16}. Neither large thermal power plants nor steel mills or refineries have such heat fluxes.

And while large cities, whose average heat rejection is usually no more than 10–50 W/m^2, clearly modify the wind flow and create localized weather anomalies (commonly known as urban heat islands), they do not create persistent effects easily distinguished from natural variability. In fact, heat rejection densities of up to 100 W/m^2 over areas larger than 100 km^2 appear acceptable which means that even the most unlikely doubling of average power densities in metropolitan areas of the densely settled northern mid-latitudes would have emissions of waste heat well within that range.

With the sole exception of CO_2 build-up and its potential climatic modification effects, better scientific understanding—often only calmer and more objective assessment of the already known facts—and sometimes also improved management and controls, have combined to do away with all of SCEP's critical problems as worrisome *global* concerns. Some of them remain unpleasant, even risky, local and regional challenges to governments, planners, and pollution control engineers but none of them is threatening global modification of the biosphere and the survival of industrial civilization. This great demise should obviously be kept in mind when looking at the new concerns dominating the global environmental assessments of the 1980s.

4.2 New concerns

'There's more evidence to come yet, please your Majesty,' said the White Rabbit, jumping up in a great hurry.

Lewis Carroll, *Alice's Adventures in Wonderland* (1865)

During the mid- and late-1970s environmental concerns continued to run strongly but they were overshadowed in media reporting and in public interest by energy affairs. This big new preoccupation was initially focused almost solely on the predicaments of rich nations but it led eventually to a rapidly increasing interest in the energetics of the poor nations and hence, inevitably, to fuelwood combustion and tropical deforestation. Concurrently, advancing studies in tropical ecology started to note the dangers of losing very large portions of this uniquely rich biome without even a satisfactory opportunity to understand its composition and functioning, and thus benefit from its resources and possibly aid in its eventual regeneration.

A new concern took off swiftly and it was instrumental in promoting a more general worry about the loss of species diversity in all kinds of receding ecosystems. And the droughts and regional famines, conversion of grasslands to grain fields, shortened rotation periods in shifting agriculture in the tropics, and rising demand for grain imports which led to cultivation of almost all previously idle American cropland and, in turn, to greater erosion, renewed the concerns about soil degradation—erosion, salinization, desertification. When Nicholas Polunin, a founder of the Foundation for Environmental Conservation, spoke about conceivable eco-disasters at the Second International Congress of Ecology in the late summer of 1978 he listed build-up of atmospheric CO_2 in first place, followed by the 'disappearance of more and more of the life-support system ... an insidious danger—particularly from such scourges as soil erosion, deforestation and other devegetation,' then followed by water shortage and salt build-up with continuing irrigation, and loss of genetic diversity.

When the Rättvik Conference set out, in November 1982, its list of ten research priorities it included the same entries. After considering forty environmental issues the following critical items of global importance requiring greatly increased research efforts were agreed upon: the depletion of tropical forests, the reduction of biological diversity, the cryptic spread of mutant genes, droughts and floods, acid deposition, CO_2 build-up and climate change, the impact of hazardous substances on ecosystems and man, the loss of productive land owing to salinization, the impact of urbanization, and the meeting of current and future energy needs.

I have already noted (section 4.1.2) that this new list, and the SCEP's appraisal a decade before, share a single entry, CO_2 build-up. A rather detailed disussion of the items on the first list tried to demonstrate that all those critical problems of the early 1970s which do not rate highly any more in the early 1980s were not dropped because we solved them—but rather because of changed scientific consensus about their gravity.

Once they were appraised in calmer circumstances, in more detailed ways, and without bending the evidence to fit some 'crisis' preconceptions, they either faded away completely—stratospheric ozone depletion from supersonic transport and the climatic effects of jet condensation trails are the two best examples—or were downgraded to where they belong, from critical global problems to difficult but manageable local or regional degradations; mercury in aquatic ecosystems, oil on the ocean and nutrients in coastal waters are obvious examples. I should stress again that even in the case of DDT, where the problem of further contamination was solved by simply banning the pesticide, rapid degradation and sequestration of residues clearly proved that global pollution dangers from continued applications would have been much smaller than originally anticipated.

Better understanding changed a litany of global crises into an array of ordinary problems or non-problems. Why CO_2 made the list again is not hard to explain: above all, no satisfactory conclusions could be made about its future releases and effects during the intervening decade. As will be seen in the next chapter, CO_2 research has been proceeding at a greatly accelerated rate since the early 1970s but so many fundamental uncertainties remain that room for the big worries—global warming with drastic consequences for food production and ocean level rise— remain undiminished, at least for those who favour belief in the worst indications from the still broad and uncertain array of possible outcomes.

As these uncertainties will not be removed not only for many years but most likely at least not for several decades, CO_2 build-up must, and will, get its share of attention in this book—as will acid deposition. In fact, these two processes are the most important environmental consequences of fossil fuel energy use, as the first is the most massive interference with a global cycle, and the other is of a potentially severe consequence to large parts of northern hemisphere continents. Both phenomena have also considerable implications for food production and so their examination within the framework of global biogeochemical cycles is absolutely unavoidable in any sensible analysis of effects and dependencies involving the triad of energy, food, and environment.

This will take care of two of the ten items on the Rättvik list of research priorities. Among the remainder there are several items which, I would strongly argue, should have been placed in the second list agreed upon at the meeting, that of management priorities. While more research about CO_2 build-up and acid deposition is clearly imperative, such needs are hardly critical as far as the impact of urbanization or droughts and floods are concerned.

In the first case both the causes and the consequences of the process have been studied abundantly for over a century and even the most liberal amounts of additional research will not deliver what is needed most to moderate the process in the poor countries where it is reaching such disagreeable proportions: balanced, sustainable ways of economic development making the life in villages and small towns more attractive. Here the solutions involve the whole structure and activity of national policies, bureaucracy, and management. Technical fixes can ease some of the strains accompanying the transformation but none of them can be singly so decisive as, for example, a desulphurization of all large power plant emissions would be in controlling emissions of SO_2.

Combatting the destructive effects of floods and droughts is, above all, a similarly multifaceted management problem and the randomness and frequent severity of these natural disasters still makes even the richest countries vulnerable: controls, by various means of run-off regulation (from impoundments to water transfers), are simply too costly to be installed everywhere in ways effectively precluding damage. While there is still no scientific basis for the prediction of floods and droughts just one year or a few months ahead it makes little sense to speculate, as did the Soviet contributor to the Rättvik meeting, that 'advances in research can create means and methods for actively influencing the atmospheric processes that cause these natural disasters'.

And while greater understanding of complex energy systems and ecosystemic effects of hazardous substances is undoubtedly necessary, these two sets of problems belong, once again, largely to the realm of management, to the applying of abundant current knowledge to ease what are mostly local or regional difficulties. Rural energy shortages in most of the poor countries are perhaps the most outstanding exception: as already related (section 2.2) their ubiquity has become a problem nearly blanketing the three poor continents—and creating one of the critical intersections of the energy, environment, and food triad as the search for fuel is reponsible for a great deal of deforestation whose environmental consequences seriously undermine future food production capabilities.

But this deforestation is just a part of tropical forest depletion, an item heading the Rättvik priorities list, at which I will look in global terms in some detail. The destruction of forests, usually the richest climax ecosystem is, of course, the leading cause of reduction in species diversity, a concern which has been receiving great attention since the late 1970s and which appears second on the Rättvik list. Diversity loss deserves a closer look even if its implications were not so obviously alarming: all highly fashionable research subjects should be probed critically.

In one of its unexplained inconsistencies the Rättvik consensus put loss of productive land owing to salinization among the top research priorities, and desertification caused by overgrazing among the management priorities. Both are reducing food production capacity and both are parts of a broader phenomenon of the loss of agricultural land—but not the most important ones: the conversion of farmland to non-agricultural uses and soil erosion are the main, and truly global, ingredients of this worrisome deterioration which must get more attention.

The cryptic spread of mutant genes is then the only outstanding research priority on the Rättvik list—and one whose inclusion leaves me most puzzled. Of course, ionizing radiation and many chemicals can increase the mutation rate and the wider use of radiation and constant introduction of large numbers of new chemicals constitute valid grounds for concern that the mutation rate in man may be accelerated. But the way in which this concern was presented in the background paper at the Rättvik meeting is clearly misleading.

Although the first sentence under the subheading of causes and effects acknowledges that 'the causes of increased mutation include radiation, man-made chemicals, and natural compounds' nothing is mentioned later to evaluate the relative contributions of man-made and natural factors. The paper just notes that medical exposure accounts for most of the man-made contributions to the genetically significant population dose—but once one appreciates that, for most of the global population, x-rays contribute no more than about 10 per cent of the total dose, with most of the rest coming from the natural background (cosmic radiation, the ground, structures, food, water, and air), the whole perspective changes.

Even in the United States where x-rays are used liberally in everyday medical and dental diagnoses and where the use of radio-pharmaceuticals and the possession of consumer products emitting low-level ionizing radiation are much higher than in the rest of the world, typical radiation exposure is about equally split between natural and man-made sources, and dental x-rays will be found to be typically the single largest exposure (Fig. 4.4). However, even if a person decides to forgo such an

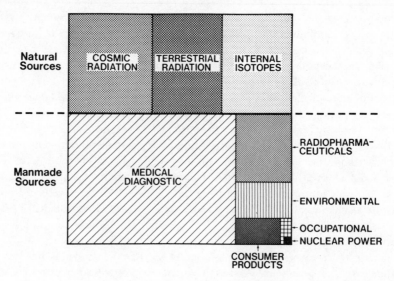

Fig. 4.4 Total annual radiation exposure in the United States in the early 1980s averaged 1.863 millisievert of gamma rays per capita. One sievert is equal to 100 rads, one rad to 100 ergs/g of tissue. Perhaps the best standard to use is that of a typical dental x-ray which delivers about one milisievert to the centre of the cheek.

examination for, say, the next three years he will still receive an equal or larger dose if he changes a wooden house for a brick one, moves from Miami to Denver, or takes four or five transcontinental return flights every year.

Clearly, very large numbers of people do all of these things without worrying about the risks of mutated genes—as they should not. Even when it is assumed that risks vary as a linear non-threshold function of the dose down to the level of the natural background radiation (and this is by no means certain) the best estimates show that perhaps only one and no more than 3 per cent of all cancers are attributable to these exposures. Moreover, long-term studies have shown no detectable increase in genetic abnormalities among the children of people who survived the Hiroshima and Nagasaki bombings; their parents received on the average 0.5 sievert to their gonads, a dose two orders of magnitude above the current total of typical natural and man-made exposures.

Arthur Upton's appraisal, written in the same year as the Rättvik meeting (1982) is worth quoting in full:

Since radiation can cause cancers and genetic defects, the question arises of exactly how hazardous the low-level radiation from background and man-made sources is. The question has been intensively investigated for many years, and the evidence assembled so far indicates that the hazard, if it is real, is too small to be detectable. Hence compared with other hazards to which people are regularly exposed, the overall hazard from low-level ionizing radiation would not appear to be any cause for concern.

As for the mutagenic chemicals, the Rättvik paper states that hydrocarbons, amines and amides, azo, amino azo, nitroso and nitrocompounds 'enter the environment through pollution of air, water or food'—ignoring again any appraisal of the relative contribution of natural and man-made mutagens. To parade a litany of such compounds all one has to do is to look at common foods—and not the ones adulterated by additions of assorted preservatives, but the traditional ones commonly considered to be quite wholesome.

Numerous amines are present in fruits, legumes, cheeses, and fermented beverages. Bananas have especially high content of serotonin (an amine causing cardiac lesions of myocardial fibrosis), avocados are high in tyramine and dopamine, sauerkraut may have lots of histamine. Histamine levels may also be high in both soft mouldy cheeses (Camemberts) and in hard, aged varieties (Emmenthaler and Gouda), as well as in red wines where numerous other amines are also present.

Nitrates are present in often very high concentrations in many vegetables and they will accumulate in some species even with little, or no application of nitrate fertilizer; spinach, broccoli, beets, radishes, lettuce, and celery are the leaders. Sodium nitrite has been traditionally used in the curing of meats, and so the consumption of vegetables and processed meats can provide a daily dose of nitrates and nitrites which can be converted in the body by enzymatic reactions to nitrosamines. Even common spices such as pepper and paprika have been implicated in the formation of nitrosamines.

The United States National Academy of Sciences Committee on Nitrite estimates that an average American consumes about 75 mg of nitrates a day with 87 per cent coming from vegetables, 6 per cent from fruits and juices, and the rest from water, cured meats, and baked goods, while vegetarians may ingest nearly four times as much. In contrast, even if all of the inhaled NO_2 in polluted urban air were converted to nitrate the total daily adult intake would not, even in Los Angeles, surpass 3 mg a day, a small fraction of nitrate in vegetables.

A survey of natural mutagenic compounds in natural food could continue at length: carcinogenic hydrocarbons have been found in dried corn; aromatic polycyclic compounds are relatively high in lettuce, molluscs, and seaweeds; harmful glycosides and toxic amino acids are common in legumes; hormones and antihormones in meats, milk, and vegetal fats; antithyroid compounds in nuts, tea, coffee, chocolate, and some vegetables. There are still fungal toxins (aflatoxin in peanuts is the best known case), ichthyotoxins, anti-enzymatic compounds and toxic minerals—altogether enough to justify a starvation diet if all these minuscule risks were to be translated into food avoidance.

I hope the point I try to make is clear: human actions have undoubtedly been increasing the exposure to low level ionizing radiation and introducing often considerable quantities of chemical mutagens into the environment, but the best quantitative assessments show that for the general public the risks arising from these contaminations are extremely low, if at all measurable and, even more importantly, that natural levels of both ionizing radiation and chemical mutagens are usually considerably higher than man-made contributions. Controls and standards should be in place to protect individuals or groups facing higher risks (ranging from farmers applying pesticides to patients using certain drugs) but the problem is not one threatening the health of the global population and the capability of the biosphere to provide irreplaceable goods and services.

With all of the Rättvik research priorities accounted for I will now turn to the three concerns discussion of which would not fit so well into the later detailed treatment of biogeochemical cycles: losses of farmland, tropical deforestation, and the decline of species diversity.

4.2.1 *Losing farmland*

> Below that thin layer comprising the delicate organism known as the soil is a planet as lifeless as the moon.

> G. Y. Jacks and R. O. Whyte (1939)

Simple comparison of the best available global arable land statistics does not tell the story: according to the Food and Agriculture Organization there was about 2.5 per cent more farmland under annual crops in the early 1980s than in the 1970s. Grain crops are by far the most important sources of food and feed, and a study by Katherine Prentice and Jerry Coiner found that between 1950 and 1975 their areas grew by 1.25 million km², an equivalent of almost 1 per cent of the Earth's ice-free land surface. Clearly, the world has been gaining farmland.

Yes—but only when considered in purely quantitative terms. Conversion of forests, shrublands, grasslands, and swamps into arable land is still proceeding at a faster rate than farmland losses—but there is little doubt that most of these recent gains have been in areas less suitable for the cultivation of the principal food crops than are those provided by the major established regions. The study just cited of grain area changes shows that more than two-fifths of the gains were between 10 °N and 30 °N, in the arid and semiarid subtropical zone where most of the fields must be irrigated to yield good harvests.

The second highest gain, about 20 per cent of the total, came in the equatorial zone, the region of abundant rainfall but of otherwise relatively inferior grain crop farming potential: in comparison with temperate latitudes, the leading producers of food, it has not only poorer soils (see the following section for more on this) but also less insolation (see section 2.4.2 for discussion of this little appreciated fact) and its high precipitation and concomitant erosion are inimical to the regular application of synthetic fertilizers as well as to widespread cultivation of annual row crops.

In contrast, the temperate and boreal zone of the northern hemisphere (40 °N–60 °N) contributed less than 10 per cent of the reclaimed total. Gains have been thus concentrated in zones which are either marginal or outrightly inferior for highly productive grain cropping while the losses, for which we have no reliable global aggregate figures, have affected all farming regions but have been especially disquieting in densely settled areas of natural high yielding capacity, in environments best suited to the cultivation of major food crops.

The Chinese experience, so representative of the global situation in terms of causes, is certainly the best illustration of this process. In 1957, the last year of the decade for which China published reliable statistics, the country had 111.8 million hectares of cultivated land. For the next two decades regular official statistics disappeared but there was no shortage of reports about large-scale reclamation efforts being carried out during the winter months by armies of peasants totalling nationally in excess of 100 million people during the early 1970s.

But the resumption of regular statistical reporting in the late 1970s brought no news of welcome gains. True, the gains from reclaiming swamps, grasslands, lakes, and forests (gains mostly dubious as they too often opened the way for desertification and erosion), and from terracing barren slopes, came to 21.2 million hectares between 1957 and 1977—but the losses totalled 33.33 million hectares, a truly incredible 30 per cent of the 1957 total! The net loss of 12 million hectares during a single generation when the country's population grew by about 300 million people means that per capita availablity of arable land dropped by 40 per cent and that China's farmland is now no more abundant than Bangladesh's—a mere one-tenth of a hectare per capita!

As anywhere else, the causes are manifold but can be split into two large categories: conversions to non-agricultural land and deteriorative processes resulting in eventual loss of productivity. Conversion of suburban farmland looms large in the first category. In 1949 there were 0.3 ha of vegetable plots for each of Beijing's residents—today only 0.05 ha are left as the cabbages and onions have been displaced by factories and apartment buildings. Similar declines happened around all large Chinese cities but the rural losses are even more extensive—and certainly more difficult to control.

Before 1978 the establishment of numerous small industrial enterprises was largely to blame, since then it has been new housing: the new rural prosperity has been expressed so exuberantly by the widespread building of houses on previously farmed land that the Communist Party daily admitted, incredibly enough, that the state had lost control over the use of land in villages. The other main reasons for large farmland losses are a host of degradative processes including reduction of organic matter, waterlogging, salinization, desertification and, above all, soil erosion.

In their northwestern provinces the Chinese have the world's largest loess formation, up to 300 m thick deposits of easily erodible fine-grained particles. Severe erosion, now extending over nearly half a million km^2 of the Loess Plateau, makes the region the area with the lowest grain yields and the poorest standard of life. Recent decades have seen such an acceleration of erosion rates (owing largely to the indiscriminate conversion of grassed surfaces to fields) that the Huanghe now carries 1.6 billion tonnes of silt to the sea, about 25 per cent more than just three decades ago.

And faster erosion has been spreading beyond the traditionally much afflicted valley of the Huanghe. No less than 15 per cent of China's territory is now affected and, a most worrying change for the Chinese, soil erosion is advancing in the previously little affected basin of the Changjiang (the Yangzijiang) with 2.4 billion tonnes of topsoil now being removed annually from the region's slopes. The Changjiang basin also has most of the country's waterlogging problems (lower yields, difficult tilth) caused by improper irrigation and inadequate wet–dry crop rotation.

Desertification in the arid northern regions has taken over 6.5 million hectares of grassland and farmland between 1949 and 1980; salinization affects irrigated land mainly in the eastern provinces and the declining recycling of organic wastes and unbalanced fertilization (too much nitrogen, too little phosphorus and potassium) is degrading soil quality in most major farming regions.

In the absence of meaningful global figures it is impossible to say how much greater, or smaller, have been the Chinese losses in comparison with worldwide, or continental, means. More reliable statistics on farmland conversion losses are available for most rich nations and urbanization is always the leading cause. In Japan during the 1970s new housing and urban transportation were responsible for about half of all the conversion losses averaging more than 50 000 ha a year. In Western Europe these conversions have been adding up to 1.5–2.5 per cent of all arable land per decade.

Curiously, American estimates vary rather substantially: the lowest values are less than 100 000 ha a year, the highest a bit in excess of 700 000 ha. But perhaps more importantly, none of these totals can reflect the critical qualitative differences: the loss of one hectare of poor soil planted to one crop of wheat or potatoes in New England is difficult to compare with a new building development taking over prime, multi-cropped farmland in California's San Joaquin valley.

This difference is brought out even more acutely when one compares a loss of one hectare of fertile land in the Nile Delta with the conversion of one hectare in North Dakota. In the first case the country's arable land corresponds to a mere 0.06 ha per capita, in the second the state has about 25 ha of farmland per person. More importantly, in North Dakota's Red River valley a suburban farmer losing some land can switch to a more lucrative crop (sunflowers, instead of wheat), an option closed to the Nile farmers whose country already has to import about half of its grain requirements. And, furthermore, a slight intensification of North Dakotan farming could easily push up yields, while Egypt's cultivation already ranks among the highest half dozen worldwide.

The loss of a North Dakotan hectare is thus inconsequential for the global food output while the loss of an Egyptian hectare can be equated with a permanent loss of 4 t of grain a year—a shortfall to be made up by higher imports. But these imported 4 t will carry not only a dollar tag: more likely than not they will lead to yet further, albeit gradual, farmland loss since most likely they will come from the United States where much of the land brought into crop production since 1972 (and then again abandoned by 1983) to meet rising export demands was much more vulnerable to erosion than the previously cropped fields.

Indeed, in the world's most productive farming nation it is not the conversion of suburban fields to shopping centres and bungalow housing which causes the greatest concern about farmland loss—it is erosion by water and wind, from the larger area of row crops planted on the slopeland and from newly opened farmland which replaced grasslands in the dry West and trees in the East. These increases have been surprisingly large for such a mature farming system: the wheat area, after years of stagnation, rose 2.5 times (!) between 1971 and 1981, with most of the gains in the Dakotas, Montana, Colorado, Oklahoma, and Texas, all—with the exception of North Dakota—being states with highly erodible land; cornfields grew by about 30 per cent and soybeans, the other main row-crop often rotated with corn, were planted in the early 1980s on nearly 60 per cent more land than a decade before. (However, as noted in the opening part of Chapter 3, by 1983 the planted area was reduced to its lowest extent in this century.)

Coinciding with this planting explosion was the most extensive study of soil erosion in the country, completed by the Soil Conservation Service in 1977. Based on 200 000 random sampling units, the study provided a reliable quantitative base from which to assess the extent and seriousness of erosion by comparing the calculated loss rates with soil-loss tolerance value. This value, denoting the maximum soil erosion losses compatible with a high level of indefinite and economic crop productivity, is assigned to virtually all mapped soils in the country and it never exceeds 11.2 t/ha · year.

In 1977 sheet and rill erosion exceeded soil tolerance on 45.4 million hectares of American cropland, or 27 per cent of the total with about 10 per cent of all cropland losing more than twice the maximum sustainable loss (it must be noted that these estimates do not include wind and gully erosion). Naturally, for row crops, grown largely on the best soils of the Corn Belt, excessive erosion rates are especially widespread: one-third of all cornfields and more than two-fifths of all soybean fields had water erosion losses surpassing 11.2 t/ha · year.

But before considering what can be done to control these losses a detour for a few paragraphs on soil loss tolerance will be necessary. It turns out that, in contrast to erosion assessment, only limited research has been done on this important topic. Knowledge of soil formation rates is clearly critical in this context but—one of the most significant lapses in our scientific achievements—embarrassingly little is known about them. Of course, no single value can suffice as differences in climate, substrate, vegetation, and soil management result in very disparate rates, but research is so sparse that wide-ranging assumptions have to be based on an uncomfortably small body of evidence.

Topsoil formation estimates range from a mere 0.025 mm to over 0.8 mm in a year, with 0.25 mm considered a typical annual gain in farmed soils. As 1 mm of topsoil weighs about 13 t/ha the highest formation rate would be compatible with the peak soil-loss tolerances of around 10 t/ha · year, whilst the mean could outweigh annual losses of no more than 3.25 t/ha. Consequently, it would appear that virtually all sloping land under cultivation is now losing its topsoil and major productivity losses will occur on these soils if erosion continues. In contrast, even some relatively high erosion rates will do little to change rooting depth and water storage on soils developed from thicker loess materials or uniform loams whose physical characteristics may be nearly identical for a depth of several metres.

Quantitative information on productivity losses caused by erosion is somewhat more abundant than on soil formation. Figure 4.5 is a good summary of one of the yield-loss studies showing a close dependence of corn productivity on organic matter declining in thinner soils. Average losses according to Fig. 4.5 would be about 50–70 kg of corn for each 1 cm of topsoil loss. Van Doren and Bartelli gathered corn yield reduction data for many different soils and their figures range from 27 to 88 kg (average around 50 kg) per each cm of soil lost from the A horizon.

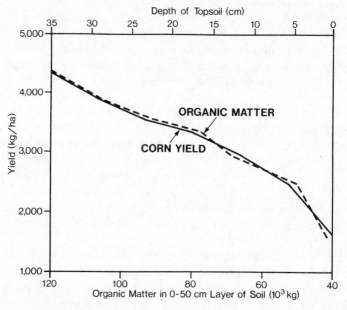

Fig. 4.5 This graph charts the relationships between organic matter in the top 50 cm of soil and corn yields in the Corn Belt and Eastern United States and between the yields and depth of topsoil.

However, Hagen and Dyke put the mean value much higher in the Corn Belt— 188 kg/cm. Loss of plant nutrients is another environmental and economic penalty. Larson, Pierce, and Dowdy calculated that in the United States annual losses of available nitrogen (1.74 million tonnes), phosphorus (34 000 tonnes), and potassium (1.15 million tonnes) add up to $(1982) 1.08 billion.

While nutrients are replaceable—although at increasingly higher cost—loss of the rooting zone is irreversible on the human time-scale. The first manifestations of declining rooting depth may not only be absolute yield reductions but also the much greater year-to-year variance of yields caused by 'pseudo-drought', a loss of moisture-holding capacity following soil loss. This consideration is especially critical in all drier, poor, subtropical countries where most of the new land reclamation occurred after 1950, as well as in the Soviet Union, where crop production is now chronically below expectations and where plans for agricultural development include extensive reclamation of new land.

The first mapping of erodible lands in the USSR, completed in the late 1970s, showed that such soils occupy 15 million km², or about two-thirds of total Soviet territory. The conclusions of the pedologists involved in the project were predictable: 'failure to allow for this factor may lead to very grave undesirable consequences', especially as most of the most sensitive soils are not, as previously believed, in the European part of the country but in the drier Asiatic regions targeted for the bulk of future reclamation.

Fortunately, the process is reversible and recent American developments show the possibility of successful management. Keeping surfaces covered is obviously the best remedy: grassland may have erosion rates easily up to ten times those of mature forests, croplands up to twenty times those of grasslands. Dead biomass will do the trick almost as well and so leaving the stubble, straw, stover, stalk, or vines on the field is the best way both to prevent the beating raindrops from detaching soil particles and to reduce surface run-off.

In practice this means turning to conservation tillage, a broad group of practices aimed at reducing soil or water losses. Unlike during conventional tillage the soil is not inverted, an appreciable amount of crop residue is left on the surface, or the field is left rough, porous, and clodding, rather than smoothly clean. No tillage is the most extreme form of the practice employing various stirring and mixing machines (discs, chisel ploughs, field cultivators, mulch treaders) and subsurface tillers (sweeps and rodweeders).

Data compiled annually by *No-till Farmer* (the publication of the journal being itself a sign of the method's diffusion) show that between 1973 and 1981 use of conservation tillage in the United States grew from 15.8 to 27.1 per cent of planted area, with no-till going from 2.0 to 2.9 per cent. With the much larger cultivated total this meant that conservation and no-till methods more than doubled from 17.8 million ha in 1971 to 39. 13 ha in 1981. Most land in conservation tillage is in the Corn Belt and Northern Plains (together about 60 per cent of the American total) but the highest relative share (just over 40 per cent) is in the Southeast, a rainy and hilly region with high erosion-loss potential.

The new ways have drawbacks. The least rational but emotionally very important one is traditional pride in the clean, trash-free field ready for seeding, a signature of work well done, a seal of tidy farming. Undoubtedly, reduced tillage is not so neat, and no-till is outright ugly until the new crop is tall enough to hide the old mess.

But there are more fundamental drawbacks: the accumulation of crop residues on the surface, the lower soil temperature, the need for higher use of pesticides and, in no-till farming, the heavy application of herbicides. No-till farming was made possible only in the 1960s when the Chevron Chemical Company introduced a powerful contact herbicide, paraquat, that kills both weeds and sod cover and then is inactivated by binding with soil clays—but the cost of these applications goes a long way to offset the savings arising from fewer trips over the field (i.e. no discing, harrowing, cultivation—just split-planting combined with herbicide and fertilizer application, and then harvesting) and makes the whole proposition less economically appealing.

The decisive factor which will determine the eventual diffusion of conservation tillage is the one on which there is little reliable information: how much more soil is suitable for conservation tillage without reducing yields. One study for the Corn Belt suggests that 64 per cent of the region's soils would be suitable, nearly double the early 1980s figure. Still, even extensive conservation tillage must be accompanied by other anti-erosion measures and in this respect the recent record of the world's richest economy is surprisingly weak.

Pavelis assembled data on annual gross and net (after depreciation) investments in individual, on-farm conservation measures since 1935, and when the annual totals are expressed in constant prices the decades after 1950 emerge as one long period of astonishing neglect (Fig. 4.6). The downward investment trend, with the late 1970s' annual expenditures only about a third of the late 1940s' peak, combined with growing depreciation allowances, has resulted in net disinvestment since 1957 with the late 1970s mean being $(1977) 230 million a year—and this decline would look even worse if the early removal or abandonment of control infrastructure were to be added. To enlarge the planted area during the last decade's expansion drive,

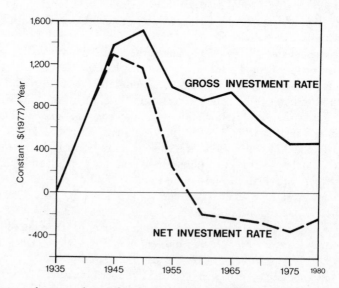

Fig. 4.6 Annual gross and net (after depreciation) investments in on-farm soil conservation measures in the United States since 1935. Neglect after the relatively brief spurt in the 1930s is stunning.

thousands of American farmers actually removed terraces, grass waterways, and windbreaks! The post–1982 slump in the planted area will go some way to restoring this damage.

If the American record is so poor what can one expect in the impoverished countries of Latin America, Africa, or Asia? The fundamental consideration is the high cost of lasting anti-erosion measures. For example, terraces are excellent on steeper slopes but they take up land, make farming more difficult (by limiting the width of field implements and making manœuvring awkward), and less profitable (more time and fuel to farm each ha), and are often quite costly to build. A cost-benefit study of a 2.1 million hectare area in Southern Iowa found that the producer's immediate costs to reduce erosion to tolerable levels were three times as much as the immediate benefits—a trade-off difficult to reverse without some outside support.

And yet even in the United States some very impressive reductions could be achieved with only slight increases in production costs largely as a result of changed cropping practices and removal of land susceptible to erosion from row-crop planting. Earl Heady and Gary Vocke simulated the trade-offs between erosion control and production costs in American agriculture by placing a penalty (from $0 to $(1978) 20 per ton) on soil loss. This would, of course, favour progressively the more widespread use of proper crop rotations (instead of continuous row monocropping), conservation tillage, and terracing, and cause regional changes in crop cultivation (for example, less corn in hilly southern Indiana, more of it in flat western Kansas).

With what surprising results! Annual soil loss could be more than halved from a national average of 12.4 t/ha to 5.5 t/ha, while farm-level supply prices would increase only modestly (by about 6 per cent). However, a further erosion reduction to 4.4 t/ha would entail price rises of almost 40 per cent, a familiar recurrent situation of increasingly costly additional benefits. Paying an additional 5–10 per cent for more than halving the erosion rates would certainly be a very sensible way for a rich society to recognize the often irreplaceable value of its lost soil.

The bill of Senator William Armstrong (Colorado) which makes crops grown on easily erodible land ineligible for subsidized crop insurance, loans, price supports, and disaster payments is an outstanding step toward recognizing this cost. Judging by the situation in the world's most productive farm economy and the largest exporter of foodstuffs, any critical appraisal of soil losses must be poised somewhere between deep resignation and strengthening optimism: poor farming practices still dominate, erosion advances, investment in conservation is inadequate—yet new ways are spreading, an encouraging degree of control could be achieved at far from crippling cost, a recognition of the necessity to act is growing.

As the basic solutions are universal, and technically fairly simple, this assessment is fully transferrable worldwide. Proper crop rotation, minimized row crop planting on slopeland, conservation tillage, strip-cropping, terracing, shrub and tree windbreaks can make the difference between an erosion-induced productivity slide and sustainable farming.

Lester Brown estimated in 1984 that the world is losing about 25 billion tonnes of soils in excess of new formation; he also assumed an average topsoil mass of about 2500 t/ha so there would be about 3.5 trillion tonnes of soil under the world's crops, and the loss of 25 billion tonnes would represent a loss of some 0.7 per cent a year. Continuation of such a rate for the next fifty years would mean a removal of one third of the world's topsoil, a catastrophically rapid rate of depletion.

Yet such calculations are unrealistic, simplistic and misleading on several counts. First, Brown's global total is derived simply by doubling the estimated excessive losses in the USA, the USSR, India, and China (and these four values are themselves fairly uncertain) and it has unknown error margins in both directions. Second, as I have just discussed, much can be done, and will be done, to check erosion. Third, the assumption that all excessive erosion means inevitable net productivity loss is incorrect as much of the eroded soil is deposited on the land elsewhere, thus actually improving crop productivity in the new location. We have no sensible figures on these transfers and hence we can say little about the real net effect of erosion.

Undoubtedly, there will be much unprecedented local and regional soil degradation in the decades ahead, but the net effects of soil erosion during the next two generations will not be translated into declines of global food output and runaway costs of staple foods. Peter Buringh's estimates, based on a wide variety of sources, are for global erosion losses of 50 million hectares of cropland between 1980 and 2000, or about 3 per cent of the current total (he also adds 15 million hectares for grasslands and 10 million hectares for forest erosion). This is about 10 per cent more than Canada's total farmland, or half of all farmland now feeding one billion Chinese; a loss of 100 m² of arable land for each person living today, or the disappearance of almost 5 ha a minute.

This may well come to pass and together with farmland conversion to non-agricultural uses (averaging annually as much, or perhaps twice as much, as erosion loss), desertification (now affecting about one-fifth of the world's cropland), and salinization (the total extent of saline soils, difficult to estimate, is now anywhere between 400 and 950 million hectares, and at least one million hectares of arable land may be abandoned each year owing to intolerable salinity levels), this planet will then have lost a minimum of 100 million hectares of farmland between 1985 and 2000.

Even so, the global cultivated area will be almost certainly larger than today as reclamation will continue: 3.2–3.4 billion hectares are potentially arable, more than twice the total now under cultivation—but, as noted repeatedly before, much, if not most of the reclaimed land will be relatively poor or it will be located in environments marginal for farming, and it will be added at the cost of further deforestation or grassland destruction. More of the same is perhaps the best way of characterizing farmland losses during the coming generation: conversions, yield declines, and field abandonment owing to local and regional erosion, desertification, or salinization that will be made up by higher productivity and reclamation elsewhere.

Global food production capacity will continue to grow concurrently with deepening local and national insufficiencies. No universal crisis—but coming ever closer to the limits of sustainability. Altogether a problem which deserves much more attention than many high-profile pollution releases for which there are good technical solutions (albeit expensive ones) and which are mostly localized in their impacts.

And a problem so old (all those tedious references to Greece, or to the Nepalese slopes ending in the Ganges) and so ever-present that recurrent calamitous reports do little to make it a big favourite with the media (and thus with the public)—or with the scientific establishment. In this respect, at least tropical deforestation has fared better—although the great attention paid to it recently will hardly change its course during the next few decades.

4.2.2 *Deforesting the tropics*

> With a large axe handle tucked in at his waist,
> straight on and deep into the ravine he goes.
> At the sound of a single blow in the empty grove,
> Hidden birds call one another to flight.
> Trees felled, tigers are driven off;
> Nests overturned, forest goblins wail.

> Pi Rixiu (833–83) *A Woodcutter's Axe*
> (W. H. Nienhauser translation)

To some observers only the superlatives spanning long eras will do to describe the loss. Norman Myers labels the tropical deforestation a biological débâcle which surpasses virtually any other change in the 3.6 billion years since the emergence of life on our planet. Thomas E. Lovejoy, Vice-President of the World Wildlife Fund, believes that the unprecedented destruction of biota involved in the massive and accelerating decline of tropical forests is 'the most critical issue of our time, indeed of all human history'.

Such catastrophic appraisals gain urgency when supported by the estimates of recent loss rates which are as high as 250 000 km² annually. Losing so much of the primary tropical rain-forest in a year—an equivalent of a bit more than Malaysia's national total, a monthly decline equalling nearly all forests of Sri Lanka, a daily drop of almost 700 km², a disappearance of close to 50 ha of the planet's richest terrestrial ecosystem every minute—makes an incredible outcome inevitable within just a few decades. Depending on the opening total used for the remaining tropical rain-forest area, the whole biome would disappear in just 25–40 years.

The dimensions and consequences of such a loss would be both unprecedented and unimaginable. In just one or two generations the earth would lose over a third of its forest cover, containing at least two-fifths of all the terrestrial biomass and more than nine-tenths of all plant and animal species. How can we even start foreseeing the consequences which this staggering disappearance would have on the future of the biosphere, on global climate, farming, and economies? One effect about which we are certain is that the loss would be, perhaps in the great majority of instances, irreversible on any human time-scale.

Although tropical rain-forests grow on a great variety of substrates—hence the highly weathered, leached red soils commonly known as laterites are found only in about half of the tropics—tropical soils, especially when compared to those supporting the rich forests of the temperate zone, are rather shallow, with thin litter layers and low nutrient content. They are thus more of an anchoring medium than a nutritional reservoir, and the striking richness and abundance of the vegetation is due largely to the efficient and rapid recycling of nutrients accumulated in the living matter itself. With trees, climbers, shrubs, and herbs gone, a deforested tropical site is stripped of most of its nutritional reserves and the thin soils, unprotected by thick layers of slowly-decaying litter such as those found commonly in temperate climates, are exposed to heavy tropical rains and erode rapidly. Loss of the forest may thus easily become a loss of the site's potential for re-vegetation.

And even at those sites where deforestation was not complete, for example where selective logging removed all valuable large trees, damaged many more, and destroyed much of the lower vegetation storey during felling and log removal, eventual regeneration would take many centuries as the re-establishment of the original species diversity is a much more protracted process than is the renewal of temperate forests where just one or a few large tree species dominate the ecosystem's biomass.

As long as the disturbance is limited to a relatively small area—a clearing for a new field or a new group of huts to be abandoned after several years or a few decades—the surrounding forest will soon re-embrace the openings. Massive felling and bulldozing to make way for pastures, cropland, tree plantations, industries, and settlements makes the destruction permanent on any meaningful planning horizon. Must we then conclude that by the second or third decade of the next century tropical rain-forest will cease to exist except in a handful of natural reserves? While I would vigorously argue against underestimating the extent and rapidity of the recent decline I consider such a loss of the planet'a richest biome extremely unlikely—and the following brief points will show why.

First, the estimate quoted above of the annual destruction of primary tropical rain-forest is the largest plausible figure and even its author, Norman Myers, stresses that it must be seen merely as an indicator of the decline's magnitude. During the mid-1970s Adrian Sommer of the Forestry Division of the Food and Agriculture Organization came up with the most likely annual global loss rate of 112 000 km^2 and a more recent inventory by Jean-Paul Lanly of the same organiza-tion ended up with a worldwide annual loss of no more than about 56 000 km^2, a mere 23 per cent of Myer's total. To find out what share of the remaining tropical rain-forests these annual loss estimates represent is simple only in theory as we are lacking a solid base for determining the denominator—the total area of the remaining tropical rain-forests.

After more than a decade of scanning by LANDSAT satellites (the first was launched in 1972) this seems incomprehensible. When we use satellite images to forecast grain harvests, or to monitor coastal shifts, why cannot we prepare a simple inventory of areas covered by different ecosystems? The obstacles do not reside in the inadequacy of the technology. LANDSAT resolution is about 80 m (much less when dealing with highly contrasting line features such as irrigation canals or roads) but this is quite sufficient for the mapping of vegetation areas with satisfactory accuracy: many small clearings common in shifting cultivation may be missed, highly selective, careful logging will not show up, isolated tiny settlements will get lost and an overall overcount may result—but in no way could the range of possible totals span more than a fourfold difference as the recently published annual tropical deforestation estimates do.

The principal difficulty in using LANDSAT images for the mapping of tropical forests is the widespread lack of concomitant information on what kind and type of vegetation is to be found in the analysed areas. Remote-sensing specialists have a very fitting technical term for this omission: the absence of ground truth. Without appropriate ground truth the accurate interpretation of images is impossible. Distinguishing primary, natural stands of tropical rain-forest from dense but degraded secondary growth, or from lightly logged and regenerating areas, is then highly elusive and it may not even be possible to separate densely forested areas (that is those where tree canopies cover at least three-quarters of the ground) from open park-type stands where large trees are scattered amidst tall grasses, or from vigorous growths of aquatic macrophyts in wetlands. Simply put, satellite images make it rather easy to identify all areas where something grows—but a great deal of ground truth and interpretation wizardry is needed to map the extents of particular vegetation covers, and only meticulous ground surveys can establish the amount of standing biomass present in the forests and its increments.

But these careful ground surveys and aerial photography, where an image may cover only 10–100 km^2, rather than LANDSAT's more than 34 000 km^2, require large numbers of qualified personnel. As a result most of the tropical nations have never taken even a single proper forest inventory, while most of those that carried out the task just once did so several decades ago. Indonesia, the world's largest tropical archipelago, is a fine example of the first category, India of the second as its forest total cited in the early 1980s was the same as that given in 1954—obviously a useless figure. The bottom line spells out wide discrepancies in the best available estimates of the remaining primary tropical rain-forests.

In one of the most cited global surveys, Swedish forester Reidar Persson put the total area of closed forests (where the tree-crowns hide at least one-fifth of the ground when the growth is viewed from above) in the tropics at 9.6 million km^2 in the mid-1970s, when another review prepared for the FAO ended up with a nearly identical 9.35 million km^2. However, Lanly and Clement, using the best updated ground inventories and LANDSAT imagery, came up with a total of 11.32 million km^2 for the year 1980, and for the same year Ariel Lugo and Sandra Brown offered 10.44 million km^2, and Fontaine 12.01 million km^2. At the other extreme, Hans-Wilhelm Windhorst credited tropical rain forests with just 4.88 million km^2

by the mid-1970s, so that the highest estimate is roughly 2.5 times larger than this most conservative figure.

Differences of concept and classification account for this large range: Windhorst takes a narrower forester's approach and considers only densely-closed true tropical rain-forests—while the other estimates pertain to all forests within the tropics and invariably include large components of ecosystems where tree canopies cover only much smaller fractions of the ground. Terminological inconsistencies are thus critical in establishing base values but they are almost always ignored in writings on tropical deforestation. What proportion of those 50 000–250 000 km² deforested annually is primary, natural, dense, closed tropical (evergreen) rain forest, and what share is accounted for by destruction of tropical seasonal (summergreen) growth or forests, so sparsely treed that they are indistinguishable from what the FAO labels an open woodland (crown cover between 5 and 20 per cent)? To these critical questions we have no satisfactory replies and yet only this knowledge can furnish the essential qualitative understanding which is missing from even the most careful estimate of undifferentiated areal loss.

The transformation of a sparse summergreen growth on level land to a properly managed pasture may actually improve the primary productivity of the ecosystem, and it should not degrade the site in any crippling way; in contrast, the cutting of all huge commercially valuable trees in a primary dense rain forest, and the concomitant destruction of most of the ground cover on a sloping site during felling and retrieval operations, will open the way to rapid total de-vegetation and irreplaceable biomass loss. Both kinds of conversions happen everyday—and countless in-between variations—but as long as our estimates of the total area affected are quantitatively very uncertain, our appreciation of the actual destruction can be outlined only in the broadest qualitative terms according to the major conversion processes.

Shifting cultivation has always been a major pantropic cause of forest conversion but estimates of current effects can be only very tentative owing to cumulative uncertainties. How many forest farmers are clearing the land year by year, 20 or 50 million families? How much, on the average, do they clear, half a hectare or two? What portion of the cleared land is virgin forest and what is secondary growth? How much of the new cultivation occurs on land opened up by commercial logging (an important point to avoid double counting of converted forest)? Commercial logging continues to be highly selective with just two to ten species accounting for virtually all exports, but construction of access roads, felling and removal of the huge trees in dense forests, destroy surprisingly large areas of original vegetation. Poorly executed logging operations, especially on slopelands, may be thus quite ruinous, while careful selective felling may allow for natural regeneration.

Similarly, provision of fuelwood may be extremely destructive when carried on as large-scale commercial clear-cutting, or relatively inconsequential when limited mostly to the gathering of dead wood or the pruning of smaller trees or shrubs for local consumption in relatively thinly populated areas. On the other hand, clearing of true tropical rain-forests for pasture is almost always destructive of the ecosystem as it causes major nutrient loss and opens the way for rapid erosion.

Leaving these currently unquantifiable, qualitative differences aside, simple extrapolations using combinations of the cited extreme estimates for annual conversions and remaining forested areas result in a total disappearance of the biome in either as little as 20 years or during the first few decades of the twenty-third century—nearly a quarter of a millenium away, a span too long and hence too full of unpredictable changes to be contemplated in any practical way. Such simple global extrapolations, however, make sense only in calling attention to the uncomfortably large uncertainties (after all, the span represents an order of magnitude difference!) underlying our best appraisals of the disappearance of tropical forests. Future conversion rates will almost certainly change, and their increases or declines may be quite considerable, and the regional and national differences will ensure that some forests will remain permanently intact, while in numerous other instances the conversion to poorer ecosystems has already been completed.

The pressures are, expectedly, closely correlated with population growth, and Asian forests are thus especially endangered. Although current population growth rates throughout tropical Southeast Asia are not as high as in tropical Africa or Latin America, the existing populations are very large and the availability of arable land is very low. Growing numbers of shifting cultivators, more frequent conversions of forests to tree crop or timber plantations, government-sponsored transmigration of peasants from densely settled lowlands, expansion of logging in order to increase export earnings, and greater fuelwood cutting combine to put great pressure on the tropical forests of southern India, Sri Lanka, Bangladesh, Thailand, the far southwest of China, Indochina, Malaysia, Indonesia, and the Philippines and continuation of the recent trends would eliminate all but inaccessible montane growth within one or two generations. Only the Burmese, and above all the New Guinean forests, would have a chance to survive much longer.

The two other most heavily affected regions embrace the lowland forests of all the Central American nations, where a complete destruction of primary stands, save for natural reserves, now appears almost inevitable within two generations (again due to a combined pressure by landless peasants and loggers and, unlike Southeast Asia, by cattle ranching), and both the evergreen and summergreen forests of West Africa from Guinea to Nigeria.

Other heavily affected countries are in Africa—Tanzania, and especially Madagascar, and Latin America—most of the relatively more densely populated fringes of Amazonia (Pará, Mato Grosso, and Rondônia in Brazil, and eastern Colombia). But most of Amazonia in Brazil, Peru, and Ecuador is undergoing moderate conversion, and the heart of the world's largest tropical forest in the Brazilian states of Amazonas and Acre is still almost untouched. Relatively light losses have been, so far, suffered by the forests of Guyana, French Guiana, and Surinam and, encouragingly, throughout most of the Zaïre basin in Zaïre and Congo, as well as in neighbouring Gabon.

The principal causes of deforestation will not change in the decades to come, but worsening rates of conversion are not inevitable in many cases. Certainly the best example is provided by the Brazilian experience. Grandiose colonization plans of the 1960s and 1970s shrank to the level of desultory forays as the predictably low crop yields on the newly cleared land along the Transamazon Highway put an end to visions of massive population shifts into the heart of Amazonia. The establishment of large-scale ranching operations, which mushroomed after 1965 with government encouragement and financial incentives, and which destroyed nearly 7 million hectares of tropical forest before the end of the 1970s, declined greatly after the state stopped promoting these obviously senseless schemes.

The unmistakeable economic setbacks accompanying Amazonian development led not only to criticism abroad but also to the advocacy of much more careful policies at home. Robert Goodland, a tropical ecologist at the Environmental Division of the World Bank, has argued that deflection of intended development to the *cerrado*, a large (about 1.6 million km²) region of under-utilized grasslands with some trees in the states of Mato Grosso, Goiás, and Minas Gerais, would be more effective than any conceivable alternative for improved development within Amazonia—because the *cerrado* environment is much more resilient than the vulnerable forest, and more familiar to people looking to set up farms and pastures.

Clearly, there is an increasing possibility that Brazil will not engage in any massive, concentrated assault on its tropical rain-forest, a prospect very different from the opinions prevailing up to the late 1970s. Similarly, the huge Zaïrean forests do not appear threatened by widespread conversion. And so the prospects for preservation of much of the world's two largest tropical rain-forest basins are much more encouraging than would have seemed likely by just looking at the global rates of annual losses.

In some regions the losses of primary growth caused by intensive logging will be alleviated by the establishment of managed plantations of commercially valuable trees. The most famous example of such an effort, Daniel Ludwig's huge Rio Jari project (100 000 ha of eucalyptus and gmelina plantations, about 450 km northwest of Belém) did not founder solely because of the recurrent problems with stand management—pest attacks, and sluggish tree growth—but because all these difficulties once again demonstrated that the establishment of sustainable tropical tree plantations requires much experimentation and research effort over extended periods of time. Research by Pedro Sanchez and his colleagues in Peruvian Amazonia also demonstrated that it is possible to have continuous grain crop harvests in the region's typical highly acid and high aluminium soils when appropriate rotations and adequate fertilization are used, but the costs and management requirements of such an effort are far beyond the means of poor peasants.

But even the most careful conversions—such as multi-species tree plantations or multi-layered agricultural arrangements, which would not result in huge declines of photosynthetic production, runaway erosion, and the eventual total de-vegetation of a site—would greatly reduce the original species diversity so characteristic of all tropical rain-forest. And while progressively greater erosion of deforested sites and losses of water-retention capacity would have undoubted severe local and regional environmental impact, it cannot be persuasively argued that such changes would alter global climate in grossly unacceptable ways—but it has been repeatedly argued that the loss of genetic diversity could have major, though mostly unforeseeable global consequences.

4.2.3 *Preserving genetic diversity*

> It is commonly agreed that genetic diversity should be preserved.... However ... we found ourselves wondering how often policy makers ... understand the wide-ranging implications of mandates that involve the protection and preservation of species for future generations.
>
> C. M. Schonewald-Cox in preface to *Genetics and Conservation* (1983)

This is a concern of enormous latitude and great complexity. For most laymen it remains identical with the traditional conservation of some aesthetically appealing environments and unique species, the domains of redwoods and pandas. For agricultural geneticists it means an increasingly more vigorous search for as yet uncollected wild species of minor land races of cultivated plants to maintain rich stores of various characteristics for future breeding. And, since the late 1970s, when the concern started to claim growing attention, for many environmentalists it has meant more and more an acute anxiety about an unprecedently rapid extinction of a very large fraction of living species which would bring such genetic impoverishment that some critical environmental services and global food production would be endangered in the not so distant future.

The last concern has, in fact, dominated most of the recent widely publicized writings on the extinction of species and loss of genetic diversity. Some reports hedge a bit, but Norman Myers and Edward Ayensu, in their background paper for the Rättvik meeting put it quite unequivocally: 'In short, Earth is suffering from an 'extinction spasm' with species vanishing off the face of the planet faster than ever before in recorded history'. By their reckoning, one or more species becomes extinct every day in the 100 000 km² of tropical forests which are converted to other uses each year, and in one of his earlier writings Myers notes that the estimate of 1000 extinct species per year (that is nearly three a day) 'seems too low'.

And his predictions are drastically more gloomy: he sees the elimination of one million species 'a far from unlikely prospect' during the final quarter of this century—and this would mean an average rate of 40 000 species gone each year, about 100 every day. As Myers' current extinction estimates are one to three species a day, this means that the 1990s would be a decade of truly frightening slaughter with some ten species being eliminated every hour. We, of course, do not know the grand total of currently living species but, should the number be as low as two million, less than fifty years would be needed to dispose of them all and the higher estimate of five million would postpone the time of complete extinction just to the end of the next century.

The extinction of species has been an essential part of the evolutionary process, and palaeontological studies offer a mass of evidence that the overwhelming majority of organisms that have existed, in excess of 90 per cent, came and went sooner or later—but these natural events were comparatively slow, with even the famous great dying off at the close of the Cretaceous period extending over at least many millions of years. Doing away with one-fifth to one-half of all biota in just one human generation would be thus a truly incomprehensible catastrophe.

An obvious question before proceeding with a discussion of the dangers arising from the loss of genetic diversity is: what are the bases of these cited huge extinction estimates; how reliable, in fact how meaningful, are they? Even a cursory look will show that the estimate of 40 000 species disappearing on the average each year between 1975 and 2000 is a pure guess. If one grants that 1000 species were becoming extinct each year in the early 1980s (as Myers and Ayensu guess—again without any evidence), and that this rate would be about constant for the first 12 years of the 25 year period, then what kind of change will cause nearly 80 000 extinctions a year after 1987 (because only thus can one get the average of 40 000 for the whole period)? If an overwhelming number of species becomes extinct owing to the disappearance of tropical rain-forests, will the deforestation rate suddenly jump 80-fold in the late 1980's? That would imply losses of 8 million km^2 of tropical rain-forest a year (again, using Myers' base of 100 000 km^2 deforested each year in the early 1980s), and the whole biome would be gone in just about one year! Clearly, the numbers do not make sense.

Julian Simon, some of whose observations about the recent vogue of bad news will be noted with approval later in this book, while others will be criticized, is quite correct when he dismisses the floating of such huge but empty numbers which will be invariably reported by newspapers, read by hundreds of millions of people, and understood as scientific statements. There is little doubt that tropical deforestation has been responsible for numerous species losses and that these extinction rates will almost certainly increase in the decades ahead, but these qualitative statements are the best we can make.

Thomas Lovejoy, who used Myers' prediction as inputs to the *Global 2000* report, defended his use of specific numbers during a discussion at the Resources for the Future by claiming that 'people think of endangered species as individual things rather than as part of a continuing process of biotic impoverishment'. This is a feeble excuse for putting down 40 000 a year: why not 35 000, 25 000, or 5000 when no evidence can be offered in preference for any of these guesses? 'I would contend,' Lovejoy closed his defence, 'that—as important as it is to determine precise rates of tropical deforestation, if only to get a better grip on this problem—in a sense all we are arguing about is the date of extinction.'

Wrong again, because such a stance presupposes the unchecked continuation of a trend. Analogically, taking the past rates of European or North American deforestation during the periods of rapid agricultural expansion, one would have to conclude that it would be only a matter of time before all the forests of the two continents were gone, and with them also all vertebrate and invertebrate species.

But both Europe's and North America's forests have been expanding, many wildlife species have been increasing in number.

Naturally, because we have no meaningful quantitative base to evaluate the current and future rates of species loss, it does not mean we should not be concerned about the undoubted diminution of species diversity, but such concerns are not served by baseless guesses. Nor are they served by repeatedly stressed fears that future extinctions will be inevitably catastrophic and grossly detrimental to mankind's welfare. I shall try to present a calmer assessment and will open it by discussing perhaps the most notable omission from the depressing accounts of species extinction—man's role in diffusing, rather than exterminating, numerous exotic species.

Plant and animal species deliberately introduced by man to new environments must number tens of thousands and most of these introductions have involved exotics from distant continents. When the wild species succeed much beyond the original intent some of these introductions became warning examples—rabbits in Australia, and water hyacinths (*Eichhornia crassipes*) choking Florida waterways, come first to mind—but it should be remembered that virtually all of our domestic animals, nearly all of our principal grains, legumes, and oilseeds, and most fruits are also exotics.

Near-Eastern wheats (*Triticum*), Southeast Asian rices (*Oryza*), and Meso-american corns (*Zea*) are now grown on every continent, and each of these crops is planted in at least 100 countries to provide the bulk of global food energy and protein. Other major food plants whose cultivation is now much more extensive outside their areas of domestication include not only potatoes (from highland South America with the Soviet Union being the number one producer), soybeans (from China, with the US at the top), or sugar-cane (from Indonesia, with Latin America now in the lead), but also bananas, barley, cacao, coffee, cucumbers, lentils, onions, oranges, peanuts, sunflowers, tomatoes, etc., etc. Similarly, numerous varieties of poultry, cattle, and pigs diffused worldwide from domestication centres on three continents (four if the turkey is added as the only American contribution to the group).

The introduction of exotic species also follows ways much less orderly than the adoption of a new agricultural crop. Wholesalers, retailers, commercial breeders, and owners of exotic fish, birds, and mammals have all been implicated in the purposeful or inadvertent releases of new species. In the United States, with its large imports of fish, birds, and mammals destined for domestication, as pets, and for research, new species introductions can be cited for every state, but it is Florida which has become a new host to more than twenty species of exotic fishes, to colourful tropical birds such as ibises, parrots, parakeets, bulbuls, and Brazilian cardinals, as well as to two species of monkeys (squirrel and rhesus), coyote, ocelot, sambar, and axis deer.

Among these scores of releases are, obviously, many species with a potential for serious economic damage—for example, the rose-ringed parrot (*Psittacula krameri*), and the monk parrot (*Myiopsitta monachus*) are destructive to the state's valuable fruit farming—or ecosystemic disruption (the walking catfish, *Clarias batrachus* is perhaps the best example of an aggressive species). However, other introductions face competition from both the native organisms and from the earlier established exotics, so their numbers will not escalate; or they simply fill previously unoccupied niches. In both cases their presence will enrich the ecosystem.

Undoubtedly, deliberate introductions have not always worked as intended and some accidental releases have been no less troublesome. However, such cases attract a disproportionate amount of negative comment and it is then easily forgotten that most introductions have not had any destructive influence on their new ecosystems but they have been either beneficial to human welfare (not just crops and domestic animals, but also many commercial timber species, ornamental flowers, game-fishes, birds, and mammals are in this category, as are many invertebrates brought in to control pests), or are a welcome and innocuous diversification of ecosystems. How could one not admire the daring of sparrows who weather the rigours of Canadian winters and are chirping in my backyard while a rising chill factor drives well-dressed humans inside?

But are not sparrows and crops two entirely different categories of species introductions? While the nearly global diffusion of those little undemanding birds must be seen as a clear extension of that species' diversity, are not the substitutions of grasslands and forests by crop or tree plantations the truest examples of genetic impoverishments, and are not these temporarily so successful and productive man-made ecosystems threatening the long-term capacity for food and wood production? Indeed, such is the standard argument. For example, Paul Ehrlich, a few years after his predicted end-of-the-ocean did not happen, turned his attention to extinction and substitution, and his writings gathered some well-known arguments about the unsuccessful performance of substitutions.

Tree plantations have reduced energy flows, cycle less minerals, lose more nutrients, have less resistance to pest attack, and are not able to provide genetic diversity comparable to that of the original forest. Annual crops put less biomass into underground production and thus contribute less to soil generation than perennials, lead to nutrient loss, and cannot be maintained without continuous external inputs of energy and resources, above all fertilizers and pesticides.

Gabor Vida gives a revealing quantitative illustration of genetic impoverishment accompanying the substitution of natural ecosystems. A 100 km² area of deciduous forest may contain about one million individual trees of a single dominant species. If this natural forest is replaced by a highly productive plantation of a single, vegetatively propagated genotype, the gene-pool value of this man-made eco-system will be only about 2 per cent of the original vegetation. The inevitable proportional reduction in the diversity of other tree, shrub, herb, and animal species may easily depress the overall gene diversity to a few tenths of 1 per cent of the original richness.

And with farm crops the loss is even greater: not only plants, but also pests and weeds are fairly uniform and the overall diversity may be less than 0.1 per cent of the eliminated natural ecosystem. Genetic uniformity of crops sown over large areas is obviously conducive to crippling pest attacks, and the probabilities of losing a significant portion of a harvest inevitably go up. In the long-run, continues the scare-mongering argument, the genetic base may be narrowed so much that the plants may become commercially non-viable.

An unexpected infestation of southern corn-leaf blight (*Helminthosporium maydis*), which destroyed about 15 per cent of the American corn crop in 1970, is often cited as the consequence of instability inherent in monocultural uniformity. The Council on Environmental Quality concludes: 'Thus by relying on food resources that can fluctuate significantly, humankind has adopted an opportunistic strategy of existence which is not in keeping with its nature as an equilibrial species.' And episodic widespread hunger, boom–bust cycles in the global economy, and geopolitical destablization are seen as the ultimate consequence of continuing such a risky strategy.

The Council's conclusions, echoed in most writings on crops and diversity, make two highly arguable generalizations. First, historical crop-yield statistics available for many countries for periods exceeding one century, show that annual outputs fluctuated with weather and pests no less, and often more, before the advent of widespread farming with genetically more uniform seeds, than after its diffusion. Looking at the United States, corn yields during the past half century clearly demonstrate the continuing decisive influence of weather (Fig. 4.7). During the dry 1930s, harvests were down in comparison with the previous years: 30 per cent in 1934 and 32 per cent in 1936, at a time when only about 1 per cent of all the crop was grown from the just-introduced hybrid seeds.

During the post-World War II period, the 1947 harvest was down 23 per cent from 1946 owing to poor weather (hybrid seeds accounted at that time for about two-thirds of all plantings), and the 1974 yield (100 per cent hybrids) was down 22 per cent compared to 1973 as a result of exceptionally wet weather.

Clearly, these weather-induced yield drops are much larger than the repeatedly cited 1970 blight-induced decline, but they are lower than the numerous fluctuations of previous decades. In general, variation of American corn yields has been reduced in recent decades in both good and bad weather years, a reality contradicting any talk of an opportunistic strategy leading to more instability. Similarly, a look at historic yields of China's rice or India's wheat will show smaller yield fluctuations in recent decades with weather, rather than new cultivars, still being responsible for the greatest dips.

The second dubious generalization clouding so much of the diversity debates concerns natural equilibria—those cherished, perfectly harmonious states which man comes to destroy in irreversible manner. But this notion of paradise-like equilibria disrupted and obliterated by human actions is partly a wishful desire to see in nature the model of perfection (a lasting debt to the romantic philosophers, a point discussed earlier in section 4.1.1), partly a misinterpretation of complex scientific evidence.

Obviously, I am not using these arguments to endorse any actions leading to the reduction of any species' genetic diversity, or to its outright extinction: but it is

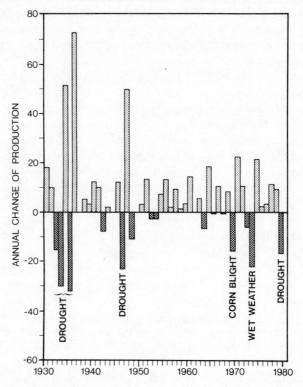

Fig. 4.7 Half a century of American corn yields is plotted in terms of annual changes: climate is still the decisive factor as it was 50 years ago and a great uniformity of seeds since the 1950s does not appear, with the exception of the 1970 *Helminthosporium maydis* attack, to make a difference.

clearly over-critical to insist on seeing the disappearance of some plant and animals, regrettable as it may be for emotional, aesthetic, or even economic reasons, as an automatic destabilization of the affected ecosystem and as an irreparable tilting towards the inevitable demise of the whole ecosystem.

The survival of many mammalian and bird species which are at the forefront of current conservation interest is definitely not essential for the continuing provision of indispensable environmental services, be it nitrogen fixing by soil bacteria, or water retention by forests (although in some cases the removal of a single species may be highly destabilizing). However, the reverse does not apply: the long-term preservation of a single species is, as Otto Frankel aptly remarks, just 'a figure of speech or a public relations exercise'. The only possibility of keeping alive a tiger or a redwood is, obviously, by the preservation of their complete ecosystems. Moreover, truly long-term conservation must be a dynamic process, nothing less than continuing evolution which means that wild species must have available a sufficiently large pool of genetic diversity to overcome environmental pressures exceeding the limits of developmental plasticity. The diversity of many natural reserves is patently too small to survive inbreeding, genetic drift, random fixation of allelles, and accidental deaths.

This limitation poses by far the greatest difficulties for the preservation of many mammals with low population densities. Michael Soulé illustrated this fact with the grey wolf (*Canis lupus*). As its adult population density is merely 5 per 100 km^2, only about one individual in three breeds, generations overlap and frequent and severe seasonal population bottlenecks interfere. The minimum effective size to avoid inbreeding would be about 600 wolves, requiring 12 000 km^2 just for short-term preservation—an area larger than Yellowstone Park.

Globally, there are now only 14 reserves larger than 5 000 km^2, the largest, not surprisingly in the Arctic (Gates of the Arctic, 30 344 km^2; Wrangell-Saint Elias, 30 129 km^2) but the two perhaps most famous animal reservations, Serengeti in Tanzania (25 000 km^2) and Kruger National Park in South Africa (19 084 km^2), are also spacious although they still may not be large enough to assure the survival of all large mammalian species.

Even with impeccable protection in the existing large parks, even with further extensions and establishment of many new natural reserves, the disappearance of many large mammalian species appears inevitable. As John Beardmore puts it: 'The text for the day I fear is that the generalists—that is, the genetically diverse—will inherit the earth.' In contrast, the outlook for the conservation of plant species in larger reserves looks much more encouraging. For example, in rich lowland tropical rain-forests there are between a few tens and several hundred individual trees of the same more common species per square kilometre, and hence very large numbers can be preserved in extensive reserves with even species having densities as low as one tree per square kilometre being perpetuated.

Although tropical deforestation rates have been destructively high, and are rising in many regions, there are large areas which have remained mostly untouched, especially, as noted, in equatorial Africa (in Zaïre and Congo), in western Amazonia, French Guiana, and Surinam. Controlled, spot-type intrusion into these huge regions and the permanent preservation of select contiguous areas larger than 20 000 km^2 could go a long way towards conserving tens of thousands of plants and hundreds of thousands of the still largely unclassified invertebrate species.

The chances of appropriate development compatible with the preservation of natural tropical rain-forest, and the setting up of larger protected areas, are not so dismal as the reading of current extinction stories would indicate. Brazilian attitudes are perhaps the best example of coming change. At the 1972 Stockholm Conference the official Brazilian position was to resist environmental protection efforts as a rich countries' plot to curb the poor world's development. A decade later Paulo Noguiera-Neto, the country's top environmental official, stated that 'we have changed 180 degrees'.

And real changes *have* been taking place. Massive financial incentives disbursed by a government agency to promote Amazonian cattle ranching (inevitably preceded by ruinous deforestation) were stopped, a shift of inestimable importance for the further protection of the region's forests. The programme of large-scale settlements along the Transamazonian Highway petered out and new development projects affecting larger areas are setting aside natural reserves: for example, the Polonoreste programme in Rondônia and the western part of Mato Grosso may eventually affect 100 000 km^2 of natural forest, but eight reserves and ecological stations totalling 19 400 km^2 are being set up to protect a variety of ecosystems ranging from flooded *várzea* forest to transitional dry growth.

And before leaving Amazonia I must note that the oft-repeated concerns about reductions of genetic diversity in the Vavilov centres, a dozen areas of unusual concentration of variety to which the origins of all the principal food crop plants can be traced, would appear to be actually least applicable to the world's largest region of tropical rain-forest. Obviously, with some 100 000 species of plants the possibilities of discovering new food, fibre, medicinal, and industrial crops (producing starches, oils, waxes, rubbers, and resins) are quite considerable—but the historical contribution of Amazonian plants to global food supply has been surprisingly negligible in comparison with the major Eurasian centres of domestication.

The only contribution which has made any real difference on a pantropical scale was the Amazonian domestication of cassava (*Manihot esculenta*), although the plant was also probably domesticated independently in Central America. About 120 million tonnes of cassava were harvested annually in the early 1980s with Brazil, Thailand, Indonesia, Zaïre, and Nigeria accounting for two-thirds of the total. Two other notable plants originating in Amazonia provide luxury foods rather than principal nutritional needs: wild *Ananas microstachys* gave rise to the cultivated pineapple (*Ananas comostus*), while *Theobroma cacao*, originally an Amazonian species domesticated a long time ago in Central America, provides a raw material for a rich drink and for the even more universally appreciated chocolate.

Gustatory pleasures aside, the second most important Amazonian contribution to major cultivars is an industrial crop: the Pará rubber tree (*Hevea brasiliensis*) is now the source of about 4 million tonnes of natural rubber each year. Nine-tenths of the harvest comes from Asia (Malaysia, Indonesia, and Thailand) while Brazil produces less than 1 per cent of the total. Once again, I have to stress the thrust of the argument: I am not recalling Amazonia's relatively small genetic contribution to global food production in order to belittle the dangers of destruction of the region's rain-forest, merely to temper recent extremely pessimistic opinions.

Rather than being a storehouse of genetic diversity for future staple crops Amazonia can be more appropriately seen as an extremely rich depository of biocides. To survive the incessant attacks of numerous heterotrophs, long-living tropical rain-forest trees have had to develop complex chemical defences and hence their saps, barks, and leaves could yield efficient biodegradable insecticides and fungicides. The extensive pharmacopoeias of the Amazonian Indians attest well to the presence of large numbers of other plant chemicals, ranging from ichthyotoxins to hallucinogens and stimulants, but the biodynamics of these substances have been assayed only in a small fraction of known cases.

But how can crop diversity be preserved in those key Vavilov centres which furnished us with most of our cereals, legumes, and oil plants? Many eforts will be necessary. First, a search for surviving ancestor species and rare, surviving old cultivated varieties has already led to several notable discoveries which have helped in breeding better strains of wheat, rice, and corn. In some instances it should be possible to preserve these small, isolated sites for further natural evolution; in most cases, however, the only practical way is to collect the seeds and store them, frozen at −20 °C in genetic banks, a far from perfect option but one which substantially increases the possiblities of germplasm preservation without cumbersome annual replanting.

A recent review of conservation of rice genetic resources by T. T. Chang is a very encouraging example of germplasm preservation for one of the world's two top food crops. Global rice harvests now surpass 450 million tonnes a year and they feed more than 2.3 billion people (half of the world's population). The widespread introduction of high-yielding semi-dwarf rices during the 1960s and 1970s, the increased use of nitrogen fertilizer, and more frequent double-cropping of the same variety led to some undesirable narrowing of the genetic base, opening a way for several major yield reducers such as tungro virus, grassy stunt virus, and rice blast which affected Indian, Southeast Asian, Korean, and Chinese crops.

Releases of new varieties effectively suppressed these pest sweeps, and wild species made significant contributions in these efforts as well as in other rice improvements. Genus *Oryza* has twenty wild species and two cultigens (Asian *Oryza sativa* and the much less important African *Oryza glaberrima*) and Chang estimates that slightly more than 100 000 cultivars of the first species now exist worldwide. National and international germplasm conservation programmes have already succeeded in assembling a substantial number of indigenous rices, with China collecting about 40 000 varieties and India 20 000.

Unfortunately, these collections conserved only very few wild rices and minor primitive cultivars (land races), as did the first rice germplasm bank of the International Rice Research Institute (IRRI) established in 1961. But a decade later, in the face of the rapid diffusion of new improved varieties, IRRI launched a massive collection campaign to preserve minor cultivars in threatened areas and to expand wild species coverage. Between 1971 and 1982, 25 530 samples (including duplicates and non-viable seed) were collected of which 9000 were locally claimed to possess special characteristics (ranging from drought to cool temperature tolerance). A similar programme in Africa brought in 4000 samples.

IRRI's International Rice Germplasm Centre, which contained in mid-1983 67 000 Asian and 2600 African cultivars and 1100 wild rices, can process the seed safely and efficiently for long-term storage and it deposits a duplicate set of freshly rejuvenated seed at the National Seed Storage Laboratory in Fort Collins (Colorado). Annual expenses to operate the collection were only $1.25 million in 1983, a negligible sum in view of the large monetary and human welfare returns realizable from the availability of a wide genetic base for this crucial crop. While more work is obviously needed, especially in collecting plants with resistance to the growing number of viral diseases and insect pests, the collection of rice germplasm shows the possibility of successful preservation of a very large slice of the existing genetic diversity.

Of course, many will argue that gene banks are a rather vunerable means for the preservation of diversity. Many seeds and most vegetatively propagated plants cannot be stored for long or at all; without long-term experience we cannot say how much of the stored material will deteriorate or when; collections may be poorly managed, mishaps occur and, indubitably, the germplasm is literally frozen and its further evolution is stopped. All granted, but improvements in preservation can go far towards reducing the problems. After all, we have barely started to do this in a systematic way and on a large scale.

Finally, we are also just starting to explore the amazing possibilities of genetic engineering. While it is obvious that agricultural applications are enormously more difficult than manipulation of bacteria, new methods of DNA manipulation will definitely start bringing significant results soon. The first vector exploited for plant genetic engineering, Ti plasmid of *Agrobacterium tumefaciens* (the organism responsible for crown gall tumours), has already carried foreign genes into dicotyledonous plants and the rapid progress in understanding the basic organization and expression of plant genes will accelerate many fascinating developments in plant DNA splicing.

I do agree that of all the currently identified global environmental problems the rapid loss of genetic diversity is certainly the potentially most crippling to human welfare—not because of the disappearance of a white rhinoceros, or a brown-eared pheasant, but mainly because of the threat to environmental services and to continued highly productive farming. But the best available evidence does not support the widely publicized predictions of imminent extinctions on a scale unprecedented in the planet's history, and the rising recognition of the possible penalties will almost certainly lead to the expansion of efforts already started, ranging from the establishment of very large natural reserves to high-quality seed banks, which will combine to make the impacts of future losses less crippling and more restitutable than they may now seem—even when the promises of plant DNA manipulation are left aside to avoid all overly optimistic forecasts.

This review concludes the critical examination of the three complex concerns which became so prominent almost concurrently during the 1970s. Only now can we look at the most fundamental linkages of energy, food, and environment, at the grand biogeochemical cycles, and at human interferences in their working.

5

Energy, Food, Environment

> Because the natural power continually seeks
> and tends to Destruction,
> Ending in Death, which would of itself be
> Eternal Death.
>
> And every Natural Effect has a Spiritual Cause,
> and not
> A Natural; for a Natural Cause only seems.
>
> William Blake, *Milton* (1804–1808)

Blake's first cited sentence may be interpreted as a prescient formulation of the Second Law of thermodynamics, an unstoppable slide into chaos and homogeneity; the second sentence can be seen as predating by one and a half centuries Teilhard de Chardin's idea of noosphere 'superimposed upon, and coextensive with (but in so many ways more close-knit and homogeneous) the biosphere'. Since the middle of the nineteenth century, civilization's vastly expanded intellectual powers have been accelerating the slide into homogeneity as huge quantities of fossil fuels, have been extracted and converted to create a new industrial society—and to turn low entropy fuels into carbon dioxide and atmospheric pollutants, and to lose the generated heat into space.

This incessant quest, this grand generation of eternal physical homogeneity in order to enjoy ephemeral civilized heterogeneity, this temporary defiance of the Second Law, has brought unprecedented material comforts and (so often unutilized) intellectual opportunities to huge numbers of people, but only since the late 1960s have we started to define and to measure the price paid. Conversions of chemical energy (low entropy) in fuels to thermal, kinetic, and electromagnetic energies (all of which end up as high entropy low-level heat) needed to secure more food, more material comforts, and more intellectual challenges have been accompanied by many environmental changes, some minor and of localized impact, others of wide reach, even of universal effect: noosphere altering biosphere in ways which may greatly complicate and eventually imperil its very existence long before the natural train of destruction could do away with the amalgam of conditions sustaining life.

Survival of civilization is the ultimate goal and, while the dangers unleashed by the normal life of industrial societies do not appear to be imperilling this aim in any acute way (as does, of course, the possibility of global nuclear war), we are too ignorant to predict how long this state of affairs will last. There are no reasons for panic but there is plenty of evidence to counsel caution. This stance will recur throughout the coming discussion: there will be no sensational and invariably dubious claims akin to those exposed and criticized in the previous chapters, but prudent assessments *sine ira et studio*.

But what shall I write about? Myriads of links tie the extraction, conversion, and utilization of energies, the production and consumption of food, and the state of the environment. Many were mentioned in detail or just touched on in the three topical chapters; many others have been the centres of lively interest ever since the concerns about energy and food supplies, and about environmental pollution and protection, became so prominent in public and scientific affairs.

Since the late 1960s thousands of papers and hundreds of books have appeared to deal with the linkages of these three interrelated subjects. The most extensive recent record is on energy and the environment; land destruction by surface coal-mining; water pollution by acid mine drainage, oilfield operations, refineries and tankers; air pollution from the combustion of fossil fuels; escapes of radioactivity during normal operations of nuclear power plants and the problems of long-term storage of highly active wastes. There appears to be no end to the minutiae studied in this vast field since the late 1960s, from the role of leguminous plants in reclaiming the stripped-mined lands to the possibilities of acclimatization of largemouth bass to warm water effluents from thermal power plants, from investigations of bacterial decomposition during marine oil spills to radiometric studies of waters used in geothermal generation.

Links between food and the environment have been studied for much longer but no less systematically than those between energy and the environment. This vast realm embraces everything from climatic factors influencing the maturation of crops to pesticide residues in flour, and the studies have dealt with such pragmatic questions as rates of mulching to reduce soil erosion in row crops, as well as with such still so theoretical concerns as transfers of nitrogen-fixing genes to nonleguminous species.

Compared to these two well-established and expansive preoccupations, analyses of energy-food links are mostly of very recent origins but no small results have accumulated since the mid-1970s when the energy analysis of food systems have started to prosper. Studies in this new field have focused both on overall energy use in national farming and food production sectors, as well as on energy costs of individual crops, cultural processes, and principal farming inputs.

All too obviously, even a cursory review of most of the interesting energy-food-environment links, something amounting to hardly more than a listing of research titles, could occupy the remainder of this book, splinter the focus and mix concerns which are critical with those just important or merely marginal. Fortunately, there is a simple yet perfect solution, an angle which still subtends a lot but which illuminates the core of the concerns: a look at the effects that civilization's efforts to consume more energy and to produce more food have on the planet's grand biogeochemical cycles and how, in turn, these changes could affect the energy and food production modes and capacities and hence determine the future of industrial societies.

5.1 Grand cycle linkages

> Not wholly, then, doth perish what may seem
> To die, since from one thing doth nature build
> Another, nor will suffer aught to come
> To birth without the death of some thing else.
> Lucretius, *De rerum natura* (99–55 B C)
> (C. E. Bennett, translation)

Open to energy inflows and outflows the Earth is a closed system as far as matter goes (one can easily disregard gas losses from the stratosphere and meteoritic bombardment), and the cyclical nature of life is a truism already much commented upon in ancient poetry. Yet, except for the easily observed sequence of evaporation, precipitation, and runoff the precise working of these grand circuits started to be uncovered only when advances in chemistry identified essential elements and compounds involved in the cycling. But these developments, falling largely to the last decades of the eighteenth and first half of the nineteenth centuries, had to be followed by a long period of interdisciplinary studies culminating in the formal emergence of ecological research earlier in this century and in detailed biochemical understanding of cellular and microbial processes.

And so a satisfactory understanding of the basic modes and sequences of biospheric cycling is only a recent acquisition and, as will be seen shortly, gaps in our detailed knowledge of many fluxes, reservoirs, and effects still abound. But first things first: which cycles should be chosen for a closer survey, what are the justifications for limiting the inquiry to a few major items?

The perpetuation of complex, highly differentiated life on Earth requires the maintenance of numerous physical variables within narrow ranges, but civilization cannot do anything to alter the shape of the orbital course, the tilt of the Earth's axis of rotation, the relative placement and rearrangements of the continents, and scores of other essentials which had to come together to make a habitable planet.

The biogeochemical foundations of life are a different matter. Water dominates the composition of all biomass, being more than four-fifths of the fresh weight of green plants and more than two-thirds the mass of animal and human bodies. And even in absolutely dry tems, biomass is composed predominantly of hydrogen and oxygen, and typical proportions of these elements closely resemble their ratio in water. Virtually all the remainder, about 45 per cent of dry weight, is carbon, the single largest elemental constituent of living matter (Fig. 5.1). And, of course, industrial civilization has been altering both water and carbon cycles.

If the magnitude of intervention were the main criterion, the water cycle would have to get most of the attention. Water is an irreplaceable requirement for the

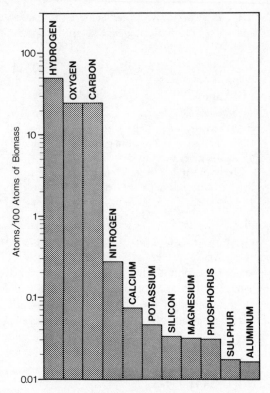

Fig. 5.1 A span of four orders of magnitude is necessary to show the most important elemental constituents of living matter. Hydrogen, oxygen, and carbon dominate, nitrogen accounts for just a third of 1 per cent of all atoms but its qualitative role is essential.

photosynthesis of flowering plants, and for the metabolism of heterotrophs, as well as an indispensable input into countless industrial processes. Order of magnitude comparisons for the world's largest economy convey best the dependence of industrial civilization on massive water flows: while the United States' energy consumption during the early 1980s amounted annually to roughly 2×10^9 t of oil equivalent, its withdrawals of water surpassed 6×10^{11} t of which about 2×10^{11} were consumed, that is evaporated or incorporated into plants, products, animals, and people. Crop irrigation represents by far the largest consumption of water, about an order of magnitude larger than water consumed in cities and by energy industries, above all in thermal energy generation. But as for the total withdrawals of water, American fossil-fuelled power plants have been surpassing the intakes by irrigation ever since the early 1960s.

Of course, all but a small fraction of this water is just used for cooling and is released back to the streams, reservoirs, lakes, or bays. Still, the location of the plants is critically dependent on the proximity of large water supplies and the released water is often appreciably warmed, causing various ecosystemic changes. And both food production, and extraction and conversion of fossil fuels result in numerous instances of water pollution.

Human interference in the water cycle is thus massive, spatially widespread, and often very intense, taking the form of transfers, delays, accelerations and pollution as local, national, or regional balances and make-ups are altered by changed land cover and land use and by farming, industrialization, and urbanization. But none of these changes affects the fundamental features of the global cycle as there are no additions to or withdrawals from the worldwide balance, the only known 'sink' being a negligible loss of hydrogen atoms to space which is readily replaced by juvenile water from volcanoes.

While the local and regional characteristics of evaporation, precipitation, runoff, surface and underground storages are often changed quite considerably by human actions, the global volume of water and the rates of its cycling are basically immune to human interferences and, as discussed before (in section 4.1.2), ocean pollution is also a serious local or regional concern but not a global degradation. To make a case for anthropogenic destabilization of the global water cycle is thus impossible, but water's second element, oxygen, appears to be a better candidate for concern—not in the waters nor in the crust (where we walk on more O_2 than is in all oceans)—but could not our combustion of fuels seriously deplete the element's atmospheric stores?

Indeed, the complete oxidation of each carbon atom requires two atoms of oxygen and the current burning of fossil fuels and biomass (including wild grassland and forest fires) sequesters each year about 20 billion tonnes of oxygen—but this represents a mere 0.002 per cent of the element's atmospheric mass, and even the complete combustion of all currently known deposits of fossil fuels would reduce the atmospheric oxygen content by only about 1 per cent (published estimates range from 0.5 to 3 per cent depending on assumptions about the ultimately recoverable fuel), a completely insignificant depletion.

With the oxygen cycle also out of contention we turn to carbon, as already noted the single largest elemental constituent of living matter. Obviously, anything interfering with the fluxes and storages of an element making up nearly one-half of dry living matter requires close attention—and the developments since the middle of the last century have combined to make this concern easily the most critical link among energy, food, and the environment.

The combustion of fossil fuels and the conversion of grasslands and forests to croplands, and for industrial and urban uses, have been releasing huge amounts of carbon dioxide into the atmosphere. The absorption of the gas by the ocean and the incorporation of carbon into new biomass have been proceeding at rates lower than these large inputs and global atmospheric concentrations of CO_2 have been steadily increasing. This rise has been seen by some as mankind's most audacious planetary experiment as the much elevated concentrations of CO_2 may change substantially the thermal balance of the Earth and the resulting warming and high-CO_2

atmosphere may have a profound effect on the future development of our civilization.

Even if this effect were much more modest than foreseen by some scientists it is undeniable that by increasing global levels of CO_2 industrial civilization is altering concentrations of the compound whose reduction forms the foundation of planetary life and whose absorptive properties have an inevitable influence on surface temperatures. Concern and caution are certainly called for. But life on this planet cannot be sustained solely by hydrogen and oxygen, whose global cycling human actions barely touch, and carbon, whose fluxes we have been influencing so heavily.

There are many other nutrient elements critically needed in the planetary biogeochemistry, elements present in the biosphere in much smaller quantities than hydrogen, oxygen, and carbon but qualitatively no less critical (Fig. 5.1). Surprisingly, most of these vital elements do not cycle. Insoluble minerals can be moved only mechanically by water and wind; the soluble ones are transported, sometimes with interruptions, from land to water—and altering both of these transfers on a global scale would not be easy.

As noted previously (at the beginning of chapter 4), several trace elements are now mobilized by the combustion of fossil fuels at rates of the same order of magnitude as natural weathering, but none of the important micronutrients (such as zinc, boron, molybdenum) are in this category. And the releases of toxic trace elements (above all of mercury) could be controlled by recycling to prevent virtually all of the existing local or regional health or ecosystem disturbances caused by elevated concentrations.

Moreover, two of the three plant macronutrients also fall into this non–cycling category: phosphorus and potassium do not really cycle and anthropogenic augmentation of their natural one–way fluxes is not known to lead to any worrying long-range, large-scale, irreversible effects on the biosphere (rich algal growth in P-rich waters is merely a local, and easily controllable, annoyance). But the third macronutrient, and the one needed in the largest quantities by plants, and filling no less an indispensable role in heterotrophs, fits the conditions of concern.

Nitrogen has permanent large reservoirs in the atmosphere (which is mostly composed of its hard–to–split molecules), in the soil (above all in organic form), and in water, and its compounds cycle readily and vigorously—oxidized and reduced by natural inorganic processes, human industries and, most of all, by micro–organisms. Anthropogenic interferences on a scale large enough to accelerate or slow-down some critical fluxes, deplete, or enlarge some reservoirs, and change the concentrations of some compounds controlling other biospheric processes must be studied closely.

Minor during the pre-industrial era, such interventions have become increasingly widespread and intensive during this century as the organic soil nitrogen reservoir shrank in newly-planted farmlands, nitrogen fertilization applications rose to rival global rates of natural nitrogen biofixation and as combustion of fossil fuels and faster denitrification put undesirable nitrogen oxides into the atmosphere.

And one more cycle must claim our attention as one more element structurally indispensable for life is doubly mobile, both water soluble and airborne, cycling vigorously on the grand scale: sulphur, the 'glue' of biomass, the element stiffening the long swirls of organic molecules composed primarily of carbon, oxygen, hydrogen, and nitrogen.

So we have three special elements sustaining life on earth by moving through the waters and through the atmosphere, passing through micro-organisms, plants, animals, men, and after these delays, moving on again in grand cycles. They must be recycled together, one atom of sulphur for 10 atoms of nitrogen and 1000 atoms of carbon being their typical ratio in phytomass. As Edward Deevey pointed out, 'it may not be an accident that all three are more reduced in the biosphere than they are in the external world. Be that as it may, they all seem to belong to the biosphere, which is otherwise mainly water'. Another interesting illustration of the importance of carbon, nitrogen, and sulphur for life is the degree to which these elements are enriched in the terrestrial biosphere compared with the lithosphere. The enrichment factors are of the same order of magnitude for both autotrophs and heterotrophs, 10^5 for carbon, 10^3 for nitrogen and 10^1 for sulphur.

But their atoms, molecules, and compounds abound also in non-living nature, as carbonates, nitrates, and sulphates dissolved in waters and deposited in the earth's crust, in the atmosphere as numerous gases, liquids, and particles, all parts of incessant transfers among their numerous reservoirs. The extraordinary mobility of carbon, nitrogen, and sulphur, their critical place in the structures and functions of life, their dependence on incessant cycling mediated by living organisms, and hence the vulnerability of their transfers and reservoirs to human intervention, thus continue to make the study of the three cycles the most revealing method of inquiry into the complex interaction of energy, food, and environment.

5.2 Carbon, the largest unifier

> The carbon cycle is the most important and most complex cycle of all, as it is the pacemaker for the other cycles which in turn codetermine flow rates in the carbon cycle.
>
> La Rivière (1979)

One may easily point out the contradiction in the statement made by the learned chairman of the Project on Biogeochemical Cycles of the Scientific Committee on Problems of the Environment: precisely because the other cycles codetermine its fluxes the carbon cycle cannot be simply labelled the most important one. The interrelatedness of life makes it impossible to rank life processes in terms of importance. The wonder of life is that everything needed for a rose, a kohlrabi, or a skunk has to click (within tighter or looser limits, of course) at the same time—that nothing needed is unimportant.

And yet the carbon cycle deserves a special place—if for no other reason than just for the magnitude of its turnovers. As already noted, carbon forms nearly one-half of dry biomass and, even in the biosphere as a whole where the enormous water mass makes hydrogen naturally the most abundant element, every fourth atom will be carbon. But carbon is also one of the most unusual elements, forming bonds which tend to be both stable and labile, creating life, still the only life which we know in this universe, ready to react and to change. No less interesting is carbon's physical diversity, ranging from graphite to diamond in elemental form, and from gaseous oxides to liquid alcohols to solid paraffins among its numerous compounds.

Of all carbon compounds it is carbon dioxide which holds the centre stage in the intricate cycle involving biota, waters, atmosphere, soils and parts of the earth's mineral crust (Fig. 5.2). All carbon atoms in complex organic compounds forming leaves, stems, blades, and crops, animal bodies and our tissues, are there only because of the reduction of CO_2 within chloroplasts of photosynthesizing plants, and as all these compounds are recycled, partially through respiration but largely through decomposition, CO_2 re-enters the atmosphere to start yet another sequence of buds, leaves, roots, fruits, and seeds.

The unravelling of the involved biochemistry of photosynthesis came only in the late 1950s and the early 1960s but quantification of global primary productivity and standing phytomass is still open to considerable errors. The International Biological Programme and other research efforts since the mid-1960s increased greatly our understanding of photosynthetic performance of all major ecosystems but averaging and extrapolation of this knowledge gained from detailed studies at a few scores of sites to biomes covering millions of square kilometres are clearly highly error-prone, especially as we do not know accurately the area of tropical rain-forest, the ecosystem which produces and stores most of the planet's photosynthate.

During the discussion of tropical deforestation (section 4.2.2) I had to stress the major uncertainties regarding the actual areas of this richest terrestrial ecosystem and their recent changes. Consequently, it would be of little help to try to compile accurate statistics for other major ecosystems such as tropical and temperate grasslands or semi-deserts which store only a fraction of the biomass locked in the tropical forests. Only for temperate and boreal forests of the rich countries, and for agro-ecosystems worldwide do we have satisfactorily reliable areal, storage, and productivity estimates but even in these cases I prefer to use heavily rounded values to survey the global distribution of phytomass.

Table 5.1 approximates the situation in the early 1980s and it is encouraging that my total of 1.06 trillion tonnes of global phytomass storage (that is 477 billion tonnes of carbon) derived by very simple procedures is virtually identical with the lower value (1.02 trillion tonnes) offered by Jerry Olson of the Oak Ridge National Laboratory as a result of his painstaking global vegetation mapping (0.5° × 0.5° cells assigned to one of twenty-five major vegetation types). Olson's mean is 1.24 trillion tonnes (range 1.02–1.46 trillion tonnes), or about 15 per cent above my estimate but this (considering the inherent uncertainties) small difference would have been much smaller if Olson's work could have included the latest consensus on grassland

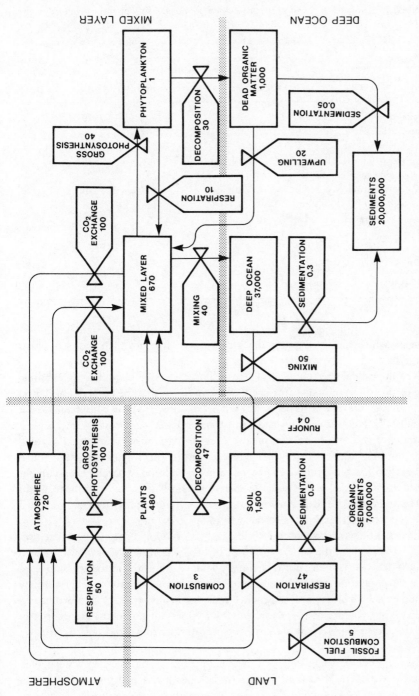

Fig. 5.2 Basic flow diagram of the global carbon cycle. Reservoirs (rectangles) carry the estimates of storages in billion t C, fluxes (valves) in billion t C a year; all values are just the best approximations.

Table 5.1 Estimates of global primary production and standing phytomass (all values are in dry terms)

Ecosystems	Average net primary productivity (t/ha)	Average phytomass (t/ha)	Total production (10^9t)	Total storage (10^9t)
Tropical moist forests	20	300	20	300
Temperate forests	10	250	10	250
Boreal forests	10	200	15	300
Woodland and shrubland	10	75	10	75
Tropical grasslands	10	20	20	40
Temperate grasslands	10	20	10	20
Cultivated land	10	10	15	15
Settlements, transportation	5	5	3	3
Tundras	1	5	1	5
Deserts and semi-deserts	1	5	2	10
Wetlands	15	75	8	40
Total	—	—	114	1 058

biomass which clearly requires the use of lower averages than those preferred formerly.

The appealingly round total of one trillion tonnes of phytomass (450 billion tonnes of carbon) thus appears to be the best global value to be offered—until most of the storage means for the major ecosystems are shown to be in considerable error (a rather unlikely possibility) and until more reliable land-cover data enable a thorough revision of areal estimates. Net primary production (that is total photosynthesis diminished by plant respiration) of this phytomass is not much higher than 100 billion dry tonnes a year and the addition of ocean photosynthesis would increase the global total by between one-half to two-thirds to about 160 billion tonnes annually. In contrast, ocean phytomass adds up to no more than half a per cent of the terrestrial total and can be easily ignored.

Photosynthesis and respiration cycle CO_2 between biomass and atmosphere, two carbon reservoirs of the same magnitude, 10^{11} tonnes. Atmospheric concentrations of all water-soluble gases must, of course, be in long-term equilibrium with their ocean levels and as the world seas contain more than 28×10^{12} t of carbon they should be able to buffer easily any conceivable rise in airborne CO_2. Unfortunately, the layered structure of the ocean causes an important mixing delay, so brief in geological terms but so critical for the fate of anthropogenic carbon and its effect on global environment.

Only the ocean's topmost layer, less than 2 per cent of all sea water, which is well mixed by winds and waves and warmed by daily heating, is in diffusion equilibrium with the atmosphere, while the equilibration with deep ocean layers cannot be accomplished in less than the several hundreds to a couple of thousand years needed for exchange between the two ocean compartments. The principal reaction $(CO_2 + CaCO_3 + H_2O \rightleftarrows 2HCO_3^- + Ca^{2+})$ will add bicarbonate and calcium ions, the deep ocean will become more acid and some of its sediments, containing some 20×10^{15} t of carbon, will slowly dissolve as the downward mixing gradually neutralizes all CO_2 newly added to the atmosphere.

Broecker and Takahashi estimated that the deep ocean calcite made available by benthic organisms stirring the marine sediments is just about sufficient to neutralize all CO_2 released by combustion of the world's known resources of fossil fuels. In reality, only a fraction of these resources will be eventually burned so there is no doubt about the ocean's capacity to accommodate any conceivable anthropogenic CO_2 releases—but only after the inevitable mixing delays and possibility of doubling, or even tripling of atmospheric CO_2 concentration.

Although the focus has been so much on the combustion of fossil fuels, there are two other major ways in which mankind has been putting CO_2 into the atmosphere: combustion of biomass and accelerated and expanded respiration owing to land-cover changes. The last process is certainly most difficult to quantify but, once again, it is satisfying to note that a simple derivation of the most likely range I will offer here is in a very close agreement with Olson's detailed account.

The difficulties of this task are compounded by the fact that inevitable errors in estimates of annual conversions are magnified by uncertainties concerning the degree of biomass loss. I think that somewhat conservative values are preferable in the case where it matters most, for losses of tropical forests: forest conversion does not always mean a complete loss of vegetation and secondary growth may often follow. Consequently, I will assume that 10 000 km² of tropical rain-forest are now completely lost each year, 100 000 km² are turned into woodlands and grasslands (storing no more than 25 t/ha rather than the original average of 300 t/ha), and that another 100 000 km² are degraded by a loss of at least one-third of its standing biomass. In extra-tropical areas the conversion of grasslands to productive (often irrigated) cropland had a much smaller effect on the overall carbon storage, with the net phytomass loss amounting to an equivalent of some 12 000 km² of good grasslands a year since the 1950s.

Estimates of desertification effects collated during the intensive research efforts of the 1970s (culminating in a United Nations conference in Nairobi in 1977) put the annual losses at about 32 000 km² of rangelands and 25 000 km² of rainfed and 1250 km² of irrigated farmland. Losses caused by new settlements, industrialization, and construction of energy and transportation projects have been estimated to be as little as 10 000 and as much as 80 000 km²; I will assume, in addition to the already listed 10 000 km² in the tropics, a 40 000 km² loss, two-thirds of it being cultivated land or grasslands, the rest forests. Finally, for vegetation losses owing to soil erosion I will use Gordon Wolman's estimate of 30 000 km² (with two-thirds of it coming from crop-fields and grassland) and another 10 000 km² will be eliminated by waterlogging, salinization, alkalinization, and toxification.

All of these estimates of areal and phytomass declines are assembled in Table 5.2

Table 5.2 Estimates of global annual decline of phytomass around 1980 (all mass values are rounded to the nearest five)

	Areas affected (10³ha)	Standing phytomass decline (t/ha)		Total phytomass loss (10⁶t)
Tropical deforestation				4 545
Development	1 000	300 →	5	295
Destruction	10 000	300 →	75	2 250
Degradation	10 000	300 →	100	2 000
Desertification				65
Grasslands	3 200	20 →	5	50
Fields	2 800	10 →	5	15
Development				410
Grasslands, fields	2 500	20 →	5	40
Forests	1 500	250 →	5	370
Erosion				275
Grasslands, fields	2 000	20 →	5	30
Forests	1 000	250 →	5	245
Farmland extension	2 500	20 →	10	25
All other losses	1 000	20 →	5	15
Total	37 500			5 335

and their aggregates add up to an annual loss of about 5.5 billion tonnes of plant mass, or 2.5 billion tonnes of carbon, roughly an equivalent of half a per cent of all terrestrial living matter stored in plants. As is clearly seen, the processes of tropical deforestation and degradation, the actual extent of which we are least certain, dominate the account and even small shifts in basic assumptions may thus bring considerable overall changes. Putting the maximum at no more than 10 per cent above the calculated total of 5.5 billion tonnes, and the minimum at half of the top rate, the most likely range of 3–6 billion tonnes must be adjusted for phytomass gains owing to successional regrowth and to afforestation.

These gains have been largest in rich countries of the northern temperate belt where the abandonment of low-productivity farmland and extension of forest preserves led to increases of forested areas and woodlands by about 500 000 km² between 1950 and 1980. Doubling this value to account for regrowth and reforestation throughout the rest of the world, and assuming average current storage in these areas at 150 t/ha, results in about half a billion tonnes of new phytomass a year for a net annual loss of 2.5–5.5 billion tonnes (roughly 1.1–2.5 billion tonnes of C).

Olson's account also includes releases of organic carbon from land-cover shifts and from burning of phytomass during land clearing on one hand, and higher storages owing to regrowth and afforestation on the other, and his most likely range is a net loss between 0.5–2.0 billion tonnes of carbon a year, while the great uncertainties in burning and regrowth estimates make him caution that the net losses may be as high as 4 billion tonnes C—or that there may be an actual gain of up to 2 billion tonnes C.

Our two most likely estimates thus overlap to a very high degree, and as the mean of nearly twenty other estimates published during the rate 1970s and the early 1980s is almost exactly 2 billion tonnes C, the best conclusion, not to be revised soon, is that the human conversions of ecosystems add annually one or two billion tonnes of carbon to the atmosphere. The disagreements over the biosphere's source or sink role in the carbon cycle which arose during the 1970s have been thus, at least temporarily, reduced to arguments about the actual value of what appears to be a major source of atmospheric carbon.

Naturally, it is vastly more difficult to reconstruct ecosystem transformations since the middle of the nineteenth century when large-scale conversions of forests and grasslands in pioneering agricultural efforts in North and Latin America, Asia, and Australia started to change vegetation cover at faster rates than at any previous time during recorded history. Relatively the easiest task is to come up with the totals for overall increases of farmland but the disaggregation of these totals according to original ecosystems, and values for initial carbon content, are largely a matter of guesswork.

As with the current biospheric carbon fluxes, there have been numerous estimates of CO_2 releases following the conversion of natural ecosystems to cropland, covering mostly the period from 1850–60 to 1950–78 and ranging from no more than 40 billion to almost 200 billion tonnes of carbon. The average of all published 1860–1980 estimates is 125 billion tonnes C, or 1.13 billion tonnes C a year. A simple check of the order of magnitude by assuming a post-1850 replacement of half of the world's closed forests (some 2.5 billion hectares) by farmlands, grasslands, and urban areas with original carbon content going down from about 125 to some 50 tonnes C/ha, results in a total of almost 190 billion tonnes C, or 1.4 billion tonnes a year, a fine agreement with the just cited elaborately derived values and yet another proof that under uncertain conditions a simple calculation works as well as complex constructs.

During the discussion of biomass energies in the poor world (section 2.2) I noted that we may never get satisfactorily accurate nationwide and global totals, a circumstance making worldwide CO_2 fluxes from combustion of wood and crop residues very uncertain. The best estimate I can offer is that at least 2.75 billion m^3 of roundwood equivalent are burned annually in the poor countries and with 150 million tonnes of fuelwood consumed in the rich nations (a mass smaller than the uncertainty in setting out the poor world's total), a global total of 1.9 billion tonnes of wood would release, upon complete oxidation, about 850 million tonnes C. Combustion of crop residues and dried dung would add almost another 300 million tonnes for a grand total of just over 1.1 billion tonnes C.

Whatever the uncertainties, the CO_2 flux for biomass fuels can be seen with

much more confidence than the best estimates of carbon releases from fires set to
clear the forest vegetation by shifting cultivators in the tropics, and the burning of
old grassland growth by tropical and subtropical pastoralists and straw by
temperate latitude farmers. My total estimate for these processes comes to about
2 billion tonnes C a year, so that the grand total of anthropogenic biomass combustion, including all fuel uses and on-site burning, is about 3 billion tonnes of carbon
a year, once more an excellent agreement with the differently derived Olson value
of 3.15 billion tonnes. Numerous uncertainties inherent in all of these estimates
mean that the actual annual value may be, most likely, between 2 and 4 billion
tonnes, the higher total being very close to carbon releases from fossil fuel
combustion.

The return of fossil carbon sequestered tens to hundreds of millions of years ago
in coals, oils, and gases is now the single largest anthropogenic contribution to
carbon's biogeochemical cycle and the one which, compared to biomass carbon
fluxes, can be accounted for with a satisfactory degree of accuracy. Fuel consumption statistics may not be perfect but the regularly published United Nations
compilations provide data which are reasonably consistent, disaggregated and, as far
as the world's largest fossil fuel consumers are concerned, sufficiently accurate for
reliable calculations of CO_2 emissions.

Differences in published global estimates thus arise largely from the choice of
carbon multipliers, an especially complicated matter in the case of coals whose
heterogeneity (from low quality lignites to outstanding anthracites), changing
quality (surface mining, now a preferred method of new extraction, has been
extracting more of the poorer bituminous and brown coals with lower carbon
content), and differences in processing (China, the world's largest producer, reports
its output in raw terms) require some care in choosing a mean carbon share.

By making the necessary adjustments for all major coal consumers I have found
out that the UNO global totals for the 1970s and the early 1980s need reduction of
about 10 per cent to get the most realistic hard coal equivalent (2.34 billion tonnes
in 1975, 2.6 billion tonnes in 1981) which, after reduction by about 5 per cent to
account for cleaning, handling, and storage losses and for a part of carbon not
oxidized to CO_2, should be multiplied by an emission factor of 0.71 to give the
global carbon flux from coal combustion.

Crude oils are much more homogeneous fuels than coals and so the only
corrections needed are the subtraction of transportation and storage losses and
petrochemical and other non-fuel uses (lubricants, paving materials); multiplication of the consumed mass by 0.835 (average C content) will give a very accurate
CO_2 emission total from the combustion of refined crude oil products. Natural gas
is the most homogeneous fuel of the three large sources of fossilized biomass and an
assumption of average carbon content of $0.53 \ kg/m^3$ will result in totals very close
to actual CO_2 releases. The combustion total must, however, be enlarged by CO_2
generated during the flaring of natural gas and Ralph Rotty's detailed studies show
that this flux would enlarge the global aggregate by less than 15 per cent in the early
1980s.

Historical statistics of fossil fuel consumption are acceptable only since the early 1920s; the global series going back to 1860 was derived by the UNO chiefly on the basis of reported United States production, a procedure open to major errors. Still, this series, together with reliable post-World War II data forms the most often used basis of historical CO_2 emission estimates. Fortunately, during the most error-prone period between 1860 and 1914 the estimated CO_2 releases from fossil fuel combustion totalled just about 20 billion tonnes C, a mass generated lately in just four years, so that the earlier under- or over-estimates cannot alter too drastically the global 1860–1980 aggregate of some 160 billion tonnes C, half of which has been released since 1960. With no more than 10 per cent error both ways, the cumulative mass of fossil fuel generated CO_2 since the onset of large-scale industrialization thus falls almost certainly between 150 and 180 billion tonnes C.

The only two other sources of anthropogenic CO_2 are cement production, now amounting to just over 100 million tonnes C a year, and the combustion of solid waste, difficult to quantify but most likely of the same magnitude. A complete disaggregated summary of all anthropogenic sources is presented in Table 5.3: the early 1980s total amounts to nearly 8 billion tonnes of carbon, with fossil fuel combustion releasing almost 5 billion tonnes, or some three-fifths of the total, and the surprisingly large remainder coming from biomass fuels (about one-eighth) and from on-site burning of trees, shrubs, and grasses (roughly one-quarter).

Table 5.3 Anthropogenic releases of carbon in the early 1980s

	Total production (10^9)	Total combustion (10^9t)	Carbon emission factor	Total carbon generation (10^9t)
Fossil fuel combustion	—	—	—	4.58
Coal	2.6	2.47	0.71	1.75
Liquid fuels	2.6	2.31	0.84	1.94
Natural gas (10^{12}m^3)	1.5	1.40	0.56	0.78
Natural gas flaring	—	0.21	0.56	0.11
Biomass fuel combustion		2.50	0.40	1.00
Fuelwood	1.9	1.90	0.40	0.20
Crop residues	2.3	0.50	0.40	0.20
Dried dung	—	0.10	0.40	0.04
Crop residue field burning	2.3	0.20	0.33	0.07
Grassland burning	—	4.80	0.33	1.60
Burning for shifting cultivation	—	1.00	0.33	0.33
Cement production	0.8	—	0.14	0.11
Solid waste combustion	1.0	0.50	0.19	0.09

Adding just a little over one billion tonnes C a year for vegetation and soil carbon losses from the conversion of natural ecosystems (including drainage of increasing areas of carbon-rich wetlands, formerly a huge sink and now in many places a large net source of CO_2), brings the grand total of carbon entering the atmosphere as a result of human interference to some 9 billion tonnes, an equivalent of about one-tenth of the flux involved in the photosynthesis-respiration exchange. But such a comparison clearly exaggerates the scale of anthropogenic 'enrichment' of the carbon cycle.

Carbon releases from combustion and decomposition of the biomass differ from those coming from the burning of fossil fuels not only because of their sources (being of very recent origin or just a few hundred years old, unlike fossil carbon) but also owing to their effects on the global cycling of the element. Much of the grassland burning for pastures management (the most uncertain figure in Table 5.3), burning of forests by shifting famers, as well as field burning of crop residues and household combustion of biomass fuels, has been done for millenia and can, in any case, be thought of as nothing else than an accelerated respiration: herbivory and decomposition would recycle that carbon a bit later anyway.

Appraisal of the post–1850 build–up of atmospheric CO_2 should be thus limited only to that part of biomass-derived carbon emissions which have been in excess of the 'normal' photosynthesis-respiration cycling, with anthropogenic 'accelerated' respiration being part of this normality. In practice it is extremely difficult to make any reliable estimate of these excess fluxes as we have no historical comparative bases regarding household and on-site combustion rates. The only sensible way is thus to limit the assessment of the CO_2 releases which have been obviously above the 'normal' pre-1850 carbon fluxes: the combustion of fossil fuels and carbon losses from conversion of natural ecosystems to cropland, urban, industrial, and transportation uses.

The already cited cumulative estimates show a rather surprising division of the two principal contributions. The most likely values for 1850–1980 carbon releases from fossil fuel combustion (150–180 billion tonnes) and from ecosystemic changes (140–90 billion tonnes) are spanning virtually identical ranges although the combinations of extreme estimates (a legitimate undertaking considering the large uncertainties of such CO_2 losses from biomass which may be as low as 100 billion tonnes) mean that each of the two sources may be considered the more important contributor!

The grand total of 'excessive' anthropogenic carbon generation since the middle of the nineteenth century falls thus most likely between 250 and 370 billion tonnes, with 300 billion tonnes being a good, single, rounded representative figure. This total represents almost exactly two-fifths of all carbon currently present in the atmosphere, and as an increase of 1.0 ppm of atmospheric CO_2 requires the addition of 2.13 billion tonnes of carbon, retention of all of this anthropogenic carbon in the air would have raised the mean background level by about 140 ppm, or by nearly 1.1 ppm a year.

Nothing like this had actually happened as a large part of the added CO_2 has been moved out from the atmosphere to the ocean and also to terrestrial sediments. The question of how large is more difficult to answer because of our merely approximate knowledge of pre-industrial concentrations of atmospheric CO_2. All pre-1870 values are dubious and most longer-term observations for the years 1870–1900 come from Western Europe, an area where fuel combustion and cyclical CO_2 releases by vegetation are not conducive to measurements of background CO_2 levels.

About sixty observations taken during the years 1882–3 are available for the South Atlantic (between 42 °–50 °S) and they, as would be expected, are 15–30 ppm lower than those recorded in Europe at the same time. The best possible reconstruction would put the background mean at 270 ppm in 1850, and 290 ppm in the year 1900. Later unsystematic measurements show concentrations rising to 310–20 ppm in the 1930s and 322–8 ppm by the mid-1950s. Finally, in the late 1950s systematic monitoring of background CO_2 levels started as a part of the International Geophysical Year.

The longest reliable record is that at the Mauna Loa Observatory on Hawaii, a site unparalleled for its near perfection for the background monitoring of atmospheric constituents: with the nearest continental shore more than 4000 km away, and with frequent thermal inversion at about 2 km above sea level, the observatory, 3600 m above the Hilo harbour, is exposed every evening to very clean air subsiding onto it from the middle reaches of the Mid-Pacific troposphere. South Pole measurements started also in 1957 and in the early 1970s two more stations were added to complete the United States National Oceanic and Atmospheric Adminsitration's network of Geophysical Monitoring for Climatic Change—Barrow, 1 km south of Alaska's northernmost tip, and Cape Matatula at the extreme northeast of Tutuila Island in American Samoa.

Consequently, only since 1957 have we had an accurate record of tropospheric CO_2 concentrations from at least two locations perfectly suited for background monitoring, and only since the mid-1970s can we compare the changes at the four stations representing polar and tropical atmospheric backgrounds of both hemispheres. The measurements, which have not been gathered without collection and analytical complications and errors, show a rise from 315.58 ppm in March, 1958 to 340.7 ppm in December, 1982 at Mauna Loa, and from 311.23 ppm in June, 1957 to 339.2 ppm in December, 1982 at the South Pole. Comparing the same months' values or annual averages is necessary as the biospheric breath causes rather large seasonal fluctuations.

As shown in Fig. 5.3, the two northern hemisphere stations are very strongly influenced by winter–summer respiration–photosynthesis fluctuations (which are just longer-wave replicates of daily oscillations with regular afternoon CO_2 minima and late night maxima). Seasonal bursts of life and sweeps of death in distant terrestrial ecosystems are strongly etched in the secular record while the much less significant amplitudes in the southern hemisphere are explainable by its predominantly oceanic character. Figure 5.4 charts this latitudinal dependence of the

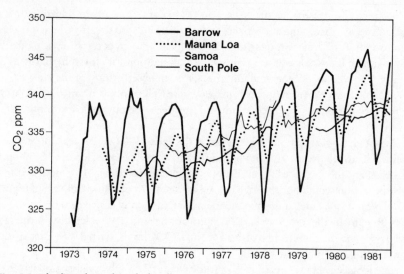

Fig. 5.3 The biospheric breath, hardly noticeable in measurements at the two Southern hemisphere stations (South Pole and Samoa), is distinct at Mauna Loa and most impressive at Barrow, Alaska as the tropospheric CO_2 levels become very low at the end of summer (maximum of the photosynthetic carbon storage) and peak at about 15 ppm higher in the spring.

seasonal CO_2 flask sampling programme which now includes 22 stations at remote locations dotted around the earth.

Calculations of annual averages or running means show steadily rising background CO_2 levels at all stations, a trend easily discernible from even a glance at Fig. 5.3, but annual increases are surprisingly irregular. For example, average increment at Mauna Loa has been about 1.1 ppm but the 1972-3 jump surpassed 2.0 ppm, setting the record, and it was followed by the next year by a rise of mere 0.58 ppm, the lowest annual gain in the series. Figure 5.5 charts these annual increments together with atmospheric CO_2 increases which would have resulted from complete retention of all CO_2 released from fossil fuel combustion, flaring of natural gas and cement production.

The discrepancies between the two means are large, unpredictable—and as yet unexplained. Deforestation, biomass combustion, and forest and grassland fires may be responsible for a part of annual CO_2 retention fluctuations but hardly for most of the erratic ups and downs. Complex and far from understood atmospheric, oceanic, and biospheric responses govern the fluctuations of annual CO_2 increases and all we can do is to note the average long-term retention rates.

Starting with the best known quantities, Mauna Loa CO_2 levels rose by 21.77 ppm between 1960 and 1980 while during the same time CO_2 emissions from fossil fuel combustion, gas flaring, and cement production amounted to an

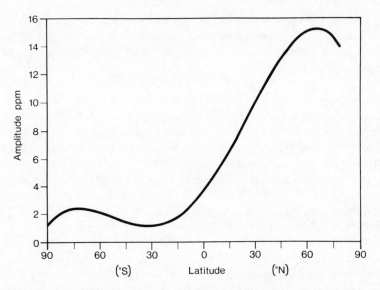

Fig. 5.4 Another illustration of the biospheric breath being barely discernible in the Southern hemisphere and increasing with seasonal vegetation fluctuations in higher northern latitudes. The smooth curve is derived from flask samplings at about two dozen sites belonging to the GMCC programme.

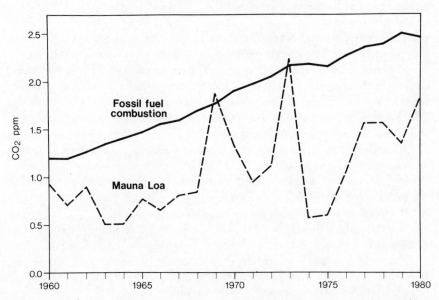

Fig. 5.5 Why are the annual concentration increments at Mauna Loa so irregular (as little as 0.5, as much as 2.0 ppm) when fossil fuel combustion has been adding fairly uniformly to the increasing mass of CO_2? The solid line shows the tropospheric CO_2 increment resulting from complete retention of all emitted CO_2.

equivalent of 37.81 ppm which means that no more than 57.6 per cent of fossil fuel-derived CO_2 remained in the atmosphere. If an average of just 1.5 billion tonnes a year is added to account for excessive CO_2 from ecosystem changes and biomass combustion, the total anthropogenic flux during those two decades would be equivalent to about 51.8 ppm and the actual atmospheric retention rate would drop to some 42 per cent

Taking a longer, and much more uncertain, perspective, if all CO_2 added to the atmosphere since 1850 by fossil combustion and eco-systemic changes had remained aloft, background concentration would have risen, as already noted, by about 140 ppm (with the most likely range between 120 and 170 ppm), so that the 1980 concentrations, assuming 270 ppm in 1850, would be around 410 ppm. Yet the actual 1980 level was 340 ppm so that only one-half (or in extremes as little as two-fifths and as much as nearly three-fifths) of all anthropogenic CO_2 was retained in the atmosphere.

Future increases depend, obviously, on the rates of energy consumption and ecosystem conversion. Upper bounds for CO_2 fluxes from both processes are not difficult to estimate but confident forecasts of actual developments are a different matter. Starting with the best available estimates of ultimately recoverable fossil fuels, Ralph Rotty and Gregg Marland calculated that burning these huge masses of coals, oils, and gases would release about twenty-five times as much CO_2 as the gas aggregate emissions since the mid-nineteenth century. Should half of it stay in the atmosphere, CO_2 concentrations would rise by 970 ppm.

Eventual conversion of all tropical rain-forests with typical phytomass declines from 300 to 100 tonnes/hectare would release some 200 billion tonnes of carbon, an equivalent (with 50 per cent retention) of 47 ppm; their complete destruction would enrich the atmosphere by 70 ppm. The conjunction of the two extremely unlikely developments—exhaustion of all recoverable fossil fuels and elimination of the earth's richest store of biomass—would thus boost atmospheric CO_2 by about 1040 ppm so that the maximum concentration would be approaching 1400 ppm, almost exactly four times the current level.

The highest conceivable CO_2 concentrations resulting from these extreme anthropogenic releases would be thus below 0.15 per cent, much lower than the Phanerozoic maxima which exceeded 0.4 per cent and still somewhat lower than the mean atmospheric CO_2 concentration of the past half billion years. Actual increments, obviously, will be lower but it is certainly possible to think of an eventual tripling of the current CO_2 levels, that is to concentrations of around 1000 ppm. How high and how rapidly the levels will go will be determined above all by the rates of future fossil fuel consumption and by the changes in atmospheric CO_2 retention. Comparisons in the last two paragraphs show clearly that the role of deforestation-derived releases will be only marginal.

Until the early 1980s, growing numbers of CO_2 forecasts put the date of doubling of pre-industrial concentrations, that is levels around 600 ppm, at no later than about 2030, an inevitable result of assuming retention rates of close to 60 per cent and long-term continuation of the 4–5 per cent post-World War II exponential rise of energy consumption. But both of these premises are no longer tenable. Growing appreciation of the rather large magnitude of biotic CO_2 releases means that assumptions of CO_2 atmospheric retention rates must be revised downward by appreciable amounts and the radically altered perception of future energy needs in the rich countries, a change discussed in detail in section 2.5, leads to much more conservative forecasts of fossil fuel consumption.

Plots of doubling times as functions of average retention and annual fossil fuel consumpton increases illustrate best the differences and inherent uncertainties of the forecasts (Fig. 5.6). Standard 1970s' assumptions bring the CO_2 doubling by 2030, typical of the early 1980s (50 per cent retention and 3 per cent global energy growth) push the doubling date to the late 2040s, and the even more conservative, yet still wholly plausible, combination of 40 per cent retention and less than 2 per cent growth postpones the time of 600 ppm close to the year 2100.

William Nordhaus and Garry Yohe's probabilistic scenarios published in 1983 put the 50th percentile in 2065, the 75th in 2050, and the 25th after 2100, expressing well the current consensus of 600 ppm coming most likely sometimes between 2050 and 2100, or roughly a century from now. The magnitude of this rise, as noted, would not be unprecedented in the biosphere's long history but its rate would be unusually rapid and the increase would be the first appreciable CO_2 shift since the emergence of complex civilizations and the introduction of intensive

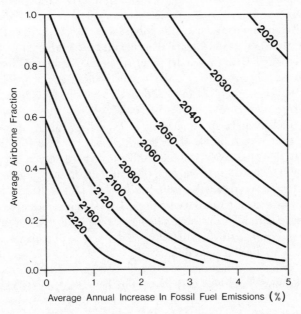

Fig. 5.6 This kind of forecasting frame makes much more sense than rigid statements: combinations of different annual increases in fossil fuel emissions and different atmospheric retention rates result in vastly different times of tropospheric CO_2 doubling.

farming. The consequences of the change must thus be taken seriously and the following two sections will examine the two critical concerns: what would CO_2 doubling do to the global climate, and how would plant productivity fare in the changed circumstances?

5.2.1 CO_2 and climatic change

> Thus human beings are now carrying out a large scale geophysical experiment of a kind that could not have happened in the past nor be reproduced in the future.
>
> Roger Revelle, Hans Suess (1957)

The simplicity with which the consequences of this grand experiment can be served up in the mass media is at the root of the lively public interest in the matter. The story is compelling: the burning of fossil fuels increases atmospheric CO_2 concentration and the gas, behaving like the glass in the greenhouse, lets in short-wave solar radiation but blocks out the heat re-radiated by the earth's surface; planetary warming, a 'greenhouse effect', is thus inevitable and only the writer's restraint and choice of adjectives determines the impact of the ensuing narration.

Cautious notes about possible changes in regional precipitation patterns and effects on crop production are not a favoured way of dealing with the consequences—it is incomparably more captivating to write about the eventual melting of polar ice-caps, the submergence of coastal regions, drastic shifts of climatic zones, catastrophic changes in grain production, widespread alterations of ecosystems, and extinction of species. Indeed, even in many scientific publications the CO_2-induced 'greenhouse effect' is portrayed as the fundamental shaper of the earth's future physique (with big coastal cities and rich farmlands under water, species extinction rivalling those of the Mesozoic time of the great dying, and altered climates)—and hence inevitably the determinator of civilization's fortunes as drastic changes of settlement and farming patterns will usher an era of stunning social upheavals.

Physical fundamentals, appreciated since Tyndall's writings in the 1860s, are convincing: carbon dioxide has three strong absorption bands in the infra-red range, two narrow (one about 3, and the other just above 4 microns), and a broad one between 13 and 18 microns, and this opaqueness to outgoing terrestrial radiation should, everything else being unchanged, lead to higher tropospheric temperatures with rising CO_2 concentrations. The effect is indeed much like that in the greenhouse—although the latter's higher temperatures are not primarily the result of the transparency of glass to UV, and its opacity to IR radiation, but rather a consequence of restricted convection.

The analogy may be functionally wrong but the image is direct, easy to grasp—planetary warming could be a very uncomfortable affair. But such a judgement betrays the bias which shows that the greenhouse simile is inappropriate even as to its final result because the temperature difference between the ambient air and the greenhouse atmosphere can easily surpass 10 °C. Such a rise in mean tropospheric temperature would obviously be most alarming because during the last million years temperatures in the Northern hemisphere have not shown fluctuations greater than ±3 °C around the mean and a relatively rapid (in a matter of a few centuries) rise by 10 °C would cetainly have profound planetary consequences.

Not surprisingly, there has been a keen interest in determining by how much the tropospheric temperatures would rise with increasing CO_2 because without a satisfactory answer to this key question one can hardly engage in sensible speculations about the further biospheric consequences. Efforts to calculate the degree of tropospheric warming attributable to rising CO_2 concentrations started with simple surface energy-balance models in the late 1950s and the early 1960s, progressed to more realistic radiative–convective exercises in the late 1960s, and reached the level of three-dimensional general circulation simulations by the mid-1970s.

By the mid-1980s there were no fewer than two scores of models to choose from but their comparisons do not show a satisfactory consensus. The average of a score of one-dimensional energy-balance models puts the mean global temperature increase resulting from CO_2 doubling at 2 °C with about a 15-fold range between the extremes of 0.25° and 3.8 °C. Extremes of two-dimensional studies range from 0.8 °C to 2.4 °C (mean 1.8 °C), and the most advanced three-dimensional models show an almost 20-fold range (0.2 °–3.9 °) and mean of 2.5 °.

Nevertheless, the three consecutive expert groups appointed by the United States National Academy of Sciences concluded in reports released in 1979, 1982, and 1983 that the best estimate of long-term climatic reponse to doubled CO_2 is an average surface temperature rise of 3° ± 1.5 °C. These conclusions were based almost solely on the results of three-dimensional models prepared by the Geophysical Fluid Dynamics Laboratory and the Goddard Institute of Space Studies—models which the GFDL and GISS researchers who headed and peopled the NAS expert panels found, not surprisingly, most to their liking.

The 'three-degree' consensus thus achieved widespread publicity and as a result it has been credited with much more solidity than it deserves. Several obvious questions have not been satisfactorily answered, some were not even raised, by the members of the CO_2 modelling establishment who endorsed their own products while virtually dismissing all other efforts. Why 3 °C when the mean of all published three-dimensional models is 2.13 °C? Should all the less complex and less interactive models be excluded from consensus-forming? Why have the calculations favouring lower temperatures increases been treated with inordinately vigorous disapproval while atypically high estimates were not attacked in comparable manner?

Following the claims and controversies of this global modelling it becomes hard not to conclude that the 'establishment', represented by the NAS panels, has been biased in favour of higher temperature response to CO_2 doubling, overconfident in the reliability of its computerized simulations, impatient with other points of view, and forgetful of the usually short lifetime which all such categoric 'consensus' positions have in the face of such extremely complex and far-from-understood phenomena. Indeed, it is the lack of humility which has been by far the most surprising aspect of this controversy since the foundations of even the most sophisticated models are so obviously simplified and limited that only a naïve faith can see in these exercises satisfactory reflections of the real world.

The computational domain of some of these models embraces just a small part of the planet (typically a wedge-shaped section of the Northern hemisphere); continents and oceans are assumed to occupy equal areas; oceans are frequently treated merely as expanses of wet land able to supply infinite evaporation (that is without any heat capacity and transport!); cloudiness is modelled in an unrealistically simplistic manner (yet we do not know even whether cloudiness changes will have positive, negative, or neutral feedbacks on future surface temperatures); small shifts in setting out key water-related parameters appear to result in large thermal differences; and for the atmospheric aerosols we cannot be sure even of the sense of their most likely effect.

I find it curious that the American CO_2-modelling establishment has been so rigorous in the defence of its 'three-degree' simulations while its two senior members, Syukuro Manabe and Richard Wetherald, have admitted that nobody should 'take too seriously the quantitative aspect of the results', and when the NAS panel has concluded that our capabilities are still not realistic enough to offer predictions fit for assessments of likely impacts.

Essentially, we are left with no more than a qualitative consensus that further CO_2 increases will lead to tropospheric warming. Other, more tentative qualitative conclusions are that such warming would be accompanied by stratospheric cooling (with rather small latitudinal variation), that the mean global evaporation and precipitation would increase (intensification of water cycle), that, at the surface, polar regions will warm up two to three times as much as the tropics and, finally, that the alterations in the water cycle would include higher circumpolar runoffs, later snowfalls, earlier snowmelts, and lesser extent and thickness of polar sea-ice.

But even cautious qualitative conclusions lose their tenuous standing once one starts looking at the CO_2-induced rise of tropospheric temperature in the only realistic way: as a part of a much broader, and enormously complex process of climatic change. Will its numerous constituents, ranging from luminosity shifts of the sun and changes of the earth's orbital eccentricity, to fluctuations of volcanic activity and variations in tidal stresses (to indicate the range of possible natural causes), and from generation of large masses of aerosols and releases of gases absorbing infrared radiation to surface albedo changes (to mention the most important human interferences besides CO_2) potentiate, neutralize, or negate the effects of higher CO_2 concentrations?

The only honest answer is to plead ignorance. True, the study of climatic change became very fashionable during the 1970s: the public media are inclined to interpret every notable departure from the mean as the onset of a new climatic era and, unfortunately, can draw on ever-ready pontifications of some scientists who prefer to hear the sound of their own voices rather than admit our utter lack of predictive powers—but at a time when a reliable local or regional one-week weather forecast remains an elusive goal it should be obvious that obstacles barring our understanding of the dynamics of global climate remain enormous.

The obstacles start well before the effects are considered. In the first place is the challenge of detecting the CO_2-induced warming, a task requiring good long-term knowledge of mean temperatures and of typical natural fluctuations around these means. Yet our records are both short (about a century for the calculation of meaningful averages) and inadequate (high latitudes and ocean data are poor even for the Northern hemisphere); moreover, as Madden and Ramanathan stress, while we are looking at the recorded temperature changes 'we cannot be certain that such changes have not masked a CO_2 effect, or conversely that they may indicate in the future that there is an observable effect of CO_2 when there is none'.

This discouraging prospect is further complicated by the growing realization of the possible thermal effects of various anthropogenic trace gases. Methane is another common carbon compound affecting the radiation balance: its long-wave opacity (at 7.66 μm) is such an important regulator of surface temperature that its disappearance would, according to Donner's and Ramanathan's calculations, cool the surface by 1–2 W/m^2 while the doubling of current concentration (averaging around 1.7 ppm in the Northern hemisphere) would lead, everything else being equal, to a fairly uniform warming of about 0.61 W/m^2.

The gas is released largely from decomposition and Sheppard's detailed global inventory of sources puts the total annual flux at 1.21 billion tonnes, with 75 per cent coming from phytomass decay, only about 5 per cent from enteric fermentation in animals, and almost 10 per cent from leakage of fossil methane from natural deposits and during distribution. Recent measurements indicate steady increases of atmospheric CH_4 concentration (between 1 to 2 per cent a year) owing largely to continuing expansion of ricefields, larger numbers of farm animals, and leakages from natural gas operations.

Moreover, future planetary warming, no matter if CO_2-driven or otherwise generated, could release huge volumes of the gas now sequestered in clathrates (methane hydrates) in continental slope sediments under cold seas. But there is no reason for intense concern: on the one hand, methane hydrate releases, if they occur, may be only a fraction of the largest published estimates (and rough guesses are all we can offer today): on the other hand, efficient removal of atmospheric CH_4, above all its reaction with hydroxyl radical, may keep the tropospheric concentrations of the gas within easily acceptable limits.

Even more attention has been focused since the mid-1970s on releases of N_2O from denitrification (to be discussed in section 5.3.2) and on chlorofluorocarbons (freons, used as propellants in aerosol cans and in refrigeration; see also section 4.1.2). They absorb infrared radiation in the critical thermal window between 7 and 14 microns, and their continuing atmospheric build-up might easily confuse any expected CO_2-induced changes. Consequently, even should a clear warming signal be detected rising above the noise of the ever-present natural fluctuations, the relative contributions of CO_2 and other anthropogenic radiation-absorbing compounds would still be tricky to apportion—and without a confident ranking of the culprits the choice of control and management strategies would remain questionable.

Carefully constructed hypotheses are thus the best we can offer and there is no more sensible way to do the guesswork on the effects of tropospheric warming than to study analogical states. The choices are many. M. I. Budyko, the leading Russian investigator of climatic change, went all the way to Pliocene to assess the change: if CO_2 doubling were to raise the Northern hemisphere's temperatures to the levels prevailing about five million years ago, then winter surface air in the Arctic could be up to 20 °C warmer and the January 0 °C isotherm would move northward by up to 15° of latitude.

As a result, the northwestern part of European Russia would have climate much like today's central France and rainfall would increase by 200–600 mm a year. Even northern Siberia would mellow considerably and higher rainfalls would improve farming's chances in currently arid Central Asia. Should Budyko's analogy be correct, Russia would clearly benefit from CO_2-induced climatic change. Similar changes can be postulated by selecting a much more recent analogy, the Altithermal period of the Holocene (about 120 000 years ago) when wetter climates prevailed throughout most of Europe and European Russia but also in the Saharan and Eastern Africa, India and Eastern China.

Another approach using climatic analogies was pursued by Wigley, Jones, and Kelly who contrasted the means of the five warmest and five coldest years between 1925 and 1977 by using temperatures of stations between 65° and 80 °N to account for the expected greater warming in higher latitudes. The principal precipitation shifts associated with warmer spells were to a large extent contradictory to the two analogies just outlined—less rain over much of the United States, most of Europe, the USSR and Japan. Useful as the analogies may be, the choice of the most fitting one in cases of contradictory evidence leaves us again only with guesses and opinions.

If the best scenarios of altered temperature and precipitation patterns are decidedly non-uniorm, what consensus can be expected in regard to the 'ultimate threat' arising from sustained CO_2-induced warming—melting of polar ice? Not surprisingly, contradictory opinions are also the norm. For example, while Hermann Flohn argues that an average temperature increase of 4°–5 °C would be sufficient to intitiate the melting of relatively thin Arctic sea-ice, and only a few decades would be needed to dispose of the whole Arctic ice-cap, others point to the inappropriateness of his paleoclimatic analogies, cite the delaying effect of increased snow cover (which must be expected with higher Arctic temperatures) on summer ice melting, and refer to sedimentary records of the past two million years which show the existence of Arctic pack-ice cover during many glacial–interglacial climatic variations.

In any case, the melting of Arctic sea ice would not trigger a new deluge, even if Greenland's glaciers also started receding the sea-level rise would be negligible, and even the wasting of the island's huge ice pack would make little difference during the first few hundred years of a long-drawn process lasting for millenia. Only when the Antarctic ice started disappearing rapidly would the global worries of flooding be justified.

But, as in the Arctic, the bulk of that ice is in the thick land-grounded ice sheet (in eastern Antarctica) which could not be melted rapidly even with extraordinarily rapid and unusually high temperature increases (such as 5 °C rise in just a century). In contrast, the thinner marine ice sheet of western Antarctica could waste much faster (weakened by warmer waters, rifting, and faster calving)—in a few hundred (rather than many thousand) years with sustained major warming. The worst case scenario associated with 3°-5 °C warming would thus see the disappearance of the west Antarctic ice sheet over a period of several centuries and a global sea level rise of 4 to 5 m.

This rise would not be unusual in biospheric terms—it would be less than that occurring during the interglacial spell 120 000 years ago—but its global human and economic implications would be obviously quite considerable, while its local and regional effects could be devastating. By far the largest population displacements would occur in Asia (in the deltas of the Ganges, Irrawaddy, Mekong, and Chang Jiang) but the highest material losses would be borne along European and North American shores. Still, Stephen Schneider and Robert Chen's detailed study for the United States indicates that a 4.6 m (15 feet) rise would flood only 1.5 per cent of the 48 continental states, displace less than 6 per cent of population, and affect about 6 per cent of the immovable assets.

As this change would be gradual, overall damage could be greatly minimized. In some areas no new investment would be allowed in preparation for their eventual surrender to the sea; in others protective dykes could easily save all immovable assets. Just to show the manageable level of the economic impact associated with such flooding I have doubled Schneider and Chen's loss estimate of $(1980) 107.5 billion arising from a 4.6 m rise and spread its cost over a century: the cost, a little over $2 billion a year, is then just about the same as the current milk price support payments transferred each year from the government budget to the nation's dairy farmers, and two orders of magnitude below defence expenditures!

Clearly, even if CO_2-induced warming were of such magnitude that it precipitated relatively very rapid melting of the west Antarctic ice the resulting gradual sea rise would prove to be a manageable, though undoubtedly locally and regionally costly and unwelcome challenge. But this transformation may never come to pass and it is patently irrational to worry about its impact at a time when we have yet to detect a clear signal of any CO_2-driven warming and when we are not at all certain what temperature increase may result from an eventual doubling of CO_2 levels (whose timing we cannot also pinpoint).

There would seem to be only one major concern associated with CO_2 build-up worthy of a serious close inquiry: the fate of plant-life. If the build-up were to lead to higher average tropospheric temperatures there would be notable ecosystemic changes ranging from extinction of many sensitive species to eventually large zonal shifts of biome boundaries. Farming would be also inevitably affected as some grain-growing regions would receive less, others more precipitation. As previously noted, both the magnitude and spatial distribution of these thermal and precipitation changes is quite uncertain but plant growth would be affected by higher CO_2 levels even if there were no climatic response, and it is this productivity effect which needs special attention.

5.2.2 Crops and forests in the greenhouse

> In the seven days in which the mint was growing in this jar of noxious air, three old shoots had extended themselves about three inches, and several new ones had made their appearance in the same time. Dr. Franklin and Sir John Pringle happened to be with me when the plant had been three or four days in this state, and took notice of its vigorous vegetation and remarkably healthy appearance in that confinement . . .
>
> Joseph Priestley, *Experiments and Observations on*
> *Different Kinds of Air* (1790)

What Priestley and his learned visitors noted—mint growing vigorously in a bell-jar in which a mouse had previously died—is the first recorded instance of the beneficial effects of higher carbon dioxide concentrations on the rates of photosynthesis. They could not formalize the causes and effects of that happening but concluded correctly that the air made 'noxious' by animal putrefaction is 'purified' again by plants. Noxious air is, of course, CO_2 and the purification is photosynthesis.

Joseph Priestley and Benjamin Franklin, who in one of the letters to his English colleague remarked that the restoration of air by 'vegetable creation . . . looks like a rational system, and seems to be of a piece with the rest', might be then rather puzzled by the recent scares about the fate of the world in which the plants would be growing in atmosphere richer in carbon dioxide. Will not this planetary bell-jar, filled with 'noxious' CO_2 from man-made combustion and land cover changes, bring more vigorous vegetation of remarkably healthy appearance—and hence bigger crops and healthier forests? In this section I shall go over the best recent evidence to assess how far conclusions from experiments with dead mice and mint sprigs are transferable to agro-ecosystems and natural forests.

During the nearly two centuries since Priestley's trials a large amount of empirical evidence gathered from work with numerous crops has confirmed beyond any doubt that in controlled conditions higher CO_2 concentrations will raise photosynthesis rates, enlarge leaf area, increase branch, fruit and seed numbers, improve germination, cause more profuse flowering, accelerate maturity, improve tolerance of some common atmospheric pollutants and, perhaps most notably, will lower transpiratory water losses. Greenhouse operators have been taking advantage of these benefits for decades and many vegetable crops are grown in atmospheres artificially enriched with CO_2.

Crop fields are not greenhouses: adequate moisture and nutrient supplies, and disease, weed, and pest controls are much harder to provide but modern crops are often well irrigated and sufficiently fertilized and do receive efficient pest and weed control so that when the whole atmosphere has turned into a CO_2 greenhouse the beneficial effects would have to come at least to modern intensive cropping; besides, higher water-use efficiency should benefit even the dryland crops. Indeed, when International Conference on Rising Atmospheric Carbon Dioxide and Plant Productivity met in Athens, Georgia in May 1982 its participants concluded that a world with 700 ppm of CO_2 would bring many desirable changes to global farming.

Although impossible to quantify on the basis of our current knowledge, there should be general improvement in drought resistance, better tolerance of air pollution, soil and water alkalinity, and high temperatures as well as significant increases in symbiotic fixation in legumes. Tuberization of potatoes may rise several-fold and, getting onto more reliably quantifiable ground, water-use efficiency (dry phytomass production per unit of transpired water) may double and the total biomass may increase by at least 20 and as much as 45 per cent in the case of C_3 plants, although much lower improvements would be expected for C_4 species. But this differential reponse should be actually most welcome because of the 15 principal crops 12 have C_3 metabolism while 14 out of 18 most obnoxious weeds are C_4 plants and high CO_2 concentration would act as an efficient weed suppressant.

Most experiments measuring actual field-crop yield increases in responses to higher CO_2 levels were done with only brief exposures which cannot be used for reliable long-term extrapolations, but out of 437 studies published up to 1982 Kimball was able to locate 81 experiments for 24 different crops (and 14 non-crop species) in which carefully measured CO_2 enrichments were applied during the daylight hours for the duration of the plants' entire growth. Average yield increases for these experiments best approximating actual field conditions were 33 per cent for doubling of current CO_2 levels, and 67 per cent for tripling of ambient concentrations.

Moreover, the results of 46 observations for 18 species where CO_2-induced transpiration changes were also noted show that CO_2 doubling cut water losses by an average of 34 per cent (from 10 per cent in wheat and barley to over 60 per cent in sorghum) and this improvement, combined with a 33 per cent boost in productivity, would push average water efficiency up by two-thirds and large arid areas could be opened to staple crop cultivation.

Sherwood Idso calls these findings 'truly amazing, and the phenomenon they portray could well prove to be a godsend in the days and years ahead' and foresees 'mind-boggling' rewards as the 'yields of plants the world over could actually double, with water usage being cut to but a fraction of what it is today'.

With deserts blooming, Idso sees the planetary greenhouse not as a disaster but as change which will take 'us back to the Eden which we left so long ago'. While I greatly approve of Idso's critique of the clearly dubious 'establishmentarian' consensus about the degree of CO_2-induced tropospheric warming, I cannot share his unrestrained enthusiasm about 'mind-boggling' plant growth. 'Days' may be easily dismissed as a rhetorical phrase but I have tried to show that the doubling of the pre-industrial CO_2 level is still generations away so the years of benefits, great or small, are far beyond any sensible planning horizon. More importantly, Idso ignored much evidence indicating that, when CO_2 doubling should come, the productivity response would fall far short of re-establishing lost Eden.

First and foremost, transferring the results of controlled small-scale, one-season experiments to actual long-term performance of field crops on milions of ha planted with numerous cultivars in various environments is a perilous exercise. Our rich experience of the huge contrasts between responses to nitrogen fertilizer on test plots and average field yields is a good analogy of how much the expectations must be scaled-down for actual everyday performance. More importantly, there are the already noted interspecific differences in response to higher CO_2 levels, with indeterminate C_3 plants doing much better than determinate and C_4 species, a fact which makes simple averages for higher global crop productivity rather misleading.

For example, soybeans, indeterminate and C_3, respond the best of all major field crops but their cultivated area is now only about two-fifths of that of corn, determinate C_4 species, whose productivity rises only very little with higher CO_2. Statistically significant means for individual species would have to be used with appropriate crop areas to determine meaningful weighted averages of CO_2-induced yield increases—but even so the estimates would apply only in the absence of common environmental stresses, that is only for well-watered, properly fertilized crops devoid of major disease and pest infestation.

In all those instances with a suboptimal supply of water and nutrients the CO_2 stimulation would still be felt, especially as higher water-use efficiency could enable cropping in previously unsuitable locations, but overall gains would be inevitably much lower. The best conclusions to be made today, in the absence of actual routine field experience, is that in the absence of stresses CO_2 doubling would boost the yields of common staple crops by about one-third, while in stressed environments the responses would vary widely, from doubling of yields where higher CO_2 would ease water shortages and salinity of soils to actual harvest declines with acute aggravation of nutrient shortages.

No matter how high the eventual crop yield increases accompanying CO_2 doubling were, any improvement in global staple food harvests would be welcome but its effects on worldwide phytomass storage would be negligible (only the additional growth of permanent woody crops would sequester more carbon for years or decades). Only when the growth of tropical, temperate, and boreal forests was stimulated by higher CO_2 levels would vegetation act as an enlarged sink of atmospheric carbon. But our predictions about responses of global forests to doubled CO_2 are nothing else but informed speculations, and the only qualitative certainties we have today are those about the differential response of principal species leading to structural shifts in ecosystems—and these changes may or may not translate into substantial net productivity and storage gains.

The summary of the Athens meeting on CO_2-induced increases in plant productivity, offered by Lemon, is perhaps the fairest assessment of the outlook: '. . . it becomes clear that after the initial metabolic uptake of CO_2 in photosynthesis, there is a hierarchy of increasingly complex processes controlling the production and allocation of end-products ... with increasing complexity, the initial advantages of more CO_2 in the photosynthesis process will be increasingly buffered.' Welcome yield increases in well-managed croplands, higher water efficiencies in arid regions, some additional storage in forests, all these changes are almost certain to accompany CO_2 doubling—but they will not add up either to a stunning transformation of the biosphere or to a huge sink of atmospheric carbon.

And what will be the plant response should the tropospheric temperatures rise as much as predicted by the CO_2 modellers' consensus of the early 1980s? From an evolutionary perspective such a change would be hardly upsetting as for more than nine-tenths of the time since the emergence of higher life forms, some 500 million years ago, the earth's surface temperatures have been appreciably warmer, and CO_2 levels have been much higher than during the past several thousand years of recorded history and spreading agriculture. We still live in what is in biospheric perspective an anomalously cold era and there is nothing disquieting in contemplating a world with 600 ppm of CO_2 in the atmosphere and with a few °C higher mean temperatures.

But could not one argue that the crops, domesticated only during the past 3000–10 000 years would respond differently than tropical rain forests which have been around for 10^7 years and that the world food prospect might be jeopardized by 3 °C $\pm$ 1.5 °C warming? Temperature and harvest records of the last century do not support this view. Surface temperature anomalies for the Northern hemisphere during the years 1881–1980 display an annual range of more than 1 °C while ranges for spring and fall, the critical times of planting and harvesting, are 1.4°–1.5 °C, which means that in those hundred years our climate has gone, in terms of extremes, through an equivalent of warming now forecast by standard models as a lower limit brought by CO_2 doubling.

As always, yields have continued to fluctuate with environmental stresses but the secular rise of staple food harvests has been unmistakable ever since improved seed, higher fertilization, and better farming practices started to change traditional agriculture. As shown in Fig. 4.7 (charting the yields of American corn), this trend transected periods of both relatively rapid warming (1915–40) and cooling (1940–70) and it has been especially strong in the northernmost, intensively cultivated areas, in North America and Europe. Moreover, principal cereal crops are grown in such a wide variety of climates that it would be no problem to adopt suitable cultivars for gradually changing conditions: wheat is grown from Kenya to Sweden, corn is a major crop in tropical Brazil as well as in cold North China, even rice is grown from the equator to nearly 50 °N.

Moreover, agronomists have been painstakingly extending cultivation of major cereal crops into colder and drier environments. In North American both corn and wheat have made spectacular poleward gains. Since the 1930s commercial grain corn farming advanced 800 km northward thanks to new, faster-maturing hybrids and corn is now a major crop as far north as the Red River Valley of Southern Manitoba where a farmer can rely on no more than ninety frost-free days in the coldest years. Red winter wheat cultivation expanded even more, its northernmost edge moving from Kansas and Southern Nebraska in the 1920s to Northern Montana, 1500 km to the northwest, by 1980.

Such rapid diffusions involved temperature and rainfall changes as great, or even greater, than those predicted by the consensus CO_2-warming models. Should the more northerly latitude experience greater CO_2-induced warming than the tropics and subtropics then these higher temperatures would only help in extending cultivation of winter wheat and corn northward and higher precipitation resulting from the warming would further help the cultivation. And even should the rainfall decline in some regions, higher water-use efficiency, resulting both from higher CO_2 levels and, as shown with the example of the Great Plains wheat farming (section 3.3.1), from better agronomic practices, may more than compensate for this change.

Should the CO_2-induced climatic change materialize during the next century there is little doubt that some of the world's major food producing areas would abundantly benefit from the combination of higher CO_2 levels, higher temperatures, and more precipitation. In others, the effects of higher temperatures and lower precipitation may be counterbalanced by higher water-use efficiency with little overall productivity change. And in all those cases where the new combination would spell harvest declines with continuation of old cultivars and old agronomic practices, the rich experience of the past 50 years shows that the introduction of new varieties, or a large-scale diffusion of new crops, and adoption of new field management techniques could be made with rapidity far surpassing the rate of environmental changes brought by a gradual warming trend.

The earth's natural ecosystems, above all the tropical and boreal forests comprising the bulk of the world's phytomass, may not become huge sinks of doubled atmospheric CO_2, but the global effect on productivity and storage should be at least mildly positive. In intensively managed agro-ecosystems the gains should be more substantial and the overall outlook for global food production should definitely improve. Obviously, new crop varieties and new farming practices will be needed to accommodate the environmental changes but the record of the past century, when highly productive farming advanced into some very marginal areas (in both the temperature and precipitation senses) makes it possible to view these challenges with confidence in their successful outcome.

Undeniably, there may be need for some relatively drastic farming changes, above all in places which could become appreciably drier, but the total effect of CO_2-induced warming on food harvests—with poleward extension of longer vegetation periods with higher water-use efficiency and with general acceleration of the water cycle—must be much preferable to the agricultural consequences of substantial cooling. With most of the world's grain output coming from the Northern hemisphere's mid-latitudes (35°–50 °N), a southward march of the principal grain-growing regions by, say, 500 km, that is from Central Illinois to Northern Mississippi, or from Northern Holland to Alsace, would entail a shrinking of cultivation zones which it is truly frightening to contemplate.

As with any aspect of complex biogeochemical changes, uncertainties abound and no reliable quantifications of eventual productivity shifts will be possible for decades. The simplest of all indubitable facts is that the present tropospheric levels of CO_2 are much below the optima for crop production. Almost as certain is the assertion that increasing CO_2 levels would have an overall beneficial effect on global food production, above all owing to remarkable improvements in water-use efficiency and to greater resilience of crops under such environmental stresses as drought, high temperatures, soil salinity, and toxic compounds in the atmosphere.

Sylvan H. Wittwer believes that the great increases in crop productivity since the early 1950s, which coincided with the substantial rise of atmospheric CO_2, may have been already much influenced by more abundant CO_2. There is no way to separate this effect, if it did really occur, from the contributions of new cultivars, better agronomy, more irrigation, fertilizer, pesticides, and machines but when atmospheric CO_2 levels have increased by another 100 ppm or so we should be able to see the beneficial effect. Doomsayers may continue seeing in more CO_2 another dangerous environmental trend; but no unprejudiced biologist can fear it and all knowledgeable agronomists must welcome it.

5.3 Nitrogen, the strongest link

> Every vital phenomenon is due to some change in a nitrogen compound and indeed in the nitrogen atom of that compound.
>
> A. F. Needham, *The Uniqueness of Biological Materials* (1965)

Nitrogen's essentiality is not surprising considering the element's presence in amino acids, nucleic acids, and enzymes. For heterotrophs the importance is both qualitative and quantitative: while dry phytomass averages only about 0.75 per cent of nitrogen (and except for the high-protein seeds the element's properties matter more than the mass), skeletal muscles giving the higher organisms the freedom to move are composed largely of protein (actin of thin filaments, myosin of the thick ones), which means that about one-seventh of their dry mass will be nitrogen.

Still, the aggregate nitrogen content of mankind—assuming 4.7 billion people, a mean weight of 50 kg and average protein content of 15 per cent of live weight—is just around 5 million tonnes. But no vertebrate can synthesize nitrogen-containing amino-acids from ammonia: they must be ingested preformed as essential dietary proteins (see section 3.1.1), and to maintain and enlarge mankind's small nitrogen pool we have to manipulate and interfere with much larger storages and fluxes of the element. The annual harvest of food and feed crops (excluding forages) contains about 50 million tonnes of nitrogen, and nitrogen in slaughtered animals amounts to almost 5 million tonnes.

However, to gather these harvests we cultivate about 1.5 billion hectares of farmland which usually holds in its top 50 cm (overwhelmingly so in organic matter), 2-10 t of nitrogen per ha. Taking 6 t/ha as a conservative average (deep soils contain more nitrogen in lower horizons) this means that we are interfering with a reservoir containing of the order of 10 billion tonnes of nitrogen. Since we started to cultivate these lands—some of them 10 000 years ago, most of them only since the last century with the opening of the vast North American, Latin American, Asian, and Australian grasslands for grain farming—we have greatly reduced their original nitrogen content and to maintain their fertility we have had to resort to further interventions by recycling organic wastes, by planting leguminous species and, in recent decades, by extensive and more intensive application of synthetic fertilizers.

This interference has become irrevocable because the existing populations

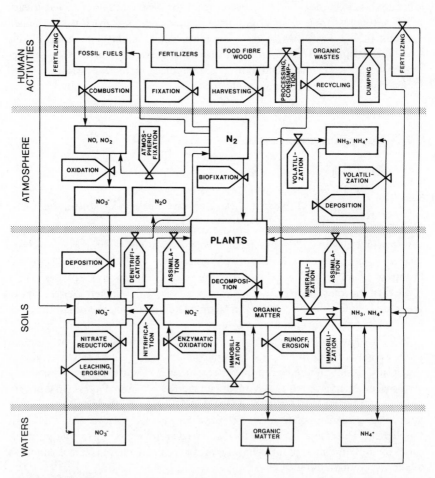

Fig. 5.7 Basic flow diagram of the global nitrogen cycle. Owing to great uncertainties in the size of the reservoirs and magnitude of the fluxes I did not put in single 'best' average estimates.

cannot be supported without highly productive farming, but it unfortunately affects the most complex and the most important segment of the planetary nitrogen cycle. As outlined in Fig. 5.7, nitrogen storages in soils are in four reservoirs connected with each other, with plants, atmosphere, and water by flows made possible largely by microbial actions. Jansson and Persson's division of nitrogen cycling in soils into three sub-cycles is an excellent elucidation of the interdependences governing the processes (Fig. 5.8).

The outermost, elemental, sub-cycle moves nitrogen from the atmosphere (where it forms nearly four-fifths of the mass, virtually all of it in tightly bound and stubbornly unreactive dinitrogen form) to plants, after their decay to soil organic matter (the biggest nitrogen reservoir involved in these exchanges), from which mineralization (reduction to NH_3 by numerous microbial species, aerobic and nonaerobic, which goes on in almost all environments but is accelerated in higher temperatures and in moister, well aerated soils) moves it into the relatively unavailable ammonia pool.

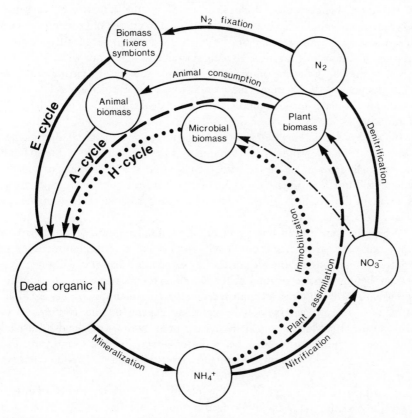

Fig. 5.8 Jansson and Persson's flow diagram of three interdependent subcycles of nitrogen cycling in soils.

Nitrification transfers ammonia nitrogen to nitrate nitrogen (two obligate aerobic bacteria, first *Nitrosomonas*, then *Nitrobacter*, carry the oxidation to NO_2^- and NO_3^-), and denitrification closes the cycle as numerous species of common heterotrophic bacteria reduce nitrate first either to N_2O or all the way to N_2. Without this recycling, current rates of primary production would deplete the biosphere's mobile nitrogen in a mere four million years.

The second sub-cycle in Fig. 5.8 is dominated by autotrophs. Fixed nitrogen is assimilable by plants either as nitrate or ammonia and both of these compounds are scarce not only in surface waters but also in soils so that primary production in general, and cultivation of high-yielding crops in particular, are more, and more often, limited by the supply of metabolizable nitrogen than by shortages of any other plant nutrient. From nitrogen comes the deep, dark green hue of vigorously growing plants, their taller stems, larger leaves, and bigger grains. From its shortages arises the yellowing of leaves and stunted growth. Directly through plant decay, and indirectly via heterotroph consumption, phytomass nitrogen returns to the large organic pool in the soil.

The third, heterotrophic, cycle is governed by the activities of microbial heterotrophs immobilizing mineralized ammonia (a process of great importance in farming where the incorporation of relatively large masses of nitrogen–poor crop residues into soils may lead to a considerable tie-up of nitrogen). Mineralization is the gate releasing nitrogen to all three sub-cycles; starting with the ammonia pool the flows are in competition which gets further intensified at the nitrate pool as plant uptake, immobilization, and denitrification deplete the NO_3^- stores.

But the key driving flux is obviously nitrogen fixation: mineralization could not continue without continuous replenishment of organic nitrogen in the soil and only a small fraction of the recharge comes from the weathering of organic sediments. More than nine-tenths come from fixation, a process which converts atmospheric dinitrogen to ammonia and other nitrogen compounds assimilable by plants.

Whenever there is a high energy flux in the atmosphere some of the dinitrogen will be split and freed: ozonization could produce up to 15 million tonnes N in oxidized compounds. Falling meteorites may fix no more than 100 000 tonnes N a year but lightning may contribute about 30 million tonnes. None of these estimates is reliable as it usually results from several averaging assumptions. For example, calculations of nitrogen fixation by lightning depend on the assumptions of nitrogen oxide production per flash and the mean number of cloud-to-ground flashes per second, both variables being obviously very difficult to pinpoint. The result is a range of global estimates from just a few to as much as 75 million tonnes a year.

What is more surprising and more perplexing is that we have to be so unsure about the contribution of biotic nitrogen fixation, the dominant, ubiquitous, and intensively studied form of providing nitrogen to terrestrial ecosystems. The reduction is done by bacteria, actinomycetes, and blue-green algae living alone or, more often, symbiotically with fungi, liverworts, and ferns as well as with thousands of angiosperm species.

Free-living heterotrophic bacteria (be it aerobic ones such as *Azotobacter* and anaerobs such as *Clostridium*) fix only small amounts of nitrogen, while free-living blue-green algae, found in almost all kinds of terrestrial environments, are much more important contributors, especially to the fertility of paddy fields. Experiments at the International Rice Research Institute in Philippines showed that the algae, fixing up to 75 kg N/ha a year, could replace all nitrogen removed by as many as twenty-four consecutive rice crops without any decline in the total soil nitrogen content. But the best performance by a green alga, *Anabaena azolla*, comes when it lives symbiotically in a large cavity of the upper lobes of a small floating fern, *Azolla pinnata*. The two organisms grow in unison and supply plenty of nitrogen to rice as well as a high-protein duck and pig feed.

Symbiotic actinomycetes belonging to genus *Frankia* are attached to the roots of about 160 species in 15 genera. The hosts are mostly trees, the best known ones including red alders (*Alnus rubra*) in the temperate zone and casuarina (*Casuarina equisetifolia*) in the tropics. Thanks to the symbiosis (annual fixation rates are between 140 and 300 kg N/ha) red alders can reach 2 m within two years and then keep adding up to 1.5 m for the next 15–20 years, providing an excellent opportunity for short rotation wood plantations in moister climates.

But it is the symbiosis between leguminous species and *Rhizobium* bacteria, an association first recognized in the late 1880s, which is nature's principal way of making atmospheric nitrogen available to plants, including many important food and feed crops. This is no surprise considering *Leguminosae* are the planet's second largest family of plants (only *Graminae* are ahead), with 650 genera and some 13 000 described species.

If an appropriate *Rhizobium* is present in the soil (unfortunately, many important species can become symbiotic with just one species of the bacteria) root nodules are formed and the plants are supplied with fixed nitrogen in exchange for some of their fixed carbon. Measurements of this exchange show extremes between 0.3 and 20 g C/gN with the best average being about 6.5 g C/g N.

In view of the great importance played by the legume-*Rhizobium* symbiosis it is surprising how inadequate is our knowledge of actual field fixation rates. Measurement methods, ranging from costly isotopic tracing to lysimetric studies, yield what might be seen as only semi-quantitative results and common differences in the growing environment further complicate a search for representative averages. Still, it is almost unbelievable that Thomas La Rue and Thomas Patterson had to conclude their 1981 review of fixation rates by writing that 'there is not a single legume crop for which we have valid estimates of the nitrogen fixed in agriculture'.

Even for the most frequently assessed legume, United States soybeans, the values are quite wide-ranging (between 15 and 300 kg N/ha) and represent just small experimental plots in specific locations rather than typical field values. Most of the measured rates for common leguminous crops would fit between 100 and 250 kg

N/ha, with maxima going up to and even a bit over 600 kg N for broad beans and alfalfa. Without understanding the reasons why, farmers have long known these prolific fertilizing properties of leguminous plants and in all traditional agro-ecosystems legumes occupied a prominent place in crop rotations.

In Egypt, planting of *berseem* clover (*Trifolium alexandrinum*) always provided essential forage to maintain the indispensable draft animals and to provide nitrogen for subsequent food crops. In China, winter planting of Chinese vetch (*Astragalus sinensis*) has been a principal way of fertilizing subsequent grain crops, and food legumes, ranging from broad beans (*Vicia faba*) and peanuts to the quintessentially oriental soybeans (*Glycine max*) were used in complex rotations with cereal and oil species. In traditional European farming clovers and alfalfa were similarly important in crop rotations.

But growing legumes as a green manure (to be ploughed-under before food crops are planted) has an obvious drawback: for a few months the fields are occupied by a non-food crop, precluding further intensification of food production, a major consideration in densely populated regions with a limited amount of arable land. Rotating food legumes with grain and oil crops would then seem to be a much better choice—except that legumes are not enthusiastically accepted as providing a greater share of everyday diets (see section 3.3.2). The presence of toxic substances and problems with palatability and digestibility make them generally staples of last resort rather than first choice.

Consequently, the most productive traditional farming has also relied to a large extent on the recycling of organic wastes to maintain soil fertility. The simplest way is, of course, to return crop residues, mainly cereal straws and stalks, and legume and tuber stems and vines, directly to the soil but in poor, populous countries this happens only in a small proportion of cases as the residues are needed as animal feed and bedding, as fuel and raw material. In the rich countries, in contrast, up to four-fifths of all residues are directly recycled, returning nutrients and, no less importantly, retaining moisture and reducing water and wind erosion rates.

Reliable nationwide figures on the extent of recycling are rarely available but a detailed USDA survey in the late 1970s showed that some 3.5 million tonnes of nitrogen are returned to fields each year, an equivalent of about 30 per cent of all nitrogen incorporated in annual crop harvests (including all residues). Approximate global calculations indicate that complete recycling of all crop residues would—when postulating losses of no more than 35 per cent—return about 10 million tonnes of nitrogen, or roughly one-third of the nutrient removed by crops.

Animal wastes are a much more concentrated source of nitrogen than crop residues but of their global output (about two billion dry tonnes annually, some 70 per cent coming from cattle, a tenth from pigs) only a part is recycled to fields. In the United States less than one-third of all manures are managed as fertilizers; in India (where dried dung is used heavily as fuel) about half of the manure produced by cattle and water buffaloes is recycled; in China, the nation with ancient tradition of manure fermentation, about two-thirds of all wastes (that is virtually all pig manures) are turned into composts or liquid fertilizers.

Manure recycling is also traditionally high in Europe but relatively low in Latin America and insignificant in Africa. For most of the world's countries there is no quantitative foundation to estimate the contribution of manurial nitrogen but such figures are anyway of only theoretical interest as large volatilization and leaching losses shrink the mass eventually available for mineralization. For example, in the United States USDA estimates these losses of manurial nitrogen at about 1.3 million tonnes, a total which prorates to some 6 kg N per capita each year, or almost exactly as much nitrogen as is actually consumed each year in plant and animal protein by an average American!

The recycling of human wastes (containing annually worldwide about 25 million tonnes N) is even more limited. In the rich, highly urbanized countries the practice is limited owing to the restrictions on feasible sludge transport distances and by the often unacceptable composition of sludges (very high concentrations of heavy metals and toxic chemicals)—and throughout the poor world, with the exception of the Chinese cultural realm, the practice is largely absent for socio-religious reasons. My best estimates for effectively (that is net after losses) recycled animal and human wastes are no more than 10 and 1 million tonnes N respectively.

Should crop residues and manures return effectively about 20 million tonnes N each year, and should nitrogen left in the soil after the cultivation of leguminous species and inputs from atmospheric deposition (nitrates and ammonia) and weathering add up to another 5 million tonnes, the grand total of roughly 25 million tonnes N would represent just about half (most likely we will not ever be able to pinpoint the share) of the nutrient removed in the annual harvest of food and feed crops. The other half comes from the human interference in this key biogeochemical cycle which appears to be the least easily replaced, an interference which also requires continuous, considerable expenditures of fossil energies and which has numerous and far-reaching environmental consequences: the application of synthetic nitrogenous fertilizers.

While it may be economically unappealing to contemplate a rapid conversion of the world's primary energy supply from a predominantly fossil-fuelled one to a mixture of nuclear generation and solar conversions, the option is technically well conceivable—and it would do away with carbon dioxide's atmospheric build-up and, at the same time, it would remove the principal anthropogenic source of acid deposition. In contrast, we cannot seriously contemplate any displacement of nitrogenous fertilizers if we want to keep the coming generations adequately fed. Moreover, even in the long-run, that is at least for the next two generations (or 40–50 years), we cannot foresee any viable alternatives to our currently dominant fossil-fuel based ammonia synthesis—nor can we, even with the best field management, effectively prevent all the undesirable environmental impacts of increased fertilization.

The next two sections will thus focus on this essential food—energy-environment link, first by detailing its energy dimension, then by reviewing its environmental repercussions. Possible control measures and better ways of management will be discussed together with measures applicable to carbon and sulphur cycles in the last section of this chapter.

5.3.1 *Energy subsidies in modern farming*

> The great conceit of industrial man imagined that his progress in agricultural yields was due to new know-how in the use of the sun. . . . This is a sad hoax, for industrial man no longer eats potatoes made from solar energy; now he eats potatoes partly made of oil.

> Howard P. Odum, *Environment, Power, and Society* (1971)

Indeed, the impressive post-World War II rise of crop yields (outlined in section 3.3.1) had nothing to do with improved efficiency of photosynthesis: until evolution develops a fundamentally different process of converting solar radiation to phytomass we cannot surpass the upper limit of photosynthetic efficiency (about 10 per cent of the total radiation reaching the ground). Yields have been going up owing to the redistribution of photosynthate (increasing the harvest index by breeding short-stalked varieties) and, above all, thanks to the provision of needed services-fertilization, irrigation, and pesticide applications. Instead of being limited by shortages of nutrients and water, and harmed by competition from weeds and attacks by pests, modern crops could come closer to reaching their full yield potential.

The price of this success has been a steady flow of fossil fuels and electricity, directly to power field machinery and indirectly to synthesize the agricultural chemicals. Every crop harvested on a modern farm incorporates not only solar radiation which reached its leaves during the growing season but also a variety of anthropogenic energy subsidies coming mostly from fossil fuels. Potatoes, bread, and rice partly made of oil now account for the overwhelming majority of global food production.

During the 1950s and 1960s when this great transformation unfolded rapidly throughout the rich world and started to spill into many poorer countries, virtually no attention was paid to its energy cost. Only in the 1970s, when the rapidly rising energy prices focussed attention on energy consumption in various products and processes, did energy accounting of agricultural production become a lively topic of interdisciplinary research. Studies of national and regional level and process analyses of individual crop cycles have since provided much valuable information on levels, distributions, and changes of energy subsidies in modern farming.

Perhaps the most misleading feature recurring in many of these studies has been the incorrect use of a simple efficiency ratio in judging the performance of the process: the energy content of the harvested crop is divided by the sum of anthropogenic energy subsidies. Not surprisingly, this ratio shows a steady decline with the intensification of modern farming but its use is clearly misleading. After all, farming is not concerned with maximizing the gross conversion efficiency of radiation into phytomass with the help of fossil fuel subsidies: if it were so then corn for silage in the temperate zones and sugar cane in the tropics would be the only obvious crop choices!

Moreover, we cultivate our crops not just for any energy but for particular digestible and palatable food energies contained in carbohydrates, lipids, and proteins and augmented by essential vitamins and minerals. One Joule of rice cannot be equated with a Joule of diesel fuel as is done in the simplistic efficiency ratio.

As long as the studied crop is a modern high-yielding cultivar the two principal conclusions of energy analyses keep recurring: nitrogen fertilizers are invariably the single largest energy subsidy, accounting typically for no less than a third and often for as much as nine-tenths of all external energies invested in production, and their utilization displays a steady secular rise which has in some instances led to optimum applications but in the overwhelming majority of cases there is still a considerable potential for more intensive fertilization.

For example, a detailed study tracing the developments of post-World War II energy inputs into American grain corn found that in 1945 nitrogen fertilizer applications averaged less than 2 kg per harvested hectare and their energy cost was less than 10 per cent of the still relatively small subsidy total dominated by fuel for field machinery—while the recent nationwide mean has been over 50 kg per harvested hectare, by far the single largest (about two-fifths) ingredient of much increased fossil fuel subsidies. And for the most productive Corn Belt farms, where all corn is heavily fertilized, the values are a good deal higher. Average nitrogen applications are basically at saturation levels of 170–230 kg/ha accounting for two-thirds of the anthropogenic energy inputs into what is one of the world's most subsidized crops.

There have been so many stunning scientific advances and engineering accomplishments during this century that applications of synthetic nitrogen look utterly unexciting but without them the rich countries could not have reached their food affluence and the poor countries could not have extended better diets to their fast growing populations—they are an irreplaceable key to planetary survival, a relative novelty whose absence has quickly become unimaginable.

For millenia the planting of legumes and the recycling of organic nitrogen were the only ways to maintain and enhance soil fertility and only the discovery of huge deposits of Chilean nitrates in 1809 introduced the first commercial source of inorganic fertilizer. The recovery of ammonium sulphate from coking ovens and the oxidation of nitrogen in electrical arc furnaces were added later in the nineteenth century. But the total consumption of nitrogen fertilizers remained negligible well into the new century. In 1913 Fritz Haber and Karl Bosch finally succeeded in their efforts to synthesize ammonia—spurred not by concerns about agricultural productivity but by a quest to develop domestic source of nitrates, the basic ingredient for explosives.

In their catalytic fixation process ammonia was synthesized from atmospheric nitrogen and hydrogen using a nickel catalyst at high temperatures (about 500 °C) and pressures (10^2 atmospheres). Later basic changes included the provision of nitrogen by fractional liquefaction of air and tapping of methane, the main constituent of natural gases, as the preferred source of hydrogen. But World War I, and the economic difficulties of the subsequent decades, and yet another war, postponed the era of large-scale fertilizing until the 1950s. The historical statistics of nitrogen fertilizer consumption in the United States best illustrate the trend: from a mere 34 000 tonnes of inorganic N in 1890 to about 800 000 tonnes by 1950 and to over 10 million tonnes by the late 1970s (Fig. 5.9).

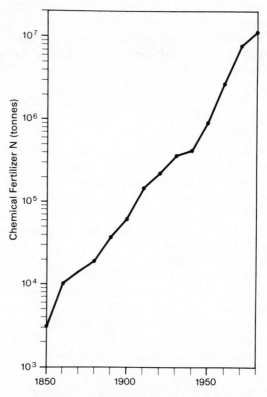

Fig. 5.9 Total nitrogen fertilizer application in the United States since 1850. The fairly straight ascent on this semi-log graph means, of course, a sustained exponential growth with the fastest rates between 1940 and 1970.

Global consumption trends have been very similar—from less than 1 million tonnes N a year in the 1920s to nearly 4 million tonnes in 1950 and over 60 million tonnes during the early 1980s—and this large post—World War II expansion was accompanied by rapid diffusion which has by now left only one continent, Africa, largely untouched by the benefits (and the problems) of nitrogen fertilization. Many countries in the Sahel and in Black Africa still do not use any fertilizers and, excluding the Republic of South Africa, the continent's mean application is negligibly low at 4 kg/N/ha · year.

The Latin American mean is just below 20 kg, the Far Eastern average no more than 25 kg, the Soviet and North American means are, respectively, about 35 and 50 kg N per hectare of farmland, and Western Europe leads the regional breakdown with some 100 kg N/ha. The global average is just over 40 kg N/ha, but national means—and even more so typical individual crop applications—show what extremes contribute to all of the just cited figures.

As already noted, most African crops receive no fertilizer while the Dutch mean stands unparalleled at more than 550 kg N/ha · year. Other very high users of synthetic nitrogen are, in descending order, Ireland, North Korea, Belgium, West Germany, and South Korea (all over 200 kg/ha), while Egypt, all Western and Central European countries, Israel, China, and a few Caribbean islands all use over 100 kg N/ha each year. In contrast, three of the world's five large grain exporters fertilize very little: Canada (20 kg/ha), Australia (5 kg), and Argentina (below 2 kg) rely on large areas rather than on high yields to produced their big harvests.

As for individual crops, the ranges are smaller than for the just listed national averages because the latter are annual applications per ha and in all intensive agro-ecosystems at least two and often three crops would be grown on the same land each year. In North America corn receives the largest amount of nitrogen, in Europe wheat is at the top, in Asia, obviously, rice with the highest applications as much as 200 kg N/ha.

These high applications make economic sense because modern cultivars respond strongly to nitrogen fertilization. Numerous studies with various crops have confirmed that higher phosphorus and potassium applications must accompany increased nitrogen fertilization but that the yield responses to P and K are much more erratic and much less universal than to nitrogen. There is an eventual levelling-off of returns but, as illustrated in Fig. 5.10 for wheat, the improved short-stalked cultivars keep on responding to much higher levels of fertilization than traditional varieties. Typical gains in the range of the strongest response are, all in kilograms per kilogram of nitrogen, 12–20 for unmilled rice, 15–20 for wheat, and 25–45 for corn.

Agronomical literature has repeatedly emphasized the importance of the whole package for achieving and sustaining higher yields. New cultivars, irrigation, pest and weed control, timely planting, good field management, and proper harvesting are all necessary but even so fertilization appears to be the single largest contributor to rising yields, accounting for at least one-quarter and in some cases perhaps for as much as one-half of all long-term gains. White-Stevens put it more emphatically in the American context:

To deny American farmers the use of artificial fertilizers . . . would require the removal of 100 million Americans, some of whom, in a democracy it would seem, are likely to vote against the idea. It would also become necessary to open up some 250 million acres of new land . . . to return some 20 million nonrural people back to the toil of the soil and stoop

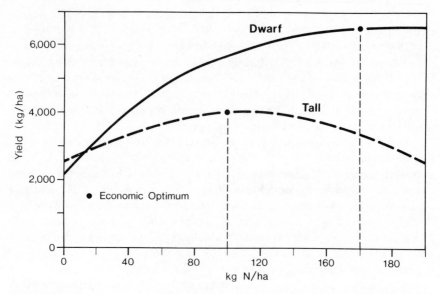

Fig. 5.10 Different responses to fertilizers by traditional (tall) and modern (short-stalked) cereal cultivars account for much of the post-World War II rise in grain yields. In this case the responses are those of a traditional Indian wheat and a new Mexican variety bred at the CIMMYT.

labor; to reduce the present standard of living by at least 50% and to retreat to the way of life a century ago.

Even perfect recycling of all organic wastes could cover only a fraction of the nitrogen needed by the world's crops, and of all the human interferences in key biospheric cycles none is less amenable to reduction than nitrogen fertilization— and none is in need of such an extensive expansion. Although an approximate accounting shows that some 40–50 million tonnes of nitrogen are removed each year by the world's farm crops (that is some 15–30 per cent less than the total applied in nitrogenous fertilizers) the surprisingly large losses of fertilizer nitrogen from agro-ecosystems (the next section will detail these processes) and the great disparities in regional, national, and individual crop applications noted earlier in this section mean that, in spite of the positive global application/removal ratio, the world's nitrogen use is still too low and that its major expansion is a key condition of higher agricultural productivity needed to support large populations.

Here, then, is the tightest link between energy and food production, a link whose inevitable maintenance also has multiple environmental implications. Minimum tillage practices may be reducing the need for energy subsidies in terms of field machinery and fuel needed to run it, higher CO_2 levels may improve water-use efficiency and hence lead to lower requirements for pumping subsidies, new pest-resistant cultivars and biotic controls may cut down the needs for pesticide applications, but even with the best field management future fertilizer applications will have to keep on rising steadily.

And so will energy costs. As anywhere in the realm of energy conversions there are undoubted possibilities for efficiency improvements—but definitely of a lower order of magnitude than in many other common processes of worldwide importance. This is the result of the past impressive technological improvements which have lowered the energy cost of ammonia synthesis from about 400 MJ per kilogram of nitrogen in the early 1920s (when the first commercial plant started operating) to less than 40 MJ/kg N in the early 1980s when the best M. W. Kellogg and Haldor Topsøe processes, each installed in more than 100 large plants worldwide by the early 1980s, needed just below 30 MJ of the synthesis of 1 kg of NH$_3$ (Fig. 5.11).

Many technological changes contributed to this improved performance, ranging

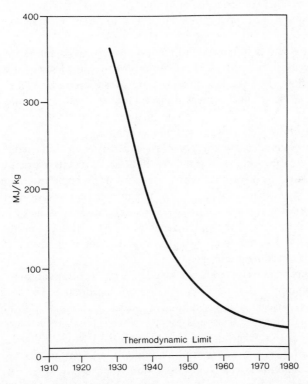

Fig. 5.11 Impressive decline of gross energy requirements needed to synthesize a kg of ammonia by the best available processes between 1930 and 1980. An order of magnitude improvement means that a surprisingly small portion of global natural gas extraction goes into NH$_3$ synthesis—but it also means that there is little room for further improvement of the top efficiencies.

from the introduction of low-pressure gas reformers and the hot carbonate process in CO_2 removal during gas purification to the increased number of refrigeration stages to reduce liquefying power requirements, but it was the replacement of turbine-driven (reciprocating) compressors by centrifugal ones which brought the greatest energy savings and also forced the adoption of large size plants working with higher load factors, thus further improving the production economies. While the small (less than 500 t NH_3/day) pre-1963 plants used up to 700 kWh to synthesize a tonne of ammonia the newest consume a mere 40 kWh, a huge saving of 95 per cent.

Although inputs of natural gas as feedstock and gas consumption for heat and power are hardly influenced by a plant's size, investments for a 1000 t NH_3/day plant are some 25 per cent below those for a small unit, and operation costs are reduced by a similar margin. As a result American ammonia prices declined nearly 50 per cent during the 1960s and in spite of large price increases during the 1970s they are now still cheaper (in constant monies, of course) than before 1973! Experience in other countries and with other nitrogenous fertilizers has not been so encouraging and most poor nations have had to resort to subsidies to raise the applications.

While it would be impossible to replicate the energy saving experience of the 1950–70 period in the NH_3 synthesis of the next two decades (a glance at Fig. 5.11 will show how close the best processes are to the thermodynamic minimum), further engineering adjustments will shave a few more MJ/kg, but in global terms the best energy conservation option lies in a still wider adoption of the two most concentrated nitrogen fertilizers, ammonia and urea.

Ammonia is not only the cheapest nitrogen fertilizer, in both energy and money terms, but also the most concentrated one (82.2 per cent of its mass is nitrogen) and so it is hardly surprising that where special storage, application, and handling equipment is affordable (NH_3 is a gas at atmospheric pressure hence the bulk storage requires refrigeration and field application must be done by injecting the compound into the soil through a series of tubes) it has become the dominant fertilizer: more than half of all nitrogen in North America now comes from NH_3 and its use is spreading in Europe.

But in Europe, as well as in Asia with its high nitrogen applications to rice, urea, ammonia's simplest derivative and the most concentrated solid fertilizer with 45 per cent of nitrogen, is fast becoming the leader. For example, Italy now puts out about half of its nitrogen as urea, as does Japan, while the shares are over two-thirds in South Korea and three-quarters in both India and Pakistan (in the United States urea now covers about one-fifth of all applications). Various nitrogen solutions such as ammonium nitrate and ammonium sulphate are other, less common, options.

The average energy cost of ammonia coming from a mixture of plants is, of course, much higher than the cited best performance for modern large units and various national or company-wide studies put it at 55–65 MJ/kg N, with 60 MJ/kg being a good worldwide average. All other fertilizers for which NH_3 is the

basic feedstock must cost more: nitrogen in urea needs 70–5 MJ/kg, in ammonium nitrate up to 90 MJ/kg, and inefficient small coal-based fertilizer plants in
China producing unstable ammonium bicarbonate may need much over 100 MJ/
kg N.

Consequently, in the United States in the early 1980s I estimate the average
energy expenditure at no more than 63 MJ/kg of applied nitrogen (NH_3
dominance tells) while in China, where large modern urea plants bought from
abroad and built domestically during the 1970s account for less than one-half of all
ouptut, the nationwide average is at least 85 MJ/kg N and the Chinese producing
nitrogen fertilizer in the same mixture and with the same efficiency as Americans
would thus save at least 25 per cent of feedstock and process energy.

As the nitrogen fertilizer production is now almost equally split between rich
and poor countries, and as even many rich nations continue to use large amounts of
fertilizer more energy intensive than ammonia and urea, worldwide savings from a
virtually complete conversion to the two cheapest fertilizers would be certainly
comparable to the difference between American and Chinese means. Although we
have fairly good worldwide data about the output of major fertilizers we have no
breakdown of global fertilizer production according to the size, age, and technology
of the plants and hence it is impossible to calculate global energy consumption in
nitrogen synthesis.

Nevertheless, a proper order of magnitude estimate is easy by assuming, rather
conservatively, an average of 80 MJ/kg N. The worldwide energy cost of nitrogen
fertilizer production in the early 1980s would be then close to 5 EJ, or a mere
1.5 per cent of global energy consumption. Even for the United States the estimates
differ, with totals ranging from as little as 600 PJ to as much as 750 PJ, or between
0.75 and 1.0 per cent of the country's total energy use.

Clearly, energy savings arising from future improvements of ammonia synthesis
will be very small in a nationwide context, even in the countries with heavy
fertilizer applications, and the worries about the dangerous dependence of modern
farming on fossil energies, so often and so dramatically publicized during the 1970s,
also appear in a different light with such a comparison. The dependence is critical,
the processes are highly energy intensive (for comparison, it takes only 10 to 25 MJ
to produce 1 kg of phosphorus, a mere 4 to 9 MJ for 1 kg of potassium) but the
overall burden, especially when considering the returns (say, on the average, at least
15 kg of grain for each kilogram of nitrogenous fertilizer) is surprisingly small.

Efforts to make ammonia synthesis more efficient should, and will, continue but
an equivalent of the total energy needs for nitrogen fertilizers production could be
rather easily gained by simple conservation measures elsewhere in a nation's
economy. For example, in the United States it would take no more than an
improvement in the average gasoline consumption of the nation's 125 million cars
in the early 1980s by a mere 8 per cent (that is a performance leaving the fleet still
much behind the current West European and Japanese performance) to gain about
700 PJ. In China 900 PJ going into nitrogen fertilizer synthesis could be gained by
improving the nation's dismal average fuel conversion efficiencies (see

section 2.3.3) by a mere 5 per cent, a trivial gain considering that it is still nearly four times as energy intensive as the Japanese economy.

No substantial changes in the currently dominant Haber–Bosch synthesis of ammonia, based largely on hydrocarbons as the feedstock, can be foreseen for the remainder of this century. The possibility of rapidly rising production costs in the case of a new cycle of ascending energy prices is thus obvious. The very high cost of plant conversion and the unavailability of low-priced coal in many countries would work against any rapid switch to coal-based synthesis. But as the hydrocarbons (above all natural gas) used as feedstock and fuel in nitrogen fixation are only such a small fraction of the total gaseous and liquid energies, and as the additional food they help to produce could not be obtained in any more efficient, more economical, and more ecosystemically sound way (except, of course, by not feeding so much of the grain to animals), there is no doubt that priority allocation, short-term rationing, and long-term subsidies would take the industry through any temporary supply shortages or rapid price rise of essential feedstocks.

Alternative ways of producing ammonia show little promise for early commercialization, in the case of new processes, or for the widespread adoption of well-tested procedures. Thermochemical dissociation of water to produce feedstock hydrogen is the least promising option—as yet there are no commercial processes available and the need for a source of high heat (nuclear energy would do well if it were economically viable), pollution potential from chemicals required for sequential reactions, and high capital and operation costs do not make this method appealing. Electrolysis of water is a well-established procedure but it makes sense only with cheap hydro-electricity, an option by now foreclosed for most of the world's nations (although China and Brazil could do well).

The replacement of hydrocarbon-based synthesis by coal is thus the best possibility, either by using large-scale gasification and turning the fuel into an equivalent of natural gas (a less likely option), or proceeding by the well-known process of partial coal oxidation in small multiple units matching the needs of ammonia plants. The relatively high capital cost and limited capacity of individual gasifiers and the necessity to cope with variabilities of coal quality and sulphur content are some of the major difficulties but the higher hydrocarbon prices of the 1970s helped to improve the economic outlook of coal-based ammonia synthesis to the point where it may be seen as an interesting long-term choice.

Regardless of the future ways and energy efficiencies of ammonia synthesis there is little doubt about a continuing increase of nitrogen fertilizer applications. Even when the global total by the end of the century will be very likely lower than forecast throughout the 1970s (a development parallelling moderated outlooks for energy consumption) the total applications will be near 100 million tonnes a year (that is definitely of the same order of magnitude as *Rhizobium* fixation), and their numerous environmental impacts will require even greater attention than they have been receiving since the late 1960s.

5.3.2 *Environmental consequences: from wells to the stratosphere*

> In sum ... we have ... thrown the nitrogen cycle seriously out of balance. ...
> Clearly, corrective measures are urgent.
>
> B. Commoner (1975)

The scope of environmental concerns associated with the application of nitrogenous fertilizers is truly enormous, from nitrates in well water (a worry which helped to launch the era of environmental consciousness in the United States; see section 4.1) to nitrous oxide in the stratosphere (a pontentially troublesome build-up which could result in a reduction of the Earth's protective ozone layer). Of course, there are environmental disturbances arising from releases of nitrogenous compounds during combustion, the second largest source of human interference in the nitrogen cycle, none of them being more important and better investigated than the role of nitrogen oxides in the formation and duration of photochemical smog.

However, these environmental degradations (including various uncertain effects of nitrogen oxides on plants and animals and on human health) can be reduced to tolerable levels by relatively simple engineering adjustments. For example, while the mid-1960s nitrogen oxide emissions in the United States averaged 2.5 g NO_x/km, the combination of combustion modification techniques and three-way catalytic conversion brought them to 1.9 g/km between 1972 and 1976, 1.25 g/km between 1977 and 1980 and 0.63 g NO_x/km in the early 1980s, a reduction of 76 per cent.

Japanese carmakers have done even better: Toyota, Honda and Fuji Heavy Industries (makers of Subaru) rely simply on ingenious combustion modifications to comply with the country's strict emission goal of a mere 0.25 g NO_x/km and Nissan's exhaust gas recirculation together with fast burn arrangement produces a mere 0.13–0.16 g NO_x/km, a 95 per cent reduction compared to pre-control emissions! Similarly, a combination of control techniques (tangential firing, two-stage combustion, low-excess-air firing and flue gas recirculation) can bring dramatic lowering (up to 85 per cent) of NO_x releases from stationary, mainly power plant, combustion.

Obviously, no amount of research and the best conceivable management could bring such reductions of fertilizer losses. Once spread on the field, or incorporated into the soil, ammonia or nitrate nitrogen in synthetic fertilizers is subject to microbial transformations powering the element's cycle as well as to inorganic processes. Volatilization, denitrification, leaching and erosion move the nutrient, mostly in gaseous compounds and in solutions, out of the agro-ecosystems, into the waters and the atmosphere. Brief summaries will outline the extent of individual loss fluxes before a presentation of an overall balance sheet for the world's leading agroecosystem and a discussion of the most worrying environmental consequences.

The volatilization of ammonia is ubiquitous and rapid in all animal manures and sewage sludges, especially when applied to alkaline soils where 80 per cent of the newly added nutrient may be lost in just two weeks—but volatilization of synthetic ammoniacal fertilizers may be also very large. Typical losses for ammonia nitrate and sulphate are between 5 and 20 per cent (the extreme range being 3–50 per cent), while for ammonia they may be negligible (with injection of anhydrous NH_3 into soils) as well as surpassing 50 per cent (with surface application of aqueous solutions).

No representative range can be quoted for urea volatilization. Some experiments found this loss to surpass those of other solid ammonia fertilizers after surface applications (up to 77 per cent wasted), others found rather low (12–16 per cent) losses. As noted in the preceding section, urea is now the preferred fertilizer in the rice-growing regions of Asia but its broadcasting onto paddy fields may result in rather large losses. The alkalinity of water is of major influence and in those paddies where algae abound (including the nitrogen fixing ones) photosynthetic with-drawal of CO_2 may push pH up to 10 by the early afternoon and direct volatiliza-tion losses will thus soar and one-fifth of all applied nitrogen can be lost within just four days after spreading!

Without denitrification there would be no return link between soil and atmospheric nitrogen but, as essential as the process is, it is still hardly welcome in some agro-ecosystems where, especially in wet and warm soils, plenty of nitrogen can be lost in a relatively short time. Longer-term monitoring of denitrification in various agro-ecosystems shows a wide range of 1–200 kg N/ha · year, representing losses anywhere from a fraction of 1 up to 50 per cent, with values of 5–25 per cent accounting for most typical reductions.

Leaching losses are almost solely due to easily soluble nitrates moving through the soil into surface and groundwaters—they are dissolved and carried into the soil during the first few minutes of rain, so by the time of eventual surface runoff they are sequestered in the soil. Then it may take months, and even years and decades, before the nitrates—moving downwards in a characteristic bulge (Fig. 5.12)—reach groundwaters or streams. The outcome is governed by the kind, rate, and place-ment of fertilizers, a soil's water-holding capacity, and the amount and distribution of precipitation and plant uptake—enough variables to make a search for average values a hopeless undertaking.

Nevertheless, the most often encountered losses in rather rainy, intensively farmed areas of Europe and North America are between 10 and 25 kg N/ha · year, with extremes ranging from negligible values to more than 50 kg. The worst time for leaching losses in temperate latitudes is fall, and spring: between May and October evapotranspiration surpasses precipitation, and crops assimilate nitrogen so there is a general upward movement of water and nutrients.

Erosion causes the worst nutrient losses where its rates have been found to be incompatible with permanent agriculture—wherever row crops are continuously planted on sloping land in rather rainy climates. In the United States virtually all of the East and Southeast and rolling parts of the Corn Belt are in this category. A detailed 1983 study by Larson, Pierce, and Dowdy put the total United States annual nitrogen loss in eroded sediments at 9.5 million tonnes, almost identical to the amount applied each year in synthetic fertilizers, while the best estimates for

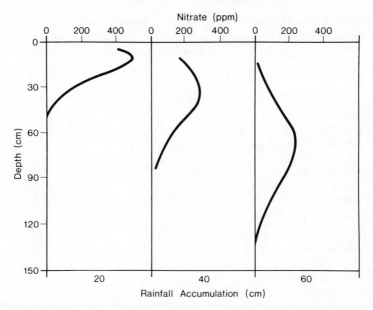

Fig. 5.12 Typical travelling bulge of soil nitrate concentrations with higher precipitation. Eventually ground waters can be seriously contaminated.

China show the loss surpassing the mass of nutrient in the nationwide fertilizer applications.

Nitrogen balances have been studied in a wide variety of agro-ecosystems ranging from small experimental plots to regional reviews and nationwide fluxes and the data presented in Table 5.4 summarize the best justifiable estimates for the United States farming. The approximate nature of all the values, except for the highly accurate total of fertilizer applications and the fairly reliable amounts of crop removal, is inevitable but it does not invalidate a few basic conclusions.

Firstly, chemical fertilizers have already become the single largest nitrogen input to this highly productive agro-ecosystem—or their contribution is just a little behind biotic fixation. Second, when nitrogen from animal wastes and from symbiotic fixation of legume crops is added, the total man-managed input of the nutrient definitely forms the bulk (about two-thirds) of all additions. Third, the country's preference for animal protein requires large nitrogen fluxes for a small final food consumption. In the early 1980s Americans were consuming less than half a million tonnes of nitrogen in plant foods but they fed crops containing some 4 million tonnes N to domestic animals (and exported grains containing roughly 3 million tonnes N). Hay and grazing provided another 5 million tonnes N in animal feeds which were converted to some 800 000 tonnes of nitrogen in meats, dairy products, and eggs.

Table 5.4 Annual nitrogen fluxes (all values in million t annually) in the United States agro-ecosystem

Fluxes	The single 'best' estimate	The most likely ranges
Inputs	33.2	26–43
Fertilizers	10.5	10–11
Fixation	8.2	5–12
Animal wastes*	6.2	5–7
Mineralization	3.0	2–4
Crop residue recycling*	2.8	2–3
Atmospheric deposition	2.5	2–6
Removals	30.6	22–40
Harvested crops and residues**	12.0	10–13
NH_3 volatilization	6.5	5–10
Hay and grazing	5.4	4–6
Denitrification	2.5	1–4
Leaching	2.0	1–3
Erosion	1.7	1–3
Immobilization	0.5	0–1

* Animal wastes and crop residues are just internal transfers rather than inputs and their totals are incorporated in the input category to show their importance. Double counting is avoided by listing theoretical totals for field and pastureland nitrogen removals. In reality only 8.4 million t rather than 17.4 million t N are taken away in field crops and grassland phytomass converted to meat, the difference being the sum of animal waste and crop residue recycling.

** Nitrogen in harvested food and feed crops totals about 8 million t and half of it was consumed by domestic animals, more than a third exported (mainly in feed and food grains) and only about one-twentieth was consumed directly in plant foods within the country.

To consume some 1.2 million tonnes of nitrogen in foodstuffs Americans thus had to remove nearly 12 million tonnes of the element from their agro-ecosystem (leaving withdrawals for exported food and feed aside) for an overall transfer efficiency of a mere 10 per cent. Or, to put it in another, and perhaps even more impressive way, no less than 25 kg of nitrogen must be cycled through the American agro-ecosystem (in natural and artificial inputs) for each kg of the nutrient contained in the nation's foodstuffs.

Conversions of phytomass by domestic animals are responsible for a large part (about two-fifths) of this cycling loss but volatilization, denitrification, leaching, and erosion take away most of the nutrient, and the environmental effects of this huge waste, which is the price paid for the meat-oriented agro-ecosystem, range from acute local problems to long-term global concerns.

Ammonia volatilization is by far the least worrying: the compound is readily absorbed by soils, water surfaces, and even plant leaves and the formation of ammonia nitrates and sulphates, and wet and dry deposition return it rather rapidly to nitrogen's huge soil pool as a most desirable fertilizer. Denitrification would be also of little concern if it merely returned dinitrogen to the atmosphere: nutrient would be lost from soil but is, after all, just an inevitable flux in the element's cycle. Unfortunately, micro-organisms often do not carry the reduction of nitrate all the way and nitrous oxide escapes from soils and waters into the atmosphere. A high nitrate level in soils favours higher N_2 releases, with N_2O nitrogen accounting for as little as a few and as much as 25 per cent of all denitrification flux (an average around 5 per cent might be a typical fraction but fluxes even within a single field will differ widely).

The gas is present in the troposphere in concentrations between 250 and 350 ppbv and regular, consistent measurements done since 1977 as a part of the Geophysical Monitoring for Climatic Change have shown a gentle rise of background N_2O levels, averaging about 1 ppbv between 1977 and 1981. A continuation of this rate of increase would double the atmospheric concentration only in the course of many generations but before these measurements were started it was feared that the gas build-up presented a considerable environmental risk.

Advancing studies of complex stratospheric chemistry identified N_2O as one of the trace gases influencing the levels of stratospheric ozone: owing to its low solubility the oxide diffuses from the troposphere to the stratosphere where it is transformed to nitric oxide, one of the principal ingredients in the ozone-destroying sequence of reactions. Models of the far from reliably understood reactions have, not surprisingly, resulted in substantially different outcomes with rising N_2O.

Some simulations for doubled atmospheric N_2O showed a mere 2 to 4 per cent ozone decrease but other models predicted 20 per cent O_3 reduction by the time synthetic nitrogen applications have reached 200 million tonnes N a year (forecast in the mid-1970s to occur as early as the year 2013). In the first case there would be barely noticeable change, in the other there might be a rapid decline likely to provoke some emergency responses.

The problem is obviously analogous to the CO_2 situation: predictions based on models built with inadequate understanding of the real world and driven, until very recently, by clearly too high future growth rates—and not integrated with the effects of other trace gases. CO_2 is not alone in absorbing outgoing infrared radiation and N_2O is not the only gas controlling stratospheric ozone levels. As already noted (section 4.1.2), the three successive NRC reports on the effects of chlorofluorocarbons, compounds which have been certainly the most studied part of stratospheric ozone chemistry, resulted in considerably lower estimates of eventual O_3 reductions.

Moreover, to suggest that the risk of possible O_3 reductions warrants substantial modifications of our way of farming appears dubious when one considers accumulating evidence showing that the worldwide burning of biomass and fossil fuels may be as large a source of N_2O as nitrogenous fertilizers (maybe even larger), and that N_2O flux from oceans and non-agricultural soils may be surpassing denitrification in agro-ecosystems by an order of magnitude! In any case, there are large uncertainties concerning both sources and sinks of the gas and any simple quantitative conclusions about the eventual consequences of fertilizer denitrification for stratospheric ozone must remain suspect.

Recently measured global background increases of N_2O would double the atmospheric concentration in about two centuries. Considering this slow increase in combination with the relatively small (and uncertain) proportion of fertilizer-derived N_2O to the total (highly uncertain) emissions of the gas, and with fundamental gaps in the knowledge of its sinks, it must be concluded that the threat of ozone reduction caused by the denitrification of nitrogenous fertilizers has been considerably overplayed during the 1970s. A better understanding of the whole N_2O segment of the nitrogen cycle and better appreciation of stratospheric chemistry are the best responses. Any recommendations for fundamental changes in modern farming are out of place.

And it also turns out that the original big nitrogen cycle upset can be now seen with much less apprehension than was the case during the late 1960s and the early 1970s when Barry Commoner and his colleagues made it seem about the greatest challenge to environmental management: nitrogen in waters is undoubtedly a problem but of a much smaller magnitude than claimed by the earlier studies.

The worry was about nitrates, leached (and also eroded) from farm soils and causing increased eutrophication (mainly algae growth) of stationary waters and presenting methemoglobinemia hazard to babies. Commoner claimed that most of the rivers in Illinois, with the Corn Belt's most intensively fertilized fields, did show an 'appreciable rise' in the average nitrate concentrations and that the new 'critical' levels of nitrate were largely caused by fertilizers. His findings were prominently reported, starting public debates and economic studies about the possible limitations of nitrogen fertilization. Some weaknesses in Commoner's arguments were spotted almost immediately but it was not until 1980 when the publication of a detailed study on nitrogen by Samuel R. Aldrich, at that time Assistant Director of the Agricultural Experiment Station at the University of Illinois, systematically refuted virtually all the earlier conclusions.

First, reviews of all available nitrate measurements for 11 Midwestern rivers show no big increases during the mid-1950s and the mid-1970s when fertilization rose rapidly. Perhaps most notably, Aldrich pointed out that Commoner's much-quoted claim of a large (about three-fold) temporal NO_3^- rise in an Illinois river (the Kaskaskia) is a chance result due to incorrect sampling at two different locations in the two periods: correct comparisons at the same site show no significant change (Fig. 5.13).

Elsewhere in the country only three among thousands of readings at nearly one hundred stations over a period of forty-three years exceeded 3 mg NO_3^-/L (that is one-third of the drinking water standard). Second, even where temporal increases occurred, correlation of NO_3^- levels with fertilizer applications is far from obvious. For example, samplings of the Illinois River at Meredosia showed declining nitrates during the period when fertilizer nitrogen applications in the drainage basin

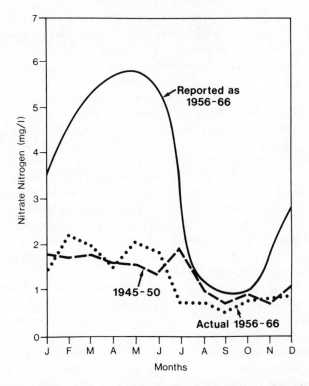

Fig. 5.13 The misinterpretation which made the news and the reality which only few noticed. Commoner's claim of huge fertilizer-associated nitrate increases in the Kaskaskia river in Illinois were set straight by Aldrich. Clearly, nitrogen cycling in Illinois soils is far from being 'out of joint' as Commoner hastily and incorrectly concluded on the basis of improper evidence.

tripled! The impossibility of relying on tracer techniques to quantify the share of NO_3^- coming from fertilizers, and inability to pinpoint denitrification losses mean that no precise quantitative apportioning of NO_3^- origins is possible.

But perhaps the most important evidence about the long-term effects of intensive fertilization on nitrates in water comes from the Netherlands. Kolenbrander's comparisons of NO_3^- levels in drinking water over a forty year span showed that just one-third of the studied waterworks had only marginally higher (by a mere 0.57 mg/L) nitrate content, and the rest remained unchanged in spite of the fertilizer application increases totalling about 150 kg N/ha!

The only common problem spots are wells in farming areas. A detailed survey of NO_3^- levels in over 6000 Iowa wells between the 1930s and the 1970s found low nitrate levels and no change of concentrations in deep public wells (although average fertilization rates rose during that time by more than 100 kg N/ha), and declining amounts of nitrate in public shallow wells—but continuing high levels in private shallow wells with nearly 30 per cent of them having NO_3^- above the maximum recommended standards.

This Iowa study did not look for specific sources of contamination but surveys from other places in North America and Europe trace most of them to improper disposal of animal manures or to domestic waste-disposal systems. Shallow wells dug too close to these sources of nitrate, and often improperly lined and maintained, are then an obvious health risk. And it was water from such wells which was repeatedly responsible for cases of methemoglobinemia in infants.

While direct poisoning with nitrates is possible only with huge doses, their conversion to nitrites in the gastrointestinal tract of babies younger than three months (they still lack acidity which prevents colonization by bacteria reducing NO_3^- to NO_2^-) leads to change of ferrous to ferric iron in the blood and the rise of methemoglobin (from a normal 2.7–4.0 per cent) accompanied first by cyanosis (blue-baby syndrome), then by hypoxia and eventually even death (with methemoglobin at 50 per cent). Fortunately, recovery is rapid and complete after administration of ascorbic acid or methylene blue.

Several thousand cases of methemoglobinemia were reported in North America an Europe between the 1940s and the 1970s, more than 95 per cent of them caused by water from private wells. There were less than 100 deaths, but since the early 1970s awareness of the possible problem in rural areas of the rich countries has made this acute disease very rare. In poor countries the hazard must still be much higher: nitrogen fertilizers are used in larger quantities, water is often drawn from streams, ponds, and shallow wells and animal and human wastes are frequently disposed of improperly. But, curiously, there have been no reports of high methemoglobinemia frequency in China or India.

In general, there is little doubt that improper disposal of organic wastes and poor wells are much more to blame than nitrogenous fertilizers *per se*. Nor can nitrates in drinking water be blamed for possible carcinogenic effects in adults. This concern arose in the late 1960s when it was realized that carcinogenic nitrosamines can be formed endogenously when nitrites react with amines in the stomach. Nitrites have been traditionally added to cured meats to provide protection against lethal botulism, to fix the characteristic pink colour of the products, and also, especially in hams, to impart specific flavour. The average daily per capita intake of nitrite in North America is just 0.77 mg (with about two-fifths coming from cured meat, a third from baked goods, the rest from vegetables; nitrite content of drinking water is usually negligible)—but ingested nitrates can raise the exposure significantly.

Most of the ingested nitrate is excreted in urine but about one-quarter of it is secreted by salivary glands and then roughly one-fifth of this amount (or some 5 per cent of all ingested NO_3^-) is reduced by bacteria in the mouth to nitrite. Nitrate concentrations in saliva appear to be directly proportional to the mass of ingested nitrate—and leaching of nitrogenous fertilizers to drinking water can be then seen as a risk factor in stomach cancer. But a National Academy of Sciences committee studying the problem found that less than 2 per cent of the average daily per capita intake of nitrate (75 mg) comes from drinking water and that, except in the local cases of high nitrate waters, the bulk of the compound (nearly 90 per cent) is ingested from vegetables. Vegetarians would have, obviously, much higher intakes.

So drinking water is hardly a problem but the link may still hold if it can be proved that higher rates of fertilization lead to higher nitrate concentration in vegetables. Indeed, they usually do but the correlation is highly species-specific and even among the susceptible species some parts of the plants accumulate the compound disproportionately. The *Brassica* genus (beets, kale, radishes, spinach, cabbages, broccoli) is the top absorber while cucumbers, peppers and onions have the lowest levels, and petioles and stems have the highest conentrations, followed by leaves, roots, and, always the lowest, fruits and floral parts.

Comparisons of nitrate levels in vegetables grown with plenty of synthetic fertilizer with concentrations measured in many species in the early decades of this century show no uniform trend, a hardly surprising finding in view of the marked variability of nitrate content (samples taken over a few months may show up to eightfold differences for a given species) determined by genetic and environmental factors and, notably, by the time of harvest (NO_3^- levels are lower on sunny days as light activates nitrate reductase needed for the compound's assimilation).

Timing of the harvest may reduce nitrate levels by 25–60 per cent and similar reductions are achievable by choosing low-nitrate cultivars, by discarding spinach petioles or cabbage wrapper-leaves, or by fertilizing with ammoniacal, rather than with nitrate, fertilizers. Consequently, high nitrate levels in vegetables appear to be a property of certain species and even more of some of their parts, and combinations of easy management methods can reduce their levels very substantially. Moreover, as it has been clearly established that vitamin C prevents the formation of nitrosamines and slows down conversion of nitrate to nitrites, the eating of vegetables, especially those higher in absorbic acid, will actually lower the risks of nitrosamine exposures.

As on so many previous occasions the whole case fits an annoying pattern: an initial, widely publicized, scare, followed by a closer look which identifies valid concern but which in no way supports any radical conclusions about intolerable risks and needs for fundamental change. This, too, is the case with the last nitrogen-related concern to be mentioned here, that of eutrophication of waters.

The process is certainly unwelcome: surface algae, normally limited by the shortage of nutrients, start expanding, soon their generation of dead phytomass leads to production of toxic substances by bottom-dwelling anaerobic decomposers and eventually to reduction of the photosynthetic output of oxygen; the water turns anaerobic and heterotrophs die. Phosphorus is most frequently the limiting nutrient preventing this sequence—but nitrogen has that role in shallower eutrophic lakes where its annual influx surpasses 2–3 g/cm.

Nitrogen leached from fertilizer may undoubtedly be a major contributor to local eutrophication but the whole process is governed by numerous environmental variables, ranging from the availability of micronutrients to the depth of the photosynthetic layer (large sediment loads in farming regions usually greatly limit light penetration). Eutrophication can be unsightly and locally outrightly harmful but there is no evidence that fertilization has been turning waters into swamps on a regional scale. Urban waste releases have been certainly a greater contributor as were, until very recently, phosphates.

Both principal environmental concerns arising from growing application of nitrogenous fertilizers have thus undergone important transformation since their formation between the late 1960s and the mid-1970s. The leaching of nitrogen, bringing the acute risks of infant methemoglobinemia and the long-term possibilities of contribution to stomach cancer, and nitrous oxide from fertilizer denitrification, part of which may be eventually participating in a lowering of stratospheric ozone levels, are not seen any longer as major worries necessitating basic adjustments in modern farming but rather as manageable complications requiring, to be sure, further study.

And the need for deeper understanding is perhaps most important where the one so far omitted environmental input of anthropogenic nitrogen compounds is concerned: acid deposition, whose origins and effects will be the subject of the following section.

5.4 Sulphur, the troublesome ingredient

> ... after this air was burned, sulphur had at the same time been precipitated: this must have been caused by the volatile spirit of sulphur which had separated from the part of the sulphur that had taken fire.
>
> Carl Wilhelm Scheele, *Stinking Sulphurous Air* (1777)
> (L. Dobbin translation)

Sulphur in the biosphere is not of such obvious importance as carbon and nitrogen: its total mass ranks only tenth, with the same order of magnitude as phosphorus or aluminium, an order below nitrogen and three orders below carbon. But, as stressed before, sulphur is qualitatively as important for living organisms as carbon and nitrogen. These elements form the unique doubly mobile trio whose cycling maintains life and sulphur's role as a bonder and fortifier of proteins makes it structurally indispensable.

The detailed outline of sulphur cycle presented in Fig. 5.14 carries values both for the reservoirs and the fluxes but this does not mean that our knowledge is so satisfactory that single figures can be assigned with confidence. The assurance is relative: storages and flows involved in the sulphur cycle are known with about the same degree of reliability as those for the carbon cycle, which means in most cases there is a much better understanding than in the case of the nitrogen cycle (whose graph in this book remains unquantified).

The lithosphere is (again as for carbon) by far the largest reservoir of sulphur, ocean's stores are about an order of magnitude smaller (roughly the same as the soil storage) but, in contrast to carbon's relatively substantial atmospheric presence

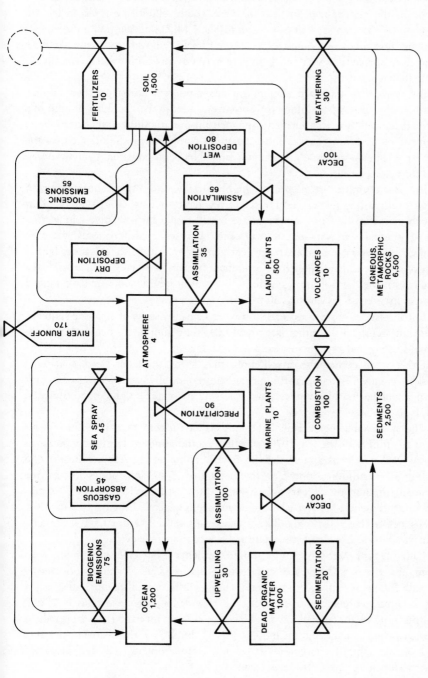

Fig. 5.14 Sulphur cycle shown in the same way as its two predecessors. Reservoirs (rectangles) and fluxes (valves) also contain single 'best' averages in, respectively, million t and million t per year.

(now approaching 800 billion tonnes) only less than 4 million tonnes of sulphur reside in the atmosphere. In fact, the verb 'reside' should be replaced by 'are momentarily in transit'. As can be seen in Fig. 5.14, about 300 million tonnes of sulphur enter the atmosphere each year but efficient removal processes make their average residence time very short, often merely a number of minutes, typically no more than a few days.

But these few days are enough for tropospheric winds to move sulphur compounds thousands of kilometres away from their sources, making sulphur one of the three doubly mobile elements and focusing our attention on the atmospheric segment of the cycle. First, a look at natural contributions. Sulphur enters the atmosphere in volcanic gases, in sea spray, and in biogenic emissions from the ocean and from anoxic soils.

Volcanic eruptions are, naturally, a highly variable source of sulphurous gases. A single event lasting just several hours to a few days can emit 10^5-10^6 t of sulphur (much of it into the stratosphere where it will reside for a relatively long time before gradual removal)—but it may send up gases containing only traces of sulphur. Averaging emissions from gently smoking, quiescent volcanoes (about fifty of them worldwide now have prominent plumes) is equally difficult and, as a result, estimates of volcanic contributions to atmospheric sulphur have ranged from as little as 1 to as much as 30 million tonnes a year.

In contrast, sea spray is easier to estimate, a steady source of airborne sulphate, with nine-tenths of its rather large total mass returning to the ocean by gravitational settling and precipitation and the rest deposited on nearby coastal regions. But the ocean contributes much more than spray sulphate. *Desulfovibrio* and *Desulfotomaculum*, two heterotrophic, anoxic genera of *Eubacteriales* are continuously reducing marine sulphates and releasing hydrogen sulphide and dimethyl sulphide (DMS).

The first reliable measurement of these gas fluxes in remote marine atmospheres were taken only in the late 1970s and the accumulating evidence now points to global DMS emissions equivalent to 35–55 million tonnes S a year and those of H_2S adding 40–50 million tonnes S so that a total of 75 million tonnes S of marine biogenic sulphur now appears to be a conservative estimate. Results of representative measurements of terrestrial biogenic emissions were published only in the early 1980s: before that spotty data did not enable any sensible estimates and the total flux was derived as a simple residuum to balance the whole cycle.

Only the four-year programme of measurements of biogenic sulphur emissions from soils, waters, and vegetated surfaces in the Eastern United States, conducted by Donald Adams and his colleagues and including different soil groups and adjusted for the mean temperature of each site, has provided a coverage extensive enough for meaningful extrapolations. Highest fluxes were measured, as expected, in Florida's saline marshes but the most surprising finding was the relatively high flux of biogenic sulphur compounds from dry, inland mineral soils typical of the Northern hemisphere's agro-ecosystems.

Taking into account the pronounced southward increase in the measured soil sulphur emissions (caused by more biomass, higher temperature, and higher humidity) Adams' results translate to about 65 million tonnes S in global terms so that the total biogenic flux is about 150 million tonnes S a year, still considerably larger than anthropogenic emissions arising predominantly from combustion of fossil fuels and, to a much lesser extent, from smelting of non-ferrous ores. Other sources—chemical industries, wood-pulping, refuse incineration, and flaring of natural gas—are smaller on a global scale than the inevitable errors deriving from estimating SO_2 generation from coal combustion, the leading industrial contributor of atmospheric sulphur.

This rather large error may appear inexplicable considering the sufficiently accurate statistics of the production of bituminous coals and lignites—but the great variety of these fuels and the large, and changing, differences of their sulphur content (from small fractions of 1 per cent to nearly 8 per cent in mass terms), and difficulties in coming up with average emission factors, make precise calculations of SO_2 flux from coal combustion surprisingly elusive.

Average sulphur content is difficult to quote not only for a small coal basin but for even a single mine whose different seams may have very different sulphur content. The commonly available data on sulphur content of exploitable reserves are not very useful as most of the coal is now washed and cleaned before combustion and these processes remove part of coal's inorganic sulphur (largely locked in pyrite, also in marcasite, and ferrous sulphate). Unfortunately, no global data are available on the removal of pyritic sulphur nor are there any regular statistics on changing sulphur content of annual coal production, a consideration especially important in those large coal-mining countries where expanding surface extraction produces increasingly high-sulphur low-energy fuel.

Yet another elusive value is the average retention of evolved sulphur oxides in combustion ashes and fly ashes. Ignored until most recently, this process may be responsible for reducing actual emissions by up to 40 per cent in the case of highly alkaline ashes—but the effect may be negligible and too little information is available to set a useful mean. Consequently, in my global estimates I have assumed no retention and taking an average coal sulphur content of 2 per cent and 3.1 billion tonnes of the fuel burned in households, factories and, above all, electricity generating plants, the annual global emissions of sulphur from this largest anthropogenic source would have amounted to 62 million tonnes in the early 1980s. Sulphur from coal used for coking and cement production would add just about 1 million tonnes a year.

Accounting for sulphur emissions from hydrocarbons is much easier. Natural gas has usually a mere trace of sulphur and desulphurization of 'sour' gases has been a standard practice for decades so the worldwide combustion of the fuel now contributes less that 10^4 t of sulphur a year. Data on sulphur content of crude oil are much more abundant than for coals but the values which matter are averages for major refinery products (gasolines, distillate fuel oils and residual fuel oils, jet fuels and kerosene) and statistics of their annual output.

Both of these data sets are readily available in fairly accurate forms and their product adds up to about 18 million tonnes of sulphur in the early 1980s. To this must be added at least 1 million tonnes from direct combustion of unrefined crudes and some 3 million tonnes released during refining, mainly from fluid catalytic cracking units. With 22 million tonnes of sulphur from combustion and refining of liquid fuels and 63 million tonnes from coal combustion, coking and cement production conversion of fossil fuels has been recently releasing some 85 million tonnes of sulphur into the troposphere.

As noted, the other anthropogenic sources contribute only a small fraction of the industrial total. The smelting of copper and, to a much lesser degree, of lead, zinc and nickel, releases just over 10 million tonnes of sulphur a year, the synthesis of sulphuric acid adds less than 2 million tonnes S, wood pulping generates just 10^5 tonnes and this is also the order of magnitude contributed by open burning and municipal incineration of organic wastes and by household burning of crop residues and fuelwood.

The grand total of anthropogenic sulphur emissions is thus about 100 million tonnes a year. The most likely error is in selecting the average sulphur content of coals: changing the multiplier just one-tenth of 1 per cent adds or subtracts no less than 3 million tonnes of sulphur annually. SO_2 retention in the ash may lower the global emissions from coal by up to 10 per cent; absence of worldwide data on control of flue gases in smelters and refineries may have introduced errors of similar magnitude. The smallest plausible anthropogenic emissions may thus total just over 80 million tonnes S, the upper limit appears to be at 110 million tonnes S.

Clearly, this is a large flux compared to the cited natural transfers of sulphur into the atmosphere but uncertainties about biogenic releases preclude the setting of a single value. Until we have much better bases to estimate emissions from tropical soils and paddy-fields it must be reckoned that the terrestrial biogenic flux is just over 30, rather than 65 million tonnes S a year—but that it may also be as large as 80 million tonnes S. If a similar range of uncertainties is applied to marine biogenic releases the total natural flux (assuming volcanic flux extremes at 5 and 30 million tonnes S) could be as little as 115 or as much as 265 million tonnes, with 195 million tonnes in Fig. 5.14 being the most likely value.

Comparing the minima and maxima of estimates for natural and anthropogenic fluxes puts the extremes of human contribution at 25–50 per cent of the total global inflow of sulphur into the atmosphere, with the most likely range between 30 and 40 per cent. Enriching natural flows by such a margin represents an obviously large interference but global comparisons, unlike the case of CO_2, N_2O, and N_2, are not the most appropriate ones in the case of the sulphur cycle because sulphur compounds are removed from the atmosphere long before they could be global pollutants. Progressively more restricted comparisons thus illustrate better the extent of the human augmentation of sulphur flows.

SO_2 emission in the Northern hemisphere accounts for 94 per cent of all combustion and industrial releases. Taking just median values, it accounts for about 93 million tonnes S, about equal to the hemisphere's natural sulphur flux (simply proportioned on the basis of ocean and continental shares). Restricting the comparison to the main industrialized belt between 25° and 55 °N, where more than 90 per cent of the hemisphere's industrial sulphur emissions are released and deposited, the human contributions surpass the natural flux about three-fold. In the Eastern United States combustion produces, on the average, about 3.8 t S/km² or no less than a hundred times the region's bacterial releases. In Ohio, the state with by far the largest SO_2 emissions in the country, the annual combustion flux, even when uniformly prorated over the whole territory (at 116 000 km² an equivalent of an average European nation), surpasses natural sulphur emissions about 400-fold.

And while most of the SO_2 released from the tall stacks of Ohio's giant coal-fired power plants is carried by prevailing westerly winds from the state and deposited far outside its boundaries there are other industrialized regions where the terrain and climate combine to keep the deposition of a large part of combustion emissions within a small territory. Perhaps the best (worst, really) example of this kind is the North Bohemian Brown Coal Region, a coal mining basin of 1000 km² in a narrow (10 km) and relatively deep (surrounding mountains top it by 400–700 m) valley with limited ventilation and low precipitation. In this region large power plants, mines, and chemical factories now release SO_2 at a rate of 500 t S/km², a mass about four orders of magnitude higher than the area's biogenic flux—and much of this burden, especially during prolonged winter temperature inversions capping the valley, is deposited right in the basin.

This intensity of interference in local, regional, and hemispheric sulphur balances is unlike anything in the two previously discussed cycles. The CO_2 problem is fundamentally global: the rise of the concentrations is worldwide. The drawbacks and benefits of any eventual tropospheric warming, increased primary productivity, and higher water utilization efficiency would be felt on a planetary scale. So would the possibly degradative consequence of increased N_2O levels—and both of these worries are also long-term, diffuse, and still to a large degree speculative.

Fertilization can dominate a field's nitrogen balance. Indiana corn or Maine potatoes may get applications of 150 kg N/ha compared to just 25 kg N/ha provided by free-living fixers, atmospheric deposition, and mineralization—but assimilation of the nutrient by the crops (120–150 kg N/ha), denitrification, volatilization, leaching and erosion remove and recycle readily all of the added nitrogen. After the end of a growth cycle these soils may have just about the same stores of nitrogen as at its beginning and if there has been any build-up of soil nitrogen stores then so much so better.

In contrast, a good cereal or tuber crop will remove annually just between 10–20 kg S/ha—but this much sulphur is most often easily supplied by weathering and by natural atmospheric deposition. Bacterial conversion must transfer annually a few kg of soil sulphur per ha to the atmosphere and any excess sulphur (usually deposited as sulphate) will either accumulate in the soil or it will move into waters. And the excesses are substantial. Large parts of Europe and North America (areas covering several million square kilometres) affected by long-range transport of industrial sulphur receive 10–30 kg S/ha in dry and wet deposition and the total enrichment in the most polluted industrialized regions (extending over 10^5 km^2) is in excess of 50 kg S/ha a year.

To stiffen its proteins, biomass growing 10^7–10^8 years ago needed no more than 0.1–0.2 per cent of its mass as sulphur; mineralization process increased this low concentration roughly by an order of magnitude and, in the case of coal, intrusions of inorganic sulphur further doubled or tripled the fuel's sulphur content. We cannot oxidize coals' and oils' ancient carbon to release heat to energize our civilization without oxidizing these relatively small stores of sulphur. But the consequences of this oxidation have been increasingly troublesome.

Sulphur in the fuel is turned into sulphur dioxide (and also some, usually not more than 5 per cent, of SO$_3$); further oxidation converts it into sulphates, a common one being H$_2$SO$_4$, one of the classic Arrhenius acids. In water this acid is rapidly ionized, splitting into SO$_4^{2-}$ anions and H$^+$ cations. Concentration of hydrogen ions expressed in a simple logarithmic notation (log 1/[H$^+$]) is the pH of a solution; a pH smaller than 7.00 denotes acidic solutions. Rains in industrialized areas, and even those far downwind from them, may have a pH of just 5 or 4, occasionally even less while the rains acidified by just weak carbonic acid (from the current 340 ppm of CO$_2$ in the troposphere) have a pH of about 5.6.

Precipitation with a pH of 4.6 will be 10 times more acidic than the 'normal' rain, at 3.6 (100 times more acidic) it will have the acidity of orange juice. Many organisms do well in acid environments but many more do not: optima for thousands of vertebrate, insects, microbial, and plant species are much above the acidity imparted to soils and waters by anthropogenic acid deposition (Fig. 5.15). The link appears to be clear, the effects obviously worrying. But before looking at these degradative changes I must discuss various recent findings which make the story of acid deposition (commonly, but imprecisely, labelled acid rain) much less linear and much more compelx than the simple outline just presented.

5.4.1 *Acid deposition: elusive understanding*

'Perhaps the worst environmental threat ever to hit us.' . . . 'Our biggest environmental problem now and for the future.' 'The greatest catastrophe' . . . Thos are the drastic kind of terms in which the acidification of land and water in Scandinavia and North America has come to be described.

Swedish Environment '82 Committee,
Acidification Today and Tomorrow (1982)

In view of such emphatic descriptions, one would think that there was no disagreement about the basics: the rapid expansion of fossil fuel combustion after World War II has been accompanied by rising emissions of SO$_2$ which led inevitably to higher acid deposition and the cumulative effects of this environmental degradation are now threatening even larger areas of forests and fields.

In the mid-1970s, just about five or so years after acid rain first gained widespread attention in Sweden, and before its North American and European

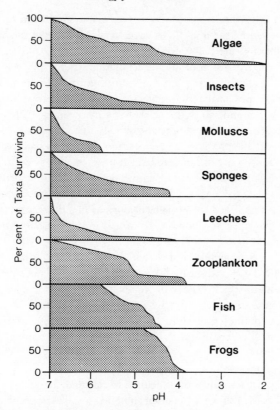

Fig. 5.15 Differences in tolerance of increasing water acidity for the main groups of aquatic biota. Individual exceptions abound but molluscs appear to be most sensitive, many insects and algae most resistant.

studies really took off, such a linear causation sounded unexceptionally correct. Since then, well past the first shocking revelations, we have entered a more difficult period of careful investigations which have been steadily revealing the great complexity of the whole acid deposition phenomenon and shrinking the space for simplistic generalizations.

Among these new findings none are more surprising than those demonstrating some fundamental uncertainties regarding the evidence for the claim of steady secular increases in the acidity of the precipitation falling on Eastern North America and on Western Europe. This still little appreciated conclusion results from careful reviews and analyses of measurements used to chart the acidity changes.

Reconstructions of rain acidities in the eastern half of the United States since the mid–1950s (largely based on measurements of major ions in wet deposition) show the area of acid deposition spreading westward and southward from the north-eastern core so that by the mid-1970s virtually all sites east of the Mississippi had average precipitation pH below 5.0. Similarly, European atmospheric chemistry network shows an expansion of high precipitation acidity (pH below 5.0) from a relatively small area covering England, northernmost France, Benelux, and southernmost Norway in the mid-1950s to virtually all Europe north of the Alps and the Loire Valley and east of the Wisla River by the mid-1970s.

These findings received widespread publicity and they have been repeatedly quoted as unassailable proofs of acid deposition spreading as a direct result of the increased combustion of fossil fuels in large sources (mainly power plants) with tall stacks. More recent studies detailing numerous uncertainties surrounding these reconstructions and puzzling facts arising from detailed analyses of the record have largely remained confined to specialized journals but their cumulative evidence upsets several principal (and now so engrained) simplistic assumptions regarding deposition of the acids.

Kramer and Tessier demonstrated that the drop in alkalinity of rain samples between the mid-1950s and the mid-1960s can be fully explained by a switch from soft glass containers (which, even with proper cleansing and rapid analysis imparted alkaline ions to the solution) to much less reactive plastics. Rain sampling in metal gauges, another common procedure during the 1950s had an even more distorting effect as their surfaces could lower the acidity of collected samples by more than one pH unit.

Many samples taken during the 1950s did not have their pH determined directly and principal solute ion values must be used to calculate the acidity levels. Yet this recalculation has been shown by Hansen and Hidy to carry errors of at least one pH unit. Effects of climate on the total ionic composition can also be considerable as the droughts in the early 1950s elevated the pH of precipitation owing to the much higher quantities of airborne alkaline dust.

A different class of problems arises from using these dubious measurements from sparse networks with changing sampling locations to represent a large area on contour maps charting pH changes. Consecutive records show that the *temporal* variability of pH data precludes any use of samples acquired at different sites and at different times to represent regional acidity trends. Interpolations across hundreds of kilometres in assembling isopleth maps from low-density samplings make the images even more misleading.

All of these considerations would be enough to conclude that historical data are both qualitatively and quantitatively insufficient to demonstrate a clear trend toward increased precipitation acidity in the Eastern United States, and such a conclusion is strongly buttressed by some reliable uniform long-term monitoring in the area of the highest assumed impact. Connecticut measurements between 1929 and 1948 show no secular trend and their high (around 4.2) acidities are identical to those now claimed to represent the current worsened conditions in the Northeast. Most significantly, seventeen years of pH monitoring of weekly bulk samples at New Hampshire's Hubbard Brook Experimental Forest show no upward trend. Between 1964 and 1981 SO_4^{2-} deposition actually declined by 25 per cent from 3.16 to 2.36 $\mu g/L$, while NO_3^- deposition rose from 0.7 to 1.66 g/L and pH fluctuated between 4.03 and 4.20.

European measurements should be much more reliable as the northwestern part of the continent has had a denser and more uniformly run precipitation chemistry network since the mid-1950s, but when Kallend and his colleagues analysed in detail the changes between 1956 and 1976, the two decades covering the rapid expansion of fossil fuel combustion, their regressions demonstrated that only 24 per cent of sampling sites showed a statistically significant trend of rising acidity and in about one-third of these instances the increases came abruptly in the mid-1960s without paralleling the secular rise in SO_2 emissions.

And even where the acidity rose, its higher annual means were not caused by generally elevated pH but rather by an increased frequency of high monthly levels: post-1965 data have a much greater scatter and some months continued to have pH similar to those before the mid-1960s. Certainly the most puzzling are the results of detailed analysis for a group of adjacent sites located within 25 km of each other near Uppsala. A common regional component in the variation of these values accounts for just 20–30 per cent of the variance—so that the bulk of the differences must be explained by site-specific measurement factors.

Undoubtedly a large part of these puzzles and uncertainties has its origin in complex relationships between SO_2 and sulphates. In general, above remote oceans the SO_2/SO_4 ratio clusters around 0.1, it rises to 0.5 over clean continental locations and ranges between 1.0–1.5 over polluted areas. Yet the particulate sulphate levels are often only poorly correlated with SO_2 concentrations. Aubrey Altshuller was the first one to note, in 1973, that in American urban locations SO_2–SO_4 relationships are non-linear to such an extent that a 90 per cent SO_2 reduction would, in a statistically average site, bring only a 53 per cent decline of sulphates.

More recently, Cooke and Wadden's 1981 analysis for Chicago found that only with low relative humidities and cold temperatures did concentrations of the two compounds display statistically significant correlations although the variance explained was still no more than 24–33 per cent. When sulphate levels were high there was no significant correlation and only two-fifths of all sulphate were attributed to local sulphur emissions. Long-range transport of upwind SO_2 emissions which get oxidized on the way is the obvious explanation of such discrepancies but uncertainties and missing links abound here as well.

Measurements taken by Pierson and his colleagues on the summit of Allegheny Mountain in Pennsylvania, a location right in the path of westerly winds carrying sulphur pollution from the Ohio Valley power plants, are a fine example of these puzzles. Two months of monitoring in July and August, the peak months for SO_2 oxidation, failed to establish a genetic relationship between atmospheric SO_2 and aerosol sulphates. No correlations were found either between SO_2 conversion to sulphate and relative humidity, wind speeds, and the presence of alkaline aerosols. The logical conclusion would be that sulphates at the site did not originate from SO_2!

And although mathematical models show the area to be near the maxima for sulphate deposition from the Ohio Valley power plants, the actually observed concentrations were four times higher than the forecasts. What then are the sources? Similarly, Reisinger and Crawford's analysis of sulphate concentration in relation to air trajectories in the Tennessee Valley demonstrated that the maximum concentrations come, as might be expected, with northeast flow from the Ohio River Valley—but the highest mass transport is associated with southwest tropical flow. As man-made emissions south of the Tennessee Valley are only about one-eighth of those north of it the only plausible explanation may be to attribute much of the sulphates to biogenic sources in Southern Louisiana and Mississippi.

The modelling of long-range transport is still very much in the category of qualitative appraisals as the inherent uncertainties and complexities of atmospheric behaviour necessitating numerous simplifying assumptions are potentiated by the tremendous variability of chemical reactions during the diffusion of gases and movement of aerosols. Conversion of SO_2 to SO_4 by photochemical gas phase oxidation is almost always very slow (often less than 1 per cent per hour) but aqueous phase reactions can convert one-fifth or one-quarter of all SO_2 in power-plant plumes within fifteen minutes after their release—or they may proceed very sluggishly.

Consequently, selecting, as the current dispersion models do, a single conversion factor to simulate the reactions extending over several days and in different environments (10^2–10^3 km can be covered during that time) is obviously a funda-mental weakness. As similar uncertainties prevail with regards to deposition processes—our knowledge of wet deposition is reasonably good but dry settling, adsorption, and impaction, the processes accounting for about half of all deposited sulphates, are very poorly known—our abilities to forecast the long-range transport and ground concentrations of sulphates remain very limited.

Undeniably, there is a great deal of long-distance transference of sulphur compounds. Their typical atmospheric residence times of 20–60 hours (40 would be the best representative value) mean that tropospheric winds can take them 500–2000 km away from the source before they are deposited. But our understanding of these transfers is quite insufficient to be sure about the effect of possible control strategies.

For example, the earliest Swedish work on acid deposition blamed sulphur transfers from British sources (above all large coal-fired power plants equipped with tall stacks) for the rising acidity of South Scandinavian rains. This appeared to be a logical explanation until simple diffusion models showed that the British share of Swedish deposition is just 10 per cent of the total, or one-half of the Swedish domestic emissions. The rest of Europe would be thus more to blame than Britain—but how to square this with actual Swedish deposition values whose variation (as noted earlier in this section) cannot be explained simply by long-distance transfer?

Long-range transport seems to be also the reason for some surprisingly high acidities of precipitation in remote locations but the explanation cannot have general validity. When the first results of the global Precipitation Chemistry

Project were published in 1982 volume-weighted pH values for all five remote locations—in Northern Australia, Southern Venezuela, Central Alaska, Bermuda, and Amsterdam Island—were mostly between 4.6 and 5.0, appreciably more acid than the pH of 5.6 for 'clean' rain.

Anthropogenic sources of these accidities are readily seen in the case of Bermudian (just 1000 km away from Eastern United States) and Alaskan precipitation. Anthropogenic emission could not be decisive either in Northern Australia or in Southern Venezuela and they must be completely excluded on Amsterdam Island, halfway between Africa and Australia at about 40 °S. A look at a global map of pressure centres and prevaling winds will demonstrate the impossibility of such an effect. The same is true of low pH values recorded at American Samoa during the GMCC measurements which also uncovered acid precipitation (volume-weighted mean of 4.9) near the top of Mauna Loa on Hawaii Island.

These Hawaiian acidities have a long-range explanation: dust from East Asia (North China) containing sulphate can be carried 10 000 km eastward and detected at Mauna Loa after less than ten days. But pH values at the observatory showed that rains coming from *every* quadrant were acid—and what man-made sources are there south or north of the island? And after ten days only a small fraction of Asian sulphates can be deposited on Hawaii but they can cause a low pH of the precipitation because at such altitude there are even fewer alkaline ions available to neutralize the acids. Hawaiian experience thus points out the necessity to consider the overall acid–base situation.

Only this often neglected acknowledgement of the complexity of airborne solutions can explain why some regions with heavy SO_2 emissions have no acid deposition problems. Perhaps most notably, Beijing's rains have a pH mostly between 6.0 and 7.0, that is *less* acid than 'pure' precipitation although the concentration of sulphate is high as must be expected in the capital of a country where raw coal is the dominant household and industrial fuel (70 per cent of primary energy consumption). The explanation of this discrepancy: high levels of suspended alkaline matter, ranging from the desert dust blown into the city to the everyday sweeping of unpaved alleys, neutralizes the acids.

A similar situation prevails in all polluted urban and industrial areas with high concentrations of alkaline terrigenic dust (India, the Middle East, Mexico). Consequently, our preoccupation with acid and acidifying substances to explain the phenomenon of acid deposition is, logical as it may seem, an unprofitably narrow one. Seen in the broader perspective could not the current acid rain problems in Western Europe and Eastern North America then be ascribed as much to the disadvantageous environment of these industrialized regions as to their large SO_2 emission?

A series of contour maps for major cations and anions in the North American atmosphere helps to answer this question positively. North America's heartland has the highest combination of alkaline cations (Ca^{2+} and Mg^{2+} from dry cultivated prairies and plains) and ammonia (NH_4^+ from volatilization of animal wastes and fertilizers) and hence a high potential to neutralize acidic precipitation. In contrast, the East, the largest producer of sulphates and nitrates, is highly deficient in these buffers and cation/anion balances are thus very different.

West of the Mississippi 75–96 per cent of all acid anions are easily neutralized–while in the Northeast 52 per cent of them must be balanced by H^+ and result in low pH. One must conclude that if the American industrial heartland with its large SO_2 were located some 1500 km to the west (its centre thus being in Kansas and eastern Colorado rather than in Ohio and Pennsylvania) the acidity of North American precipitation would not be much of a concern. And if there are such strong reasons to believe that the location of the combustion is at least as much the cause of the problem as the emission themselves, it is even more intriguing to consider a possibility that our well-intended actions since the 1950s have greatly aggravated that situation.

I am referring to a great loss of airborne alkaline material which resulted from the large-scale displacement of coal as the dominant household, transportation, and industrial boiler fuel by oils and gases, and by the widespread and very efficient capture of fly ash from electricity generating plants which became coal's main consumers. Whatever the actual, and never to be precisely known, value of the fly ash decline might have been, there is no doubt that the atmosphere above the Northeastern United States has been deprived of a large part (at least one-third, perhaps up to two-thirds) of anthropogenic alkaline aerosols—and as the natural alkaline terrigenic matter is much rarer in the region's atmosphere (three to twenty times) than in the drier and less vegetated central and western parts of the continent, this substantial decline had to have an effect on the Northeast's relatively high acidity of precipitation.

When the objectionable particulates were being eliminated by the obligatory installation of high-efficiency electrostatic precipitators nobody was suggesting that this so obviously desirable development might bring an intensification of a much more worrying pollution problem—but we must conclude that for the great success in controlling fly ash we have paid a most unexpected price by *de facto* aiding the formation and diffusion of acid deposition.

So far I have looked at five broad areas where our best current understanding is much less resolute than the earlier conclusions and where the uncertainties and unanswered questions have come to dominate simple answers: secular increases of precipitation acidity, relationship between sulphur dioxide and sulphates, deposition processes, long-range transport and rain acidity in remote places, and acid–base considerations in atmospheric chemistry (with a special emphasis on an unintended intensification of acid deposition due to efficient particulate controls). This might be enough to indicate that the title of the section, elusive understanding, is no hyperbole—but I must raise just one more conglomerate of uncertainties, the one regarding the relative roles of sulphates and nitrates.

The combustion of fossil fuels in stationary sources produces nitrogen oxides both by oxidation of the element present in the fuel and by formation of NO and NO_2 from atmospheric dinitrogen. The total global flux from these sources is now about 20 million tonnes of nitrogen a year. Nitrogen oxides from motor vehicles are equivalent to some 6 million tonnes N and airplanes, train engines, boats, ships, and various off-highway vehicles add at least another 2 million tonnes. Chemical industries, petroleum refining, production of portland cement, and coking represent only minor fluxes but the combustion of fuelwood and crop residues releases at least 4 million tonnes N as nitrogen oxides. Vegetation burned by shifting cultivators releases around 2 million tonnes N so that the total annual anthropogenic flux of nitrogen oxides is around 35 million tonnes N (a substantial mass compared to the uncertain estimates of 20–90 million tonnes N for biogenic NO_x releases).

These emissions can raise the background concentrations by up to two orders of magnitude in the most polluted urban and industrial areas but nitrogen oxides usually reside less than two days in the atmosphere and processes responsible for this fast removal involve a variety of oxidative and, above all, photochemical reactions whose study has been very intensive ever since it was realized (in the 1950s) that they are essential in the formation and maintenance of smog. Deposition of the reacted compounds is mostly in the form of ammonia and nitrates and the latter, together with sulphates, are the major contributors to the acidity of precipitation. However, much is unclear about the relationship between the precursors of these two compounds and some conclusions have been outright contradictory.

For example, Rodhe, Crutzen, and Vanderpol concluded on the basis of sulphate and nitrate monitoring in Northwestern Europe that an increase of the NO_x/SO_2 emission ratio may not be directly responsible for a higher content of HNO_3 in precipitation—while measurements in the American Northeast support a rather clear link between the two compounds. The best long-term evidence for this comes from the already mentioned Hubbard Brook Experimental Forest. There the HNO_3 contributions rose from a maximum of 20 per cent in summer 1964 to 39 per cent in summer 1979 (winter contributions going up less, from 44 to 61 per cent during the same period) and higher NO_x emissions offer the best explanation of this rise.

The NO_x/SO_2 ratio for American emissions shows a steady rise from about 0.25 in 1940 to nearly 0.6 in the late 1970s, a hardly surprising change in view of the massive post-World War II diffusion of cars and the strong shift of coal consumption to high-temperature (and hence high NO_x generation) combustion in power plants. In yet another contradiction of European and American findings, the recent reliable measurements from the Northeastern United States do not provide any support for conclusions about rising SO_4^{2-}/NO_3^- ratio with long-range atmospheric transport.

This seems logical considering the relatively rapid reaction of NO_2 with hydroxyl radical, a process faster than H_2SO_4 generation, which should produce locally more nitrate than sulphate. But American data, although showing clear seasonal variations with SO_4^{2-}/NO_3^- ratio always higher in summer (about 1.4, compared to 0.9 in winter), display a surprisingly uniform spatial pattern: stations in such source states as Illinois and Ohio have the same sulphate/nitrate ratios as in

Northern New York, a sensitive receptor area about 1000 km downwind. Oxidation of SO_2 thus appears, unlike in European experience, to be as efficient as conversion of nitrogen oxides.

If nitrates are already dominant during winter their further increase—and emissions of NO_x are expected to grow faster than SO_2 releases—may soon put them into the forefront of control efforts and such a shift would make a fundamental difference to the management of the whole acid deposition problem. Unlike sulphur dioxide, now generated overwhelmingly in a relatively small number of large power plants and non-ferrous smelters, nitrogen oxides are produced in a much more diffuse manner (for example, in North America nearly half of an NO_x emission comes from transportation, a source which contributes less than 3 per cent to SO_2 generation) and their controls are thus logistically more taxing.

But, to address the last uncertainty in this section, if the acidification effects of nitrates are not equivalent to those of sulphates then it would not be so worrying, or it may actually be beneficial to have nitrates dominating the acid deposition. This argument rests on the well-known differences in behaviour of the two compounds. While sulphate outputs usually equal inputs, only about one-quarter of nitrates deposited on monitored ecosystems leaves them as the nitrogen is extensively utilized by the plants. Assimilation consumes one mole of hydrogen for each mole of nitrate in converting HNO_3 to ammonia and CO_2—and the same exchange accompanies denitrification ending up with N_2, CO_2 and H_2O.

Consequently, nitric acid is effectively decomposed by biota, its nitrogen either absorbed by plants or recycled by denitrifying bacteria, its hydrogen incorporated into ammonia or water, in both cases an innocuous outcome. While it is true that nitrate reduction proceeds better in an anaerobic and neutral environment it can go on, at slower rates, in aerobic conditions and at acidities of pH 4.0 or less which means that nitrates can be reduced not only in organic soils but even in acidified waters. If the NO_3^- reductions are proceeding at rates equalling or surpassing those of nitrate deposition there should be no acidification of soils and waters.

On the other hand, in soils and waters which have been already acidified to such an extent that nitrate reduction is much depressed, the HNO_3 contributions to further acidification would be inevitable. Moreover, HNO_3 may be a major acidifying agent in aquatic ecosystems during spring when its winter accumulation (snow has almost always more HNO_3 than rain) is suddenly released into waters (small lakes would be especially vulnerable) where the low pH shock may kill many sensitive organisms before the acid is gradually decomposed.

The case of HNO_3 illustrates well the impossibility of simple generalizations regarding the environmental consequences of acid deposition. The changes noted have ranged from a productivity-raising fertilizing effect in coniferous forests to acute degradation of lakes and I will try to capture some of this complexity in the next section on acidity in waters, soils and biota.

5.4.2 *Acidification of the environment*

> Perhaps the first well-demonstrated widespread effect of burning fossil fuel is the destruction of soft-water ecosystems by 'acid rain' which has been caused by anthropogenic emissions of sulfur and nitrogen oxides that are further oxidized in the atmosphere.
>
> National Research Council, *Atmosphere-Biosphere Interactions* (1981)

That acidification of lakes should be the first instance of a clear link between combustion of fossil fuels and a large-scale degradation of ecosystems is rather remarkable. Localized effects were, not surprisingly, noted a long time ago. When R. A. Smith wrote in 1872 his now classical account on *Air and Rain*, he mentioned not only the corrosive damage to building materials and metals but also the bleaching of chlorophyll in aquatic plants and the interference of acid gases with the growth of crops near cities—and he already could cite several detailed field studies and tests of such effects conducted earlier in the nineteenth century.

But most of this obvious, acute urban-generated damage disappeared throughout the rich Western world with the conversion of household combustion from coal to hydrocarbons, with the location of large coal-fired power plants away from the cities, and with the construction of tall stacks. These developments were accompanied by impressive declines of urban SO_2 and SO_4^{2-} levels but they led to the much longer transport of emissions—when the buoyancy of large hot power plant plumes is taken into account the effective height of SO_2 emission in summer is as much as two to three km and vigorous tropospheric mixing can carry the gases and aerosols much higher, positioning them for transfers of many hundreds to a few thousands of km—and hence to acid deposition far away from the sources.

And so it was not the releases of fine aerosols (about whose climatic effects there is no shortage of speculation), the build-up of carbon dioxide (whose ecosystemic impacts, if any, are yet to be discerned amidst the ever-present atmospheric perturbations), or the dispersal of toxic metals (above all mercury, an earlier environmental scare discussed in section 4.1.2) but acid deposition which came to be identified as the first obvious large-scale (though not global) ecosystemic degradation arising from the combustion of fossil fuels.

The undeniably severe disruptions of some sensitive aquatic ecosystems are the best evidence of these undesirable effects but before reviewing the essentials of this by now fairly well-understood degradation, and then proceeding to discuss the much more complex case of acid deposition's effects on agroecosystems and forests, at least a few pages describing the atmospheric changes and damages to materials and human health are in order.

Even before the sulphates and nitrates are deposited they have a profound effect on atmospheric environment by greatly reducing visibility—bringing the haziness and blurred near-horizons not only to industrialized regions but increasingly also to large areas of rural countryside. Visibilities are determined by the mass of fine aerosols in the range of one-tenth to one micron and sulphates, followed by nitrates, are dominant constituents of this particulate matter. Moreover, condensation nuclei provided by these fine aerosols lead to easier formation of dense advection fogs which the recent measurements showed to be much more acidic than the average rains in the same area.

Field measurements indicate that at least 40 and up to almost 90 per cent of the light extinction attributable to particulate scattering is caused by sulphates. Seasonal differences are quite pronounced—with haze maxima in July and August and with the lowest values between November and February—and comparison of historical data shows large visibility declines in regions of expanding sulphurous emission. Most notably, July-September visibilities in the Carolinas, Ohio River Valley, Tennessee, and Kentucky declined from 24 km in the early 1950s to 13 km two decades later, clearly parallelling the region's huge expansion of coal combustion in large power plants.

The effects of sulphurous air pollution in mid-latitude atmospheres have been known for decades but the influence reaches much further afield: since the mid-1970s investigations of Arctic haze—often a surprisingly thick brownish or orange layer extending throughout the troposphere, covering an area almost as large as North America and restricting the horizontal visibility to just 3-8 km—proved its largely anthropogenic origin, with industrial sources in the central part of the USSR and in Western Europe appearing to supply most of the aerosols. The eventual influence of Arctic haze on the region's radiation balance remains a matter of speculation: the haze is seasonal (its build-up starts in late fall and its density peaks in March and April), its thickness and extent varies, as does its black soot content, largely responsible for absorption.

Of all the effects of acid deposition none is easier to study and to test in minute detail than the weathering and corrosion of materials. Comparative exposures of metals and building materials have provided a wealth of information on the rates and conditions of these surprisingly expensive degradations. Reactions of SO_2 and SO_4^{2-} with $CaCO_3$ are responsible for most of the deterioration of limestone and marble and where rains keep removing the $CaSO_4$ in solution the weathering can continue indefinitely: interesting experiments by Braun and Wilson showed that weathered limestone surfaces from a fifteenth century building of Eton College removed as much SO_2 as fresh cut facets of the same block!

Damage done to ancient monuments, so often built of or clad in marble, has attracted most concern. The deterioration of the Acropolis (built of Pentelic marble) skyrocketed since 1945 with the rapid industrialization of Athens and the shift to burning of high-sulphur fuel oils. Preservation of the badly decayed structures remains an unsolved problem as there is no consensus regarding the effects of synthetic resin coatings and as stringent controls of the city's serious air pollution will not be implemented in the immediate future. The Taj Mahal's turn appears to be next as the new refinery in Agra is just 30 km away and its SO_2 emissions (25-30 t/day) will be carried in the mausoleum's direction between October and March by the prevailing winter monsoon winds.

Metal corrosion losses amount globally to many billions of dollars as unprotected carbon steel, galvanized steel, nickel-plated steel, nickel, and zinc corrode much faster in environments with elevated SO_2 and SO_4^{2-} levels, especially in the presence of higher moisture. Buried metals can also be seriously affected by the actions of sulphur-reducing bacteria and a mere 0.1 per cent of sulphate in groundwater can initiate degradation of underground concrete structures. Paints are also frequently affected by discolouration, loss of gloss, sulphate blooms, and eventually complete destruction takes place.

Health effects of SO_2 were of great concern during the 1950s and 1960s but current annual averages in Europe, Japan, and North America are usually below 50 $\mu g/m^3$, too low to cause acute, serious changes in pulmonary function and structure which appear to arise only with SO_2 levels at 3–10 mg/m^3. However, disagreements continue in regard to sulphur dioxide's effects on general mortality and on the higher incidence of respiratory diseases.

General mortality data appear to be too insensitive for accurate evaluations of chronic low-level pollution but, in any case, much careful analysis indicates that sulphur oxides or particulates at the levels now usually prevailing in rich industrialized countries have a negligible effect on mortality. But the controversies will not be stilled soon: the room for interpretation is still wide enough to permit differing conclusions. For example, critics of one of the United States Environmental Protection Agency's major studies argued that it used 'naïve models' and that its results 'should not be regarded as reliable in any way. Their value can be doubted even as orders of magnitude'.

In response, the study's authors argue that there is no way to eliminate estimator risks in studies of this kind and define their interpretations with numerous references to the theory of methods and meanings of advanced statistics and econometrics. In any case, there appear to be no unequivocal proofs linking current typical concentrations of SO_2 and sulphates with increased mortality and at best there are only suggestive links with higher morbidity.

Perhaps the best illustration of our inability to offer meaningful evaluations of SO_2 effects on human health are the assumptions of an influential OECD study published in 1981. While the costs for its strictest control scenario were just 2–3 times higher than for the least effective option, uncertainty about the health effects forced the authors to use a 25-fold range in estimating the health benefits. As a result, total benefits for each of the three scenarios would be much below the control costs if the lowest assumptions of health gains best reflected the reality—or, in the case of the highest assumed costs of mortality and morbidity the cost-benefit ratio was very favourable. Obviously, such evaluations are of no help in deciding about management strategies.

Uncertainties surrounding the understanding of health effects give a good foretaste of difficulties encountered in disentangling the role of acid deposition in ecosystems. Even in the best demonstrated case, the acidification of freshwater ecosystems, generalizations often run up against peculiarities as the declining pH of lakes and streams is not simply just a function of acidity of precipitation but it also depends heavily on the nature of drainage basins.

In well-buffered basins with calcareous substrates the alkalinities of waters are high and inputs of acid will be continuously neutralized with no, or only negligible, pH changes even after decades of acidification. In contrast, soft, poorly-buffered waters become susceptible to acidification once the alkalinity falls below 300 μEq/L and pH can drop rapidly below 5.0, first just during spring after acid meltwater surges and later, when bicarbonate is essentially eliminated from the water and replaced by sulphate as the major anion, these high acidities can become permanent.

The other most important environmental factors determining the rate of acidification are high precipitation on impermeable ground (making difficult deep penetration into the ground where contact with neutralizing materials could lower the acidity), the small size of many basins (with little possiblity of diluting the sudden influx of acid water), small catchment areas (with less opportunity to neutralize the precipitation during its travel through the soils), and headwater location (where thinner soils and short catchment travel are, obviously, much more likely).

The combination of these influences results in a highly specific response: lakes in the same region receiving identical acidity inputs per unit of catchment area and water surface will range from heavily acidified small bodies surrounded by steep rocky slopes to considerably much less affected larger lakes with gently sloping catchment areas and soils with higher buffering potential.

In affected lakes low pH and high sulphate concentrations are accompanied by dissolution of aluminium hydroxide which neutralizes further pH declines but puts large amounts of highly toxic Al^{3+} into solution. Unaffected lakes thus have calcium-magnesium-bicarbonate waters while the heavily acidified ones are characterized by a hydrogen-calcium-magnesium-sulphate mixture with unusually high levels of aluminium and heavy metals (mercury, manganese, zinc, lead, copper, and cadmium).

The total number of lakes affected in this way has been increasing in both Europe and North America. In Sweden more than 18 000 of the country's 85 000 lakes with area larger than one ha were seriously acidified by 1982, in Southern Norway lakes covering over 13 000 km² have no fish, in Canada (mainly in Ontario and Quebec) and Northeastern United States hundreds of lakes have pH falling below 5.0, diminished fish counts or complete disappearance of sensitive species. This spreading 'dying' of lakes has been intensively studied since the early 1970s and enough is known to offer reliable explanations.

Increased acidity inhibits many biochemical processes and these changes, interfering with nutrient cycling, lead to sequential disappearance of biota. Some phytoplankton species start disappearing at pH just below 7.0 (see Fig. 5.15), but the total lake phytomass may not change much. Long before a complete loss of individual fish species there are numerous degradative signs, above all poor recruitment, failure of females to reach spawning condition, high egg and larvae mortalities, fin and gill-cover deformities. Numbers of fish then start declining drastically and their age structure changes rapidly with fish younger than one year least often present in highly acidified (pH 4.0–4.5) waters. Species totally lost in such lakes include brown trout and Atlantic salmon in Southern Norway, char and roach in Sweden, lake trout, smallmouth bass and lake whitefish in Ontario. The physiological explanation of fish demise involves the disruption of normal ionic and acid-base balances: gill membranes are highly permeable to hydrogen ion and low pH causes losses of plasma sodium and chloride as well as inhibition of RNA synthesis. Moreover, aluminium ions irritate the gills, destroy protective mucus, erode gill filaments and cause eventual suffocation.

While the decline of fish numbers and the complete elimination of sensitive species attracted most research and popular press attention a much more worrying problem from the viewpoint of human health is the acidification of groundwaters. Marked changes in hardness (doubling of Ca^{2+}) and alkalinity (halving of HCO_3^-) have been encountered in many Swedish localities but the most worrying is the leaching of metals out of the acidified soils and also out of piping systems when the pH drops below 4.0. Corrosion of copper pipes can raise the drinking water's metal content to 30 times the recommended maximum and, while in municipal waterworks treatment can easily maintain neutral pH, in rural areas water from many private wells will continue to be acidified.

Expensive alkaline filters can handle this problem—but not the removal of aluminium released from acid soils. Currently there are no public health standards on Al in drinking water and it is possible that the total amount of the metal in the body is homeostatically controlled within the range of normal intakes (1–30 mg Al/day) but it is aluminium's potential role in Alzheimer's disease which provides perhaps the most worrying, although still only very tenuous, link between acidification of ecosystems and human health.

People with Alzheimer's disease (neuro-fibrillary degeneration affecting tens of millions of older men and women worldwide) have markedly higher blood and brain Al levels and animal studies of molecular interaction of Al with the central nervous system showed pathological changes resembling those of Alzheimer's disease. Although we do not know if Al is the etiologic agent, a recent review of Alzheimer's management concluded that all 'known extra sources of entry of this metal should be eliminated insofar as is feasible'.

But a closer look at the complexities of soil chemistry will demonstrate that natural processes and changes in land use may have as much, and often much more, impact on the acidity of soils, and hence on the quality of waters, as deposition of acid compounds derived from combustion of fossil fuels. To begin with, acidity of 'normal' rain, that is one with pH 5.6, causes surprisingly large releases of $CaCO_3$ (in the order of 400 kg/ha with precipitation of one m a year) and rain with pH of 4.6 will dissolve only additional 10–12 kg of limestone, an insignificant increment especially in acid farm soils where limestone applications average 2–5 t per ha!

The second little appreciated reality concerns the direction of natural soil formation processes: as soils mature they become more acid than their substrate regardless of whether the parent material was acidic (igneous) or basic (sedimentary). Soils are living assemblages and respiration of their flora and fauna always creates acid conditions in the topmost layer of mineral soils. The third twist deals with the equivalence of increases in H^+ in acid rain and losses of nutrient cations. Their ratio cannot be, as commonly accepted, a unity because then the already strongly acid soils would be rapidly acidified owing to their low nutrient content. In reality, removal of nutrient cations by H^+ in soils with pH below 5.0 is relatively slow.

Explanation of this counterintuitive phenomenon lies in the buffering capacity of organic acids abundant in strongly acid soils. These humic acids are made less soluble by added H^+ and hence the inputs of H_2SO_4 promote sulphate output but lower the flux of organic anions. The net result is little or no change of soil pH so that even in seemingly susceptible regions measurements cannot find any excessive chemical weathering attributable to acid deposition. As for the leaching of aluminium ions there seems to be no need to point at acid deposition as the key mobilizational factor in Al^{3+} releases. Naturally acid soils have so much of soluble Al^{3+} that any rain will flush it out. Besides, the solubility and mobility of aluminium complexed by organic anions drops drastically with pH below 4.5.

The final set of arguments looks at the effects of changing land use. Edward Krug and Charles Frink who systematized the previous four points concerning complexities of soil acidification also gathered plenty of material to prove that in the Northeastern United States, where acid rain is commonly seen as *the* reason for environmental degradation, it is the recovery of the region's forests from previous burning, overcutting, overgrowing, and heavy erosion which has brought the acidification of forest floors. Similarly, studies in Norway showed that the exchange acidity of surface humus of a 90-year old spruce forest growing on an abandoned field is equivalent to the input of strong acids from about 1000 years of annual precipitation of one cm with pH 4.3.

Restating these conclusions is not superfluous as they dispel most of the commonly held 'truisms' about acid rain: excess acidity introduced from precipitation containing acids derived from combustion is just a small fraction of normal rain's acidity; natural soil formation is strongly acidifying; the hypothesis of equilvalent leaching of H^+ and nutrient cations is not supported by field measurements; Al^{3+} losses are not necessarily an outcome of anthropogenic acidification; and soil formation in previously disturbed ecosystems appears to be the greatest source of naturally rising soil acidity.

And one more common misperception regarding the acidification of soils needs correcting: frequently reprinted outline maps showing the sensitivity of soils to acidification must leave an impression that huge areas in the industrialized countries are so endangered. Figure 5.16 is a good example putting a large part of the eastern half of the United States into a highly sensitive category. How misleading is this generalized image can be easily seen by referring to a detailed map of soil sensitivity east of the Mississippi prepared by McFee.

As will be seen in Fig. 5.17 the only sensitive soils are in northern parts of Wisconsin and Michigan, parts of the Appalachian region, and a portion of the Northeast and New England. The total area of highly sensitive soils is thus relatively

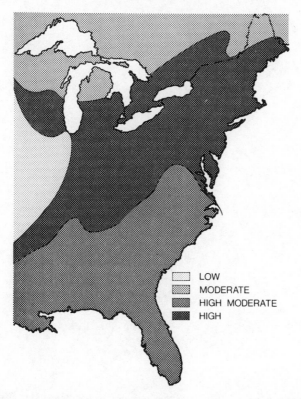

Fig. 5.16 Maps at this level of generalization have been frequently reprinted to demonstrate the widespread susceptibility of eastern North America to acidification.

small and, moreover, detailed examination of an enlargement of the Adirondacks region, generally considered the most sensitive part of the Eastern United States, shows that even there susceptible locations and resilient areas form a complex patchwork. Similarly, assignation of most of the Canadian Shield soils to the sensitivity category on the basis of granitic bedrock ignores the overlays of calcareous glacial drift providing considerable buffering capacity (Fig. 5.18), and the Scandinavian soils are also a mixture of the sensitive and resilient.

Besides, in all soils under intensive cultivation acid deposition will be greatly overshadowed by effects of nitrogenous fertilizers. Mineralization of ammonia contributes, assuming average rate of fertilization is about 100 kg N in ammonia compounds, at least thirteen times as many H^+ ions each year as does acid precipitation—and to offset this acidification relatively large amounts of liming materials (10^2–10^3 kg/ha · year) are regularly added to soils with low calcium content. Swedish experience illustrates quite well the extent of acidifying effects of nitrogenous fertilizers on a national level (Fig. 5.19).

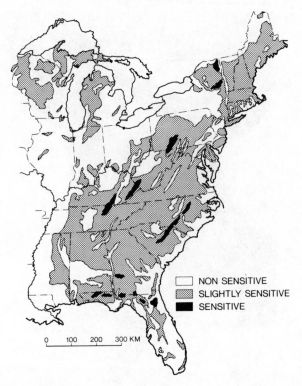

Fig. 5.17 A closer analysis of soil acidities, such as this map prepared by McFee, reveals a different picture: sensitive areas are limited to only small and disjointed patches. Even more detailed mapping shows that there is little uniformity even within the most sensitive plots which contain a good deal of non-sensitive soil.

Effects of any conceivable acid deposition on soils fertilized with ammoniacal compounds are thus almost negligible and their correction would necessitate only marginal increases in liming rates. If most farm soils were not much affected by acid deposition it is not surprising that annual crops should not be greatly hurt either. Of course, classical air pollution studies have documented in detail both acute and chronic plant injuries following exposures to higher SO_2 concentrations, but this kind of damage has become much rarer with tall stacks, and yield losses attributable to SO_2 are very low even in the areas with large concentrations of SO_2 sources.

The best such example is the Ohio River Valley, the world's leading concentration of SO_2 emissions, where SO_2-related crop yield reductions were put at just 0.6–1.1 per cent of all air pollution-induced losses (ozone being responsible for virtually all of the damage). The effects of higher precipitation acidity on yields of rain-fed species have been rigorously investigated only since the late 1970s and the conclusions are agreeably bland.

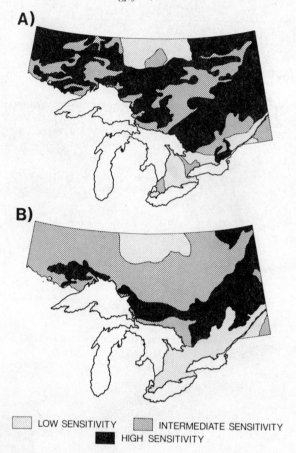

Fig. 5.18 Another illustration of differences between two evaluations of susceptibility to acidification. When judging by the substrate, most of the land north of the Great Lakes and south of 50 °N appears to be highly sensitive (A). But glacial drift put plenty of calcareous deposits over that bedrock and soils in the area are thus much less susceptible to acidification than a bedrock appraisal would indicate (B).

In experiments with 28 crops grown to harvest in field chambers exposed to simulated acid rain with pH 3.0, 3.5, and 4.0 and to control precipitation with pH 5.6, yields were lower for just five vegetable species (radishes, beets, carrots, mustard greens, broccoli), higher for six (tomatoes, green peppers, strawberries, alfalfa, orchard grass, timothy hay), ambiguous for one (white potatoes) and showed no change for fifteen crops, including staple grains which appear to be the least sensitive to acid deposition.

While the annual crops offer a picture of good resistance, many tree species are obviously much more vulnerable and the fate of forests in the regions of high

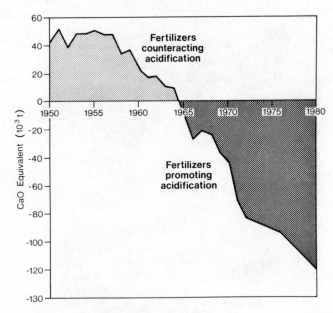

Fig. 5.19 Ammoniacal fertilizers are now the leading cause of soil acidification in Western agroecosystems. Swedish experience, charted here, shows how before 1965 the applied fertilizers were actually lowering the acidity but since then liming is needed to counteract the abundant supply of H^+.

sulphate and nitrate deposition has become the most controversial, and undoubtedly also the most critical, point of acid rain debates. After all, what are even thousands of fishless lakes in sensitive watersheds compared to the gradual decline of productivity and eventual demise of thousands of ha of both natural climax and managed commercial forests?

Yet it is precisely in this potentially most damaging case where simple causative links (and their fixed linear or exponential progression) are least appropriate. Acid deposition does not mean inevitable damage to forests and more of it does not automatically raise its eventual impact. Such effects are abundantly documented for plant injuries downwind from large point sources of SO_2 (crippling damages from smelter plumes have provided much material for detailed studies of truly stunning environmental degradation)—but they are not a result of diffuse acid deposition hundreds to thousands kilometres away from SO_2 sources.

The complexitiy of acid deposition effects on forest growth is perhaps best illustrated by comparing the best conclusions from three afflicted areas: Southern Scandinavia, Northeastern United States, and West Germany. Extensive Norwegian experiments with artificial acid rain found no decline in the growth of coniferous trees. The summary of the long-term Norwegian project on the ecosystemic effects of acidification is the best conclusion of Scandinavian experience: 'Decrease in forest growth due to acid deposits has not been demonstrated. The increased nitrogen supply often associated with acid precipitation may have a positive effect. This does not exclude, however, the possibility that adverse influences may be developing over time in the more susceptible forest ecosystems.'

American experience seems, especially at first glance, to support a strong causal link: spreading dieback and large reductions in basal area and density in high-elevation red spruce forests of New York, Vermont, and New Hampshire are there for everybody to see—and the region is seen as a principal deposition area of acid compounds carried from the Ohio River Valley. But a comparison of growth rates of various trees starts altering the perspective.

Between 1964 and 1982 red spruce diameters at the Whiteface Mountain in the Adirondacks declined by over 70 per cent—but those of younger firs expanded by up to 50 per cent, and birch diameters went up about 2.5 times. Clearly, a major change of species composition is in the making but its effect on the total phytomass of the forest may be negligible (or even positive)—and its cause may not be acid deposition after all.

Current decline may be a wholly natural process triggered and sustained by drought, or a combination of drought and subsequent infection, or a complex of drought, infection and acid deposition stresses. Drought is the best documented cause of growth disruption with historical data recording a wave of spruce diebacks during the last quarter of the nineteenth century and with the beginning of dramatic decreases of annual increments following an unusually dry period at the beginning of the 1960s. Moreover, aluminium toxicity appears to be playing no part in this decline as higher levels of Al are actually recorded in healthy root systems.

If neither the Scandinavian nor the American experience add up to a clear, worrying link, the current West German claims more than make up for this uncertainty. For decades forest decline has been linked with acid deposition in several heavily exposed regions of Central Europe but it was the survey of damages conducted in West Germany in summer 1981 and published in 1982 which revealed a surprisingly extensive *Waldsterben* (forest dying) and which attributed it to acid deposition.

According to the survey nearly 8 per cent of West German forests were damaged in 1981 and while severe losses were limited to just 6 per cent of the total (or to some 35 000 ha) 60 per cent of the country's silver firs were so affected. Updated inventories released in 1983 sharply raised the total from 8 to 34 per cent of all forests and the third survey in 1984 brought the total to half of the country's forest cover with damage to firs surpassing three-quarters of all trees, and dealing with *Waldsterben* became a matter of national urgency for forest-conscious Germans. The original report concluded that high acid deposition is the only common stress factor in all the damaged areas but it also acknowledged that environmental factors do have a contributory, if not a causal role, and recurrent droughts of the late 1960s and 1970s and severe frosts of 1978, 1979, and 1981 were hardly blameless.

The best conclusions we can offer on the basis of nearly two decades of intensifying research into acid deposition's effects on forests point to a complex genesis of damage. There are some important similarities between North American and European forest declines—inlcuding, most notably, increases of the severity of damage with elevation and droughts preceding the onset of damage—which point clearly to a major contribution of triggering function of natural phenomena. On the other hand, tree mortalities have been found in many different settings, accompanied by calcium deficiency in the roots and excess sulphur in the needles, findings consistent with the expected effect of acid deposition.

Experimental work will not solve these uncertainties. Boreal forests grow slowly, are exposed to a number of natural stresses and their soils are rather acid anyway. Cumulative changes brought to such ecosystems by acid deposition are slow and unpredictable. Sensitive species may be completely eliminated but a growing body of research on microbial responses to acidification shows that vacated niches are readily filled and that the essential processes sustaining the boreal forests, above all the turnover of organic debris covering their floors, may continue largely unaffected.

But simple conclusions must be avoided. Future research may demonstrate declines of decomposition rates in acidifying forests—and any number of other points strengthening or weakening the case for stringent controls of sulphur and nitrogen oxide emissions. Regrettable as it may be, our understanding of acidification's effects on ecosystems is still very much evolving and the basis for solid guidelines is missing. As it is with CO_2 and tropospheric warming and with N_2O and the destruction of stratospheric ozone so it is with SO_4^{2-}, NO_3^- and forests: a generation from now we shall know better and some feel such a wait would be worthwhile; others find it unconscionable and demand action now.

And if there are various relatively simple and financially bearable measures to ease, even to eliminate, the worst of human interferences in nature's grand cycles should we not embrace them readily, erring on the cautious side? But if the controls are elusively complex and cripplingly expensive (with cost measured in traditional narrow terms disregarding the value of such public good as average August temperature or pH of groundwater) should we go ahead to find a few generations later that they made little or no difference? After all, tropospheric warming, declining ozone, or withering forests have resulted from natural influences in the past—and they will be induced so again. Looking at the available control options should make it easier to decide which path to follow in particular cases.

5.5 Managing the interferences

> The implications of our findings and conclusions for choosing among possible emission-control strategies, should they be deemed necessary, are limited.
>
> Committee on Atmospheric Transport and Chemical Transformation
> in Acid Precipitation (1983)

In 1981 a National Academy of Sciences Report dealing with atmospheric pollution from combustion of fossil fuels concluded that lowering the fresh water acidity in the most seriously affected areas of Northeastern United States would require a 50 per cent reduction in deposited hydrogen ions. Almost immediately this statement was misinterpreted by environmental activists, media, and assorted politicians (recognizing acid rain as a fine popular issue to be identified with) as a recommendation for a 50 per cent reduction in emissions of sulphur and nitrogen oxides.

A new NAS committee looked again at the link between emissions and deposition and its conclusion opens this section and perfectly characterizes the problems we face in controlling our interference with the three grand cycles. I have tried to demonstrate that while our quest for energy and food leads to anthropogenic fluxes of carbon, nitrogen, and sulphur on scales comparable to important natural transfers of the three elements, there is little evidence to suggest that these interventions bring irreversible global threats to the earth's environment.

But while we should not panic or strive for drastic changes in the way our civilization secures energy and food we should manage better in order to eliminate all these obviously degradative local and regional consequences of our interferences as well as to minimize any future probabilities of unacceptable global alterations. In doing so the biggest difficulty is not the absence of immediate appropriate, effective control technologies or the lack of new long-term options aimed at gradual but radical changes of the basic offending interferences—but the choice of the kinds, degrees, and scales of these management approaches.

The admission cited in the opening of this section is an all too frequent reality: our current understanding offers only limited guidance for confident choices of what we could see, even with considerable hindsight, as optimum or at least reasonable courses of action. To stay with the subject of the cited conclusion, we have no way to recommend precise levels of SO_2 emission controls in a source region in order to obtain a desired reduction of acid deposition in a reception area 10^2–10^3 km downwind. More, we cannot unequivocally quantify what any general minimum, maximum or optimum deposition rates should be as such determinations depend on numerous site-specific conditions.

All of this, naturally, does not prevent many people from calling for very specific reductions, usually some convenient large numbers such as 25, 60 or 66 per cent, or recommending precise general deposition limits such as 15 or 25 kg of sulphate per ha. But once some such levels receive overwhelming scientific blessing, and once the necessity of appropriate controls gets a sufficient public backing, substantial reduction, even virtual elimination of SO_2-generated acid deposition would be technically perfectly feasible and the costs would be well within the capacities of rich North American and European nations which are both the largest emitters of SO_2 and receptors of acid deposition.

Unlike C_2O and NO_2, acid deposition is not a global problem and major control successes can be achieved without universal co-operation using the currently available techniques. However, regional co-operation is a must. Diffusion modelling shows that almost exactly half of all sulphur emitted in Europe is transferred from the source countries to their neighbours and beyond and this obviously makes any purely national control efforts ineffective. Similarly, exchanges of acidifying compounds between the eastern parts of the United States and Canada necessitate a joint approach to any effective future controls.

Of the three basic control choices—coal desulphurization, fluidized bed combustion, and flue gas desulphurization—the last one has emerged as the dominant practical option. Various coal-cleaning methods are effective in removing inorganic sulphur in the fuel (mostly associated with iron as pyrites) by gravity separation but removal of organic sulphur, bound with the fuel and accounting for 10–50 per cent of all S, is not yet a commercial possibility although more than a score of chemical cleaning methods is now under development.

Intensive coal clean-up is undoubtedly cost effective for most large stations but it is insufficient to remove enough sulphur to prevent undesirably high emissions. Fluidized bed combustion (FBC), where coal is burned in contact with crushed limestone which captures virtually all of the formed SO_2, is a nearly perfect choice as its low temperatures also prevent NO_x formation but intensive development of small FBC steam generators started only in the mid-1970s and it will not be until the 1990s that the now commonest generating units with capacities of 500–1000 MW will be equipped with large FBC boilers.

Consequently, flue gas desulphurization (FGD) has become the mainstay of commercial SO_2 control of the 1980s. Stripping fluctuating, but relatively low, concentrations of SO_2 from 10^5–10^6 m^3 of hot flue gases emitted each hour by a large power plant has been a difficult engineering challenge and it was only in the late 1960s, after decades of development, when the first commercial FGD units entered service at large coal-fired stations in the United States.

Wet alkali scrubbing emerged soon as the best control option and at the close of the 1970s FGDs systems were installed on more than 21 000 MW of American generating units and the mid-1980s total is 50 000 MW, or nearly 15 per cent of all United coal-fired power capacity, altogether an impressively rapid expansion of a technology which is based on simple principles but which can be beset by many operational complications.

In alkali desulphurization water slurries of lime or limestone scrub (mostly in spray towers) SO_2 from the hot gases forming, after complex reactions, calcium sulphate. Operational problems included plugging and corrosion and perhaps the greatest challenge is the disposal of large quantities of wet sulphate in economic and environmentally sound ways as the storage sites should be leakproof, close to the plant, and large enough to take in the wastes for 20–50 years of plant operation. Not surprisingly, the waste liquors exceed drinking water standards for calcium, sulphate, chlorine, magnesium, and sodium, and often also for trace metals (from fly ash scrubbed together with SO_2), and contamination of groundwater should thus be always prevented.

Costs of the disposal can be high but they can be reduced by industrial use of the sulphates or by installation of regenerable processes. However, only a small fraction of the generated total will be taken by the building industry and costs of regenerative controls are obviously much more expensive than the investment and operating expenses for the throw-away FGD. Typical wet alkali scrubbers need investment of $(1984) 170–190/kW of installed capacity and around 8 (1984) mills/kWh to operate while regenerable processes cost about 40 per cent more to build and 30 per cent more to operate.

Several detailed power plant cost studies based on a large number of data for leading American utilities conclude that the installation of a typical FGD system raises a plant's capital cost by 25–33 per cent and operating expenses by 25–40 per cent. Two fundamental innovations aimed at lowering these costs are dry scrubbing and the increased use of fly ash in desulphurization.

Dry scrubbers, now accounting for less than one-tenth of operating American installations, are particularly suitable for desulphurizing gases from plants burning coal with less than 1 per cent of sulphur, and alkaline fly ash scrubbing also works best for low-sulphur coals. Yet another simple but effective innovation is the addition of magnesium to scrubbing liquor where its high solubility (180–1000 times more than calcium) improves the performance. The eventual aim of advancing flue gas clean-up research is to achieve simultaneous control of sulphur and nitrogen emissions and Japanese companies have been at the forefront of this development but as of the mid-1980s there were no commercial applications.

But even if there were no future cost reductions and if FGD were to be installed at all coal-fired power stations in North America and Europe the costs would not be crippling. In these two continents (including the USSR) there are now about 800 GW of coal-fuelled power plants and with average installation costs of $(1984) 180/kW complete capital cost of desulphurizing this generating capacity would be about $(1984) 150 billion, a huge sum but an equivalent of no more than about one-quarter of the total these countries now spend *each* year on their military.

Needless to say, not all the plants would need such expensive FGD and the construction programme would have to be spread over a number of years so that the burden would become surprisingly tolerable: if all these retrofits were to take a decade the capital costs would average a mere $(1984) 15 a year for each person in the rich industrial nations of North America and Europe. The increase of electricity costs would be similarly bearable: about half of all the rich world's electricity is generated from coal and if FGD would push the average operation costs by 30 per cent there would be just a 15 per cent rise in the rich countries' total electricity bill, an increment which a few widespread conservation measures would make truly invisible.

While during the 1970s acid deposition control efforts focused almost solely on reduction of SO_2 emissions, since the early 1980s first the West German and later also American findings have been stressing the need for concurrent limits on NO_x emissions. As noted previously (section 5.3.2), there are relatively simple but very effective Japanesse processes for automotive NO_x controls and combustion modifications also greatly reduce stationary emissions. Again, as with SO_2 controls, the essence of the problem is to define economically the acceptable extent and degrees of controls.

In contrast to drastic cuts of acid deposition which would require considerable long-term investment in SO_x and NO_x controls, reductions of nitrogen losses from fertilizers can be achieved with only minimal or no costs and they can, at the same time, bring appreciable financial benefits as higher efficiency of utilization boosts the return ratio.

The most obvious efficiency improvement is the simple choice of doing without: reduced intake of animal foods in rich countries, a shift stopping far from pure vegetarianism, would free very large areas of land from cultivation. But such an option is open to no more than one-quarter of mankind and it is most unlikely that it will be voluntarily selected as it would run against deep-seated dietary preferences and it would cause major economic dislocations.

While there is no doubt that any improvements and extensions of biotic nitrogen fixation would be highly beneficial, the scope and rate of such developments will be decidedly modest. By far the most difficult, although immensely rewarding, breakthrough would be the transfer of nitrogen-fixing genes into all major food and feed crops. The complexities of this path (after all, during the long evolutionary span no higher plants incorporated nitrogenase into their cells) mean that, even with concentrated research, decades will elapse before any practical advances. Vincent Gutschick, after a careful review of these limitations, concluded that '100 years, if ever' is the likely prospect.

Trying to improve the nitrogen-fixing capabilities of legumes and algal systems is undoubtedly a more sensible route to tangible near-term benefits. The selection and diffusion of superior *Rhizobium* strains and the development of effective inocula for major leguminous crops would help to optimize the fixation rates but field experience has been so far unimpressive as several researchers have concluded that commercial soybean and alfalfa inoculants did little to raise the fixation activity in comparison with native strains already present in the soils. As already noted, the best average for energy cost of *Rhizobium* fixation is about 6 g C/gN— and by coincidence crops spend almost exactly the same amount of carbon in reducing 1 g of nitrate–nitrogen in the root. This means that there is no loss of yield when the fixing is done by living organisms—but we prefer to have it done in ammonia plants because *Rhizobium* symbioses work with only few crops we grow for food or feed, and using these as green manures before planting the desirable cereal or oil crops means giving up cultivation of one more food crop in a year.

Consequently, rich as well as poor countries will be further intensifying fertilization to expand their harvests, creating greater need for careful management aimed at maximizing the yield response. Of the long list of variables determining this response, the rate of application, the timing, and the placement of nitrogen are the most important (the kind of nitrogenous compound used appears to have little effect on overall fertilizer efficiency as long as the application has been done properly). All of them are critical in satisfying the key requirement of efficient fertilizing: providing enough nitrogen to cover the plant's assimilation as well as the potential volatilization, leaching, erosion, and denitrification losses.

This need is usually met by applying nitrogen in a few large batches, the largest one being commonly at the time of non-existent or very low uptake—during, or shortly after, planting. The remedies rest in finding out optimum application rates by soil testing (or plant tissue analysis) and delaying the fertilization until shortly before the time of the rapid uptake.

The first part is much easier to do. Soil testing today is fast and relatively cheap and it is thus surprising to find that even in the rich countries many farmers do not have their fields' nutrients appraised regularly. But it must be noted that soil testing for nitrogen has its complications: the simple establishment of total nitrogen is not very helpful as most of it is unavailable (bound in organic compounds); tests for nitrate may overestimate the availability in areas of higher precipitation; results are much influenced by sampling depth; to represent a field properly dozens, even scores, of samples must be taken; and, unfortunately, analytical errors may be so large that even supposedly identical testing methods show up to a threefold range in results.

Even with a perfect appraisal of soil nitrogen and the knowledge of the intended cultivar's requirements there is no sure way to optimize the outcome. Soil scientists in Nebraska had soil samples analysed by five different laboratories which recommended between 72 and 162 kg N/ha for irrigated beets (2.25-fold difference)—but the maximum yield (achieved with the least recommended amount!) was just 11 per cent above the minimum one and differences in sugar content were a mere 3 per cent. The inherent complexity of soil-water plant relationships will keep soil testing and optimized fertilizer application as much a matter of art as of science but, with its limits in mind, regular testing should be a basic step towards higher efficiency.

Delayed applications run into different problems. Too long a wait may mean that the growing crop will not permit the use of standard field equipment; of course, fertilizer distributed efficiently in irrigation water (obviously not in furrows but by expensive centre pivot systems) may be applied at any time. Splitting the applications to match the needs requires additional passes over a field increasing fuel and machinery costs. And the experiments have yet to provide (except for ammoniacal nitrogen) a clear-cut conclusion about the principal application controversy, the merits of fall fertilizing: claims continue to range from insignificant differences to large losses with fall applications.

But there is much consensus and confirmed successes, too. Foremost here is the combination of delayed application until some practical time after planting and incorporation of the material into the soil. A series of international studies with rice showed highest uptakes with application just before or during primordial initiation, that is weeks after transplanting (Fig. 5.20) and, similarly, experiments with American corn proved that the usual application rates of 170–200 kg/ha can be reduced by 25 per cent without yield losses when three-quarters of all nitrogen is applied only five to six weeks after planting. Sufficiently deep injection of ammonia is obviously essential to prevent volatilization losses and incorporation of solid ammoniacal fertilizers 5–15 cm into the soil can raise nitrogen utilization efficiency nearly twofold compared with surface spreading.

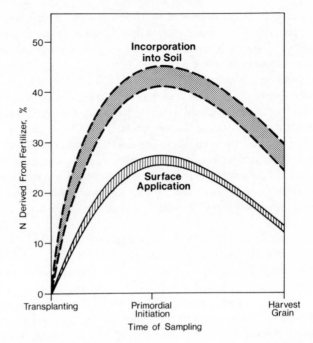

Fig. 5.20 Timing of nitrogen application is critical in any attempts to maximize fertilizer efficiency—as is, of course, the mode of application. This graph illustrates two safe generalizations in what is an extremely complex game: incorporation into the soil and application around the time of primordial initiation will maximise the grain yield.

Optimizing application rates, choosing the right time, and placing the fertilizers properly presents a powerful combination in the quest for higher efficiency but, fortunately, many other measures can bring as great or greater combined effect. Enough has been written here (section 4.2.1) about soil erosion: cutting down these losses would bring, especially in the tropics and in humid temperate regions, nitrogen savings totalling globally in the order of 10^7 t each year! Reduced tillage and no-till farming, terracing and avoidance of row crops on sloping land are the principal ways of managing field erosion whose impacts, as discussed, go much beyond nutrient losses.

Crop rotations are an excellent boost to nitrogen utilization efficienty as nonleguminous crops can benefit from previously fixed nitrogen and, in turn, deep-rooted legumes can scavenge nitrates leached into deeper soil horizons after grain crops. In poor countries with cheaper labour, intensive inter-cropping is another outstanding option. And attention to good agronomic practices every-where can reduce fertilizer losses by surprisingly large margins. The FAO estimates that the choice of inappropriate crop varieties lowers utilization efficiency by 20–40 per cent, as does the delay in sowing, while poor seedbed preparation, inadequate plant population and insufficient irrigation are typically responsible for 10–25 per cent reductions. Uncontrolled weeds and pest attacks may bring losses exceeding 50 per cent.

There is little doubt that following sound field management (from erosion to weed control) and applying optimized amounts of fertilizers properly could bring such nitrogen efficiency increases that current global fertilizer use could be reduced by at least one-quarter (and, less likely but not impossibly, perhaps by even one-third) without any loss of harvests. Future manipulation of natural nitrogen fluxes could enlarge these savings appreciably although the only such commercialized method has not met with a general success.

Inhibiting nitrification by applications of compounds blocking ammonia oxidation has been a commercial possibility since the early 1970s with development of Dow Chemical's N-Serve (nitrapyrin) and about a dozen similar Japanese products which suppress *Nitrosomonas* activity for up to six weeks following the ammonia application. The highest pay-off would be in heavily fertilized fields in humid regions, with the Corn Belt being a perfect example, but long-term tests have not been uniformly successful. Crop yield response after nitrapyrin application ranged from an increase of more than 30 per cent down to no change (4 to 10 per cent was the most common improvement).

Reduced concentration of nitrate nitrogen, higher protein content, and lower incidence of some plant disease are three welcome side-benefits of nitrification inhibitors but their application, although deserving the widest possible diffusion, will remain only a small fragment in a combinations of ways, virtually all of them relatively simple and well-tested, that are available to increase efficiency of nitrogen use.

However, there is a fundamental difference between controlling SO_2 and NO_x emissions and reducing nitrogen losses in agro-ecosystems: while it is possible to eliminate virtually all gaseous releases from fossil fuel combustion by universal, and for poorer nations perhaps still prohibitively expensive, installation of control systems, enhanced volatilization, denitrification, and leaching of nitrogenous compounds will be with us as long as we continue to fertilize our crops. Future scientific and engineering breakthroughs may make both interferences obsolete as renewable energies could supplant fossil fuels and genetic manipulation could combine all major crops with nitrogen-fixing microbes but none of these changes will dominate the next one to three generations when slower annual increases of fossil fuel combustion (see section 2.5), and spreading SO_2 and NO_x controls, could bring substantial cuts of anthropogenic gaseous releases far surpassing reductions of nitrogen losses in farming.

CO_2 controls are by far the toughest proposition of the three. Combustion of fossil fuels is nothing else but the rapid oxidation of carbon so unless (and it is just a theoretical possibility) we cut the conversion efficiency by half and stop the process with alcohols, aldehyds, and acids rather than with CO_2 (how would we dispose of those wastes?) every gram of fully oxidized carbon in coals, oils, and gases will be turned into 3.66 g of CO_2. No precombustion treatment can help and although there have been some impressively detailed proposals for scrubbing the generated CO_2 their engineering, costs, and reliability are all in the realm of speculation and, even more importantly, there are no workable solutions to the disposal of, at current generation rates, some 7 billion tonnes of CO_2 a year.

Injection into deeper ocean layers is an obvious choice but, again, today we can only speculate how it might be done. Even less realistic suggestions are pumping the gas into greenhouses (the mismatch between the CO_2 volumes generated by combustion and those absorbable by greenhouse plants is plain and, anyway, all of this would mean just a very short delay before transfer to the atmosphere), or fertilizing the ocean with phosphorus to boost phytoplankton productivity and to sequester the organic litter at the ocean floor.

As for the fuels themselves, burning more hydrocarbons would lower CO_2 emissions—but the make-up of global resources of fossil fuels, dominated by coals, makes such a switch imposssible. And because CO_2 build-up, unlike acid deposition and nitrogen losses from agro-ecosystems (except for N_2O releases) is a global affair any efforts to limit the combustion of coal would have to be an international undertaking and as the United States, the USSR, and China control about three-quarters of global coal reserves their close co-operation would be indispensable—but, setting the obvious political implications aside, why should these countries, in Zimermeyer's apt charaterization, take the second step before the first, that is act before the effects of increased CO_2 levels have been clearly determined?

Consequently, although the CO_2 problem is so different from acid deposition and from interferences in nitrogen cycle it is, once again, the lack of adequate understanding which presents the greatest management challenge. Roger Revelle summed this up well by noting that 'about the only facts available are the actual measurements of atmospheric carbon dioxide ... and some fairly reliable data ... on the annual consumption of fossil fuels'. Difficult as it would be, should we know today beyond any doubt that doubled atmospheric CO_2 would carry consequences endangering civilization's survival we could start working with the heightened urgency on a gradual transition to non-fossil energetics.

But there is no evidence of imminent disasters and after some earlier warnings about unprecedented global geophysical experiments, humanity in great peril, and the need to make immediate decisions on how to curtail CO_2 emissions, a new scientific consensus of the early 1980s has a decidedly non-catastrophic tone. For example, Rotty and Marland concluded that 'the need is not for immediate, short-range, and probably ill-conceived crash programs, but rather for long-range, carefully developed programs in the area of alternatives to fossil fuels'.

No less emphatically, the latest NAS assessment does not believe 'that the evidence at hand about CO_2-induced climate change would support steps to change current fuel-use patterns away from fossil fuels. Such steps may be necessary or desirable at some time in the future ... but the very near future would be better spent improving our knowledge than in changing fuel mix or use.'

In the case of CO_2 build-up, whose eventual effects must be assessed within the wide context of the still so poorly known natural climatic changes, ecosystemic resilience and planetary equilibria, the need for better understanding, rather than for any precipitous commitment to different ways of securing heat, electricity, and motive energy is especially obvious—but the imperative of unglamorous, incremental enlargement of a solid intellectual base to which we can anchor sensible decisions able to withstand the tests of time applies no less in all other cases of worrying environmental degradation as well as in countless other instances of managing our energy and food production.

Unfortunately, the frequency of false perceptions, misplaced worries, panicky reactions and dubious commitments does not appear to be decreasing: in spite of the rapidly growing fund of scientific knowledge we are as prone as ever to dissemble in the face of irrationally inflated threats, to cling to recurrent myths, and to chase attractive chimeras. A brief look at two well publicized appraisals of the global prospect and a fifty-year retrospective will illustrate some of these biases and close this book with a reminder of unpredictable futures.

6

Looking Ahead

What we may confidently add for the present is that between now and the end of the century we shall probably have occasion to see whether there is a chance of mankind's accession to the state of reason being accomplished soon, or whether, on the contrary, the human race, unable to take itself in hand will be relegated for long ages to the chaotic gestations of history.

Dominique Dubarle (1970)

Inevitably, we look ahead. As individuals, as nations, and ever since the Earth has become covered by the web of a rapidly communicating civilization, also as a planet. The global look, until a few decades ago solely the province of geophysical sciences, is now a focus of much thinking and speculation. One does not have to invoke the nightmare of nuclear war to inquire anxiously about the long-term prospects for mankind: the malfunctioning of any of the three critical dependences analysed in the preceding chapters can usher in a period of such deep changes that the very survival of industrial society may be in peril.

Numerous specific arguments bearing on this fundamental question of the human outlook were introduced in the previous chapters, and instead of summarizing these, I shall briefly reflect on the two disparate appraisals of the future state of the world offered by doom-saying neo-Malthusians on one side and ever-optimistic techno-fixers on the other.

The most recent, and widely publicized, instalment of this discord came as a result of the publication of the *Global 2000 Report to the President*, a gloomy assessment of the world ordered from the federal bureaucracy by Jimmy Carter, and the Report's rebuttal in *The Resourceful Earth* edited by that most inveterate techno-optimist, the late Herman Kahn of the Hudson Institute and Julian Simon, an economist with unlimited vistas.

6.1 Appraising the future

The original 1980 *Global 2000 Report to the President* is frightening. It received extraordinarily wide circulation, and it has influenced crucial government policies. But it is dead wrong.

J. L. Simon and H. Kahn, *The Resourceful Earth* (1984)

The controversy has left no common ground: the most publicized comprehensive interdisciplinary study of the world outlook produced by agencies of the United States government was found to be 'dead wrong' by its most outspoken critics led by Julian L. Simon whose earlier book, *The Ultimate Resource*, published while he was still a Professor of Economics at the University of Illinois at Urbana-Champaign, served as the intellectual foundation for *The Resourceful Earth*, an edited volume called by its publisher 'the famous report which demolishes *Global 2000* . . . a devastating indictment of all doomsday books and also the most scientific inquiry into the future ever organized'.

Choosing sides is easy. If one feels that the world around us has been gradually disintegrating, a perception inculcated since the early 1970s by the stream of media reporting on famines, energy shortages and environmental disasters, then the arguments and statistics assembled in the *Global 2000* seem to convey an expected and incontrovertibly correct bleak appraisal:

If present trends continue, the world in 2000 will be more crowded, more polluted, less stable ecologically, and more vulnerable to disruption than the world we live in now. . . . Despite greater material output, the world's people will be poorer in many ways than they are today. For hundreds of millions of the desperately poor, the outlook for food and other necessities of life will be no better. For many it will be worse. Barring revolutionary advances in technology, life for most people on earth will be more precarious in 2000 than it is now.

In contrast to this Malthusian anguish the editors of *The Resourceful Earth* concluded that

If present trends continue, the world in 2000 will be *less crowded* (though more populated), *less polluted, more stable ecologically*, and less vulnerable to *resource-supply disruptions*. . . . The world's people will be *richer* in most ways than they are today. . . . The outlook for food and other necessities of life will be *better* . . . life for most people on earth will be *less precarious* economically than it is now.

Altogether a satisfying symphony of progress and achievement, a prognosis attuned to the perceptions of all those who have seen plenty of signs of continuous gradual improvement of the human lot.

A systematic, point-by-point examination of differences between the two reports' data, arguments, and conclusions might be an interesting heuristic exercise but it is hardly needed in order to explain their unbridgeable disparities. The most fundamental difference is the initial mind-set, those grand-scale perceptions which guided the two efforts and which so obviously permeate the wording of both contentious conclusions.

The organizers of *The Resourceful Earth*, enthusiastic believers in the omnipotence of human inventiveness and the powers of the market, started from premises very unlike those shared by anonymous experts hidden in government bureaucracies whose invariably gloomy outlook is an essential part of justifying the very existence and further expansion of these enormous institutions. In this context it is most revealing to note that the study's director himself was unable to change what he considered to be a critical, incorrect statement formulated by 'an editor at the Council of Environmental Quality'!

Not surprisingly then, both studies contain many transparently dubious assertions which could have been avoided by adopting more of a *sine ira et studio* approach. Just a couple of choice examples will illustrate these often embarrassing attempts to fit the trends and numbers into the moulds of preconceived postures. The very first specific conclusion presented in *The Resourceful Earth* offers an incredibly naïve interpretation of a less crowded world in the year 2000.

Simon and Kahn do, of course, admit that there will be, barring a catastrophe, more people on earth by that time but they argue that a growing population implies no greater 'crowding' because 'as the world's people have increasingly higher incomes they purchase better housing and mobility'. Consequently, with more floor-space they have more privacy, and with higher mobility they can move more freely and benefit from access to greater area.

After presenting United States Statistics on residential floor-space (for the years 1940–74), land in public parks (1944–80), and the number of visits to public parks (1920–60), it is immediately claimed that 'these trends mean that people increasingly have much more space available for their use, despite the increase in total population, *even in the poorer countries* (the stress is mine). All this suggests to us that the world is getting less crowded by reasonable tests relevant to human life.' I would rather argue that such statements suggest an astounding absence of rigorous research, if not common sense.

The inadmissibility of global generalization on the basis of a few United States trends is obvious and the authors' ethnocentricity is abyssmal. If they had ever walked Shanghai streets—where over six million people are now living within an area of 158 km² (a staggering density of 38 000/km²), and do so without any possibility of driving away to a national park (!!!), or strolled down the muddy, manure-strewn alleys of Egyptian villages in the Delta—where the number of poor peasants per hectare of habitable land about doubled during the past three decades and where even the miraculous ownership of a car offers no escape as it can do nothing to remove the triangle's desert encirclement—they simply could not write such nonsense.

Even for the United States, a country representing a fabulous exception rather than a representative norm in our world, their conclusions are forced. While it is true that proportionately fewer households now live in crowded conditions than two or three generations ago the 'moving around more freely' argument is quite incorrect. For example, during the generation between the early 1960s and the early 1980s, United States highways were extended by only about 11 per cent but the number of vehicles grew 2.15 times and passenger · km travelled on them rose 2.1 times which means that crowding on an average stretch highway almost doubled in 20 years.

And, of course, highways leading to favourite parks have become travelled much more heavily and, upon reaching the destination, crowding in the parks has indubitably increased. For example, while land in the United States national parks rose from 4.94 million hectares in 1950 to nearly 6.39 million hectares in 1980, the total number of visits during the same period jumped from 13.9 to 60.2 million, which means that (not surprisingly) America's splendid national parks became, on the average, more than three times as crowded as in the earliest post-World War II years.

No less than *The Resourceful Earth,* the *Global 2000* forces its conclusions, often relying on quantitatively dubious estimates. Perhaps the best example of this dubious practice is the report's frightening estimate of the rate of species extinction which has been criticized already, earlier in this book (section 4.2.3). But certainly the weakest attribute of the report is not any particular estimate, indefensible as it may be, but rather its simple-minded reliance on a concatenation of numerous computer models used by various bureaucracies of the federal government and originally designed for different purposes than forecasting long-range planetary futures.

Bound by limits of scientific understanding, led astray by illusory perceptions and guided by 'primary' prejudices, is it then surprising that we have rushed into dubious decisions and pressed ahead with solutions which do little credit to any rational management?

Looking back at the discussion of changing environmental 'crises' (section 4.1.2) it is striking how, in retrospect, many fears receded as if by magic: the nitrogen cycle does not appear to be out of joint any more, mercury in fish is not going to poison the whole of mankind, DDT has been rapidly sequestered, the ocean is not dying, subsequent predictions of the chlorofluoro-carbon effect on the ozone layer are coming up with findings of declining impact. Ridiculous plans, blundering management, wasted efforts, and counter-intuitive results appear to be an inevitable tax on evolutionary complexification. Frequently harmful and always regrettable, it must be paid because the complexity of our affairs clearly precludes any faultless choice of optimized solutions. We have still much to learn and we can do much better.

6.2 A fifty year perspective

The contrast between efforts and results could not have been sharper: never before were there as many forecasts of economic performance and technological change as we have had since the early 1970s, but their life-span has been ephemeral, their utility largely as *objets amusants* or as illustrations of how not to do it. This book is sprinkled with reminders of this unimpressive performance: oceans have not died, crude oil prices have not topped $100/bbl, grain costs have been less in real terms, recession has not been a permanent feature of spent Western economies. A list of all those who have been wrong reads like a directory of top international organizations, 'think tanks', and scientists turned into media celebrities: so many groups and individuals are in this category that I shall not randomly name any more names.

Surprisingly, even very short-term efforts have fared no better. American economists, a clan now as numerous as car salesmen, missed the onset of the worst post-1945 recession in 1981-2, and right after that they misjudged the start and strength of the strong recovery. The rapid softening of crude oil prices, just after they reached their peak in 1982 and the concurrent across-the-board decline of raw metal and food prices, were also unanticipated, a fact causing great hardship to many primary producers.

And, since the early 1970s, many futurologists have been trying to forecast what is perhaps most elusive: the ways and diffusion rates of innovations, those sometimes sudden and sometimes gradual developments which bring new products and processes creating unpredictable explosions of demand and exposing the uselessness of any rigid long-term (and even worse when central) planning. The most obvious illustration from the early 1980s is also one of the best in recent decades: there were no personal computers in the early 1970s but a decade later more than 150 manufacturers served a market worth $(1985) 15 billion, and were expanding rapidly. And the story of 'Cabbage Patch' dolls proves that no more than a marketing ploy ('adoption' of an otherwise run-of-the-mill toy) can create an explosive demand.

But there is an obverse to successful innovations, one which can not be forecast any better than their emergence. Rather than being smitten by those busily peddled notions of how assorted high techologies will transmute everything we do by their magic touches, we should be reminded that many technical innovations, spectacular as they were, have made much less advantageous differences than we originally imagined. Perhaps most notably, they have not provided enough of the most needed change, more jobs for growing populations, and the latest round of this misplaced infatuation appears to provide yet another confirmation of this unwelcome trade-off as new high-tech openings will clearly not replace the jobs lost in old industries.

One of *The Economist*'s right-on-the-mark editorials put it best:

Less skill is needed to assemble computer parts than to make car components or English muffins. . . . The moral is that most future jobs will not come from nurturing costly high-tech enterprises based on computers or telecoms. They will come much more from the stuffing of new high-tech tricks into old-time manufacturing and services.

Of course, there is nothing new about these increasing failures, lack of imagination, or exaggerated expectations. Ever since the growing numbers of new machines and techniques started to enter the service of Western industrial civilization in the nineteenth century, countless experts have felt obliged to predict the course and, in especially assured tones, the limits or limitlessness of future developments. Collections of these dead-wrong predictions make amusing reading and one is hard pressed to pick the most ludicrous statement: my favourite is the appeal of the Director of the United States Patent Office urging President William McKinley to abolish the institution and his job because 'everything that can be invented has been invented'.

Why then do I want to add a coda about the unfathomable future? Because I want to end this book by stressing the necessity of accepting profound uncertainties which open up the opportunities for great choices. Our growing search for future certainties—be it expressed in doomsdayish warnings, technological forecasts, or econometric models, in government reports or 'think tank' releases—is worthless, besides being deeply anti-evolutionary. We cannot forecast the whole—yet it is the whole which matters.

When we are unable to forecast even the short-term changes of a few scores of selected and well-studied variables, how can anybody be so arrogant as to offer pronouncements about the state of the world decades from now and try to palm them off as something other than irrelevant personal, or group, opinions? So it is to show these elusive combinations and concatenations of changes whose result is the world as it is, and as it could not be anticipated, that I will step back just fifty years from the date I sent a letter to Oxford proposing to write this book and I will look at the world of 1983 through the eyes of the year 1933.

A fifty-year forecast of anything, prepared in 1933, would have been done in the midst of the most extensive, longest-lasting economic crisis of this century when the 'prevailing mood' phenomenon, illustrated so well earlier in this book by shifting global energy projections, was bound to be exceedingly strong. The best available global figures showed energy use declining by about 8 per cent in 1930, and a further 10 per cent in both 1931 and 1932, clearly not the best setting in which to forecast a nearly six-fold rise of fossil fuel consumption during the coming five decades!

Such a huge growth, about twice as fast as during the preceding half century, was made possible by many unforeseen discoveries of rich fossil fuel deposits. In 1933 only the old Caspian fields were working in the Soviet Union, and the country was producing a mere 21 million tonnes a year; in 1983 it was the world's largest oil producer, extracting nearly 600 million tonnes and discovering more under the permafrosts of Siberia. China, whose very independent existence seemed to be at stake in 1933 after the Japanese invasion, and whose output was made up of just 28 million tonnes of coal, became the world's third largest fossil fuel producer (over 600 million tonnes of coal equivalent) with further rich prospects ahead. And, of course, there was no inkling in 1933 that the sands of Saudi Arabia were covering the world's largest crude oil reservoirs as the first discoveries in the country date from 1938.

In technical terms there were no suitable steels, materials, components, and engineering skills in 1933 to give rise, during the next half century, to such innovations as offshore oil-drilling, continuous coal-mining, large-diameter transcontinental pipelines, unit coal trains, liquefied natural gas tankers, and direct current extra-high voltage transmission—as well as to a stunning growth and qualitative improvements in established techniques which brought about impressive economies of scale and declining real costs.

To mention just a few of these achievements, between 1933 and 1983 the size of the largest excavator buckets in surface mining grew nearly twenty-fold, the common depth of exploratory oil and gas wells nearly ten-fold, the size of the largest crude oil tanker increased 30 times (as did the tonnage of an average carrier), and while in 1933 the world's largest multi-unit coal-fired power plant had a total installed capacity of 200 MW, half a century later the size of an average new generating unit in rich countries rose to 300–500 MW, and typical long-distance transmission voltages quadrupled.

If the 1933 forecasters could have been *au courant* with advances in subatomic physics (a not very likely proposition), they might have known of Chadwick's recent discovery of the neutron but, as even the discoverer did not consider it at that time an elementary particle, and as the first sustained chain reaction under the bleachers of Chicago's Staggs Field was still a decade away, they could in no way foresee the emergence and commercialization of nuclear electricity, a form of primary energy which in the following fifty years was to go from a slow start to a relatively rapid growth and from forecasts of an essential role in future global supply to a much more subdued outlook. Nevertheless, the world gets today roughly every tenth kilowatt hour of electricity from nuclear fission, a technique completely outside the engineering realm of 1933.

On the energy consumption side, the explosion of car ownership (from less than 30 million in 1933 to 350 million by 1983), and the demand for convenient heating fuels led to the demise of coal for transportation (steam locomotives, dominant for a century, are now museum pieces in the Western world), and household use and the steep ascent of hydrocarbons, with coal increasingly fuelling electricity generation the demand for which expanded at much faster rates than total energy consumption. By 1983, electricity energized 10^8 TV sets and 10^6 computers ranging from giant rocket-guidance and weather-forecasting machines to cheap toy distractions—whereas in 1933 there were only a few dozen tiny experimental TV prototypes and another decade had to elapse before the first unwieldy, breakdown-prone computer (Mark I) was put into operation at Harvard in 1943.

Changes in farming add up to a no less profound transformation. Average global yields of wheat and rice doubled, and those of corn nearly tripled as new short-stalked high-yielding varieties (non-existent before the mid-1950s) supplanted traditional cultivars of the first two staples, and hybrids replaced old corn varieties (in 1933 less than 0.5 per cent of world cornfields were planted with hybrid seeds). Mechanization advanced both quantitatively (perhaps best illustrated by the worldwide tractor total rising from a mere 1.5 million in 1933 to over 20 million in 1983) and qualitatively (in 1933 less than 10 per cent of working tractors had rubber tyres, none had hydraulic hitches, or automatic lowering and lifting of implements).

Global application of nitrogen fertilizers rose from less than 2 to more than 60 million tonnes a year; synthetic insecticides and herbicides, both non-existent in 1933, became an essential ingredient of field-farming worldwide. Old crops took off in new places. For example, the United States changed from an importer of soybeans (1933 production only about half a million tonnes) to the world's largest grower (1983 harvest over 60 million tonnes) and the Chinese turned to corn (their corn harvest in 1933 was about 6 million tonnes; 50 years later it topped 60 million tonnes).

A long list of other innovations, from fast-maturing varieties pushing the limits of rice, corn, and winter wheat cultivation hundreds of kilometres northward to efficient centre-pivot irrigation, from no-till planting to micro-nutrient fertilizing, has contributed to increase the productivity and production to such an extent that social inadequacies (above all the need for a more equitable distribution), rather than physical constraints, have precluded ample food for everybody.

And comparisons of environmental matters are monotonously repetitive: no controls, not even any awareness of many problems in 1933—compared to numerous strict and constantly improving controls and extensive attention to scores of environmental degradations in 1983. What is more important than a nearly universal application of important control techniques known, but not diffused, in 1933 (such as electrostatic precipitation of fly ash, or water pollution control in refineries), and the spreading use of controls not available half a century ago (such as flue-gas desulphurization and sharp reductions in automotive emissions) is the awareness of a multitude of interrelated degradations and their effects, and the existence of sophisticated analytical methods to monitor and to evaluate their levels and trends.

Advances in scientific disciplines, ranging from ecological energetics and micro-analytical chemistry to cellular biology and remote sensing, make it possible to approach the whole challenge of environmental degradation in ways which, if applied consistently and assiduously, may succeed in preventing any irreparable large-scale insults to the biosphere: the basis for such an understanding and actions was completely absent in 1933.

All of these developments in energy, food, and environmental affairs have cut across political and cultural boundaries, and although different methods of economic management and differing national traditions do obviously produce different specific results, general changes in the ways of satisfying civilization's imperatives have affected the lives of billions of people more profoundly than the peculiarities of socio-economic arrangements.

Synthetic nitrogen from large ammonia factories fed and powered by natural gas and electricity now props up high crop yields in Iowa cornfields, as well as in Ukrainian wheatfields, or in Sichuan's paddies; coal-fired power plants are the cornerstone of electrified economies in the USA, the USSR, West Germany, or China, and their CO_2 and SO_2 releases make scientists everywhere worry about their global and regional environmental effects.

Irreconcilable political cleavages may yet lead to the obliteration of much of what industrial civilization has accomplished since the middle of the last century: we have to hope that adaptive, self-preserving solutions to these conflicts will be found instead. But even the lifting of those perils will not diminish the challenges of sustaining industrial civilization with enough energy and food in habitable environments. Even a brief glance from 1933 to 1983 is encouraging as an indicator of what it is possible to accomplish in half a century, now a bit less than an average human lifespan on the Earth.

There is no reason to believe that the next fifty years will see changes less profound and less far-reaching. But that brief glance also shows that there is no way to foresee the totality of those changes and, ignorant as we are in this critical respect, we must admit the possibilities of both painful failures and unprecedented advances. At all times our key objective should be the avoidance of costly mistakes, an approach requiring good understanding and management of all obvious and not-so-obvious links and connections, rather than a linear pursuit of narrow objectives.

Even so, the effects of the unpredictable will remain strong and omnipresent uncertainty escalates when looking ahead farther than 50 years: I will not. But I feel confident that the earth's resources and knowledge that we now have at hand, together with those inevitably large gains we shall make in decades ahead, will make it possible to secure a better life for an even larger section of an increased global population. Providing, of course, we learn enough from our mistakes and manage our affairs less by impressionistic rushes and more by adherence to long-term realities.

Or: providing we can show even better adaptive skills than ever before, because past accomplishments and our collective intellect give us unprecedented opportunities for further great advancement in civilized life. But this very complexity of the world, now including countless man-regulated interdependencies, also hides the increased changes for miscalculations and failures. Industrial civilization has vastly enlarged our powers in so many ways but it not only *did* not, it *could* not, bring any greater certainties than its predecessors—rather, it made the future even more open. This uncertainty cannot be shed: it is the quintessence of the human condition.

The setting is always different but it is not the first time that heightened uncertainty has had to be faced. On one such occasion, eighteen centuries ago, an emperor-philosopher considered the prospects. In times of war and plagues at the head of his armies defending the northern border of the empire, Marcus Aurelius put the choices squarely:

Either there is a fatal necessity and invincible order, or a kind Providence, or a confusion without a purpose and without a dictator. If then there is an invincible necessity why dost thou resist? But if there is a Providence which allows itself to be propitiated, make thyself worthy of the help of the divinity. But if there is a confusion without a governor, be content that in such a tempest thou hast in thyself a certain ruling intelligence. And even if the tempest carry thee away, let it carry away the poor flesh, the poor breath, everything else; for the intelligence at least it will not carry away.

On this our hope rests.

RECOMMENDED READING

In preparing this book I have amassed more than one thousand references—books, conference proceedings, papers from several scores of scientific journals in many disciplines, news items from leading weeklies—but even before I started to write I decided not to burden the text with them. Instead, I have selected about one hundred books and monographs which have either outstanding breadth of coverage, or contain unusually penetrating analyses, or are just delight to read. A reader, eager to find more information on any of the key topics discussed in this book, will thus find himself very busy when he tries to follow the hundreds of other leads emanating from the listed recommended readings which have been conveniently arranged first by topics, then alphabetically by authors or editors.

Chapter 1

Requirements for a habitable planet are engrossingly reviewed by Dole (1970). A fine glimpse of past energy conversions can be found in a special issue of *History Today* (March 1980) while Schurr and Netschert (1960) offer an exhaustive historical review for the United States and Adams (1982) provides a much shorter but no less interesting survey for the United Kingdom. Odum (1971), and Odum and Odum (1981) tie many energy-environment-food connections through revealing flow diagrams (although the energy circuit graphs do not make an easy inspection for an uninitiated reader) and collections from the *Scientific American* are fine introductions to the biosphere (1970), energy (1971), agriculture (1976), and nutrition (1978).

Chapter 2

Choosing from the abundance of post-1973 literature on energy is not easy. Good overall introductions are those by Cook (1976), Teller (1979), and Ramage (1983). Oil and OPEC tales have been well told by Danielsen (1982), Jaidah (1983), Lieber (1983), Odell and Rosing (1983), Riva (1983), and Schneider (1983). The rise and fall of nuclear energy can be traced in Falk (1982) and Pasqualetti and Pijawka (1984). Renewables are extolled and criticized in Lovins (1977), Nash (1979), Bach *et al.* (1980), and Smil (1983). For the study of energy-economy links, books by

Darmstadter, Dunkerley and Alterman (1977), Slesser (1978), and Gordon (1981) give a good start. A taste of energy forecasts can be had by sampling the work of Schurr *et al.* (1979), the Committee on Nuclear and Alternative Energy Systems (1980), Häfele (1981), and the International Energy Agency (1982). Finally, surveys of important new developments in the broad field of energy studies are perhaps most conveniently accessible in the *Annual Review of Energy*.

CHAPTER 3

Thorough general overviews of human dietary requirements, including treatments of individual major nutrients and of the most important nutritional diseases and risks, are available in Beaton and Bengoa (1976), Rao (1976), Hodges (1979), Irwin (1980), Scarpa *et al.* (1982), Hui (1983), and Solimano and Lederman (1983). Energy and protein requirements are discussed in Joint FAO/WHO (1973) and in Joint FAO/WHO/UNU (1985) reports. FAO (1980) is a handy introduction to carbohydrates and Burkitt and Trowell (1975), Trowell and Burkitt (1981), and Birch and Parker (1983) contain all the essentials on dietary fibre. Excellent overviews of the vast literature on dietary fats are Levy *et al.* (1979) and Perkins and Visek (1983). The best surveys of the global food situation are FAO's annual *Production Yearbook* and *The State of Food and Agriculture* and the irregular *Food Balance Sheets* series. Of many appraisals of world food outlook, Woods (1981) contains a very broad interdisciplinary overview while Johnson and Wittver (1984) evaluate the prospects of agricultural technology until 2030. Hanson *et al.* (1982) is an excellent concise introduction to wheat farming; new crops are reviewed in Ritchie (1979); non-traditional foods in Ferrando (1981); breastfeeding in Jelliffe and Jelliffe (1978); mushrooms in Chang and Hayes (1978).

CHAPTER 4

For a detailed overview of the first decade of heightened environmental concerns Holdgate *et al.* (1982) is perhaps the best single-volume source. To appreciate the changes in the perception of global problems it is necessary to look at the Study of Critical Environmental Problems (1970) and at the Rättvik conference whose papers were published in *Ambio* (vol. 12, no. 2, 1983). Farmland losses and erosion are described in the now classical volume by Jacks and Whyte (1939), in Eckholm (1976), Brown (1978), Schmidt *et al.* (1982), and Smil (1984). Tropical deforestation and the ecology of the tropics are detailed in Goodland and Irwin (1975), Ayensu (1980), Myers (1980), and Lanly and Gillis (1980). The disappearance of species has been brought to wider attention in volumes by Prance (1977) and Myers (1979) while interdisciplinary problems of conservation and loss of genetic diversity are thoroughly treated in Soulé and Wilcox (1980), Frankel and Soulé (1981), and Schonewald-Cox *et al.* (1983). For a very broad-ranging source on developments in world ecology Kormondy and McCormick (1981) is perhaps the best choice.

CHAPTER 5

Global biogeochemical cycles and human interference in their functioning are broadly treated in Stumm (1977), the Committee on the Atmosphere and the Biosphere (1981), Likens (1981), and Smil (1985). Among many recent detailed reviews of the carbon dioxide problem, those edited by Clark (1982) and prepared by the Carbon Dioxide Assessment Committee (1983) are clearly the most comprehensive. Writings on the nitrogen cycle are much more dispersed; besides the general volumes on global cycles, the indispensable sources are reviews by the committees of the National Academy of Sciences (1977, 1978), Aldrich (1980), and a volume edited by Stevenson (1982). Interdisciplinary reviews of acid deposition are available in volumes prepared by the Subcommittee on Water (1981), the Swedish Ministry of Agriculture (1982) and the Committee on Atmospheric Transport (1983). Good introductions to revelations, opportunities and problems of energy analysis of food production are Green (1978), Stout (1979), and Smil *et al.* (1983).

CHAPTER 6

Very different global views are offered by the volumes assembled by the Council on Environmental Quality (1981) and that edited by Simon and Kahn (1984). For captivating perspectives on civilization, adaptive controls, and global survival I must recommend Lovelock (1979) and Prigogine and Stengers (1984).

SELECT BIBLIOGRAPHY

Adams, R. N. (1982). *Paradoxical Harvest.* Cambridge University Press, Cambridge.

Aldrich, S. R. (1980). *Nitrogen.* University of Illinois at Urbana–Champaign, Illinois.

Ayensu, E., ed. (1980). *Jungles.* Crown Publishers, New York.

Bach, W., W. Manshard, W. H. Matthews and H. Brown (1980). *Renewable Energy Prospects.* Pergamon Press, Oxford.

Beaton, G. and J. Bengoa, eds. (1976). *Nutrition in Preventive Medicine.* WHO, Geneva.

Birch, G. G. and K. J. Parker (1983). *Dietary Fibre*, Applied Science Publishers, London.

Brown, L. R. (1978). *The Worldwide Loss of Cropland.* Worldwatch Institute, Washington, D.C.

Burkitt, D. P. and H. C. Trowell (1975). *Refined Carbohydrate Foods and Disease: Some Implications of Dietary Fibre.* Academic Press, London.

Carbon Dioxide Assessment Committee (1983). *Changing Climate.* National Academy Press, Washington, D.C.

Chang, S. T. and W. H. Hayes, ed. (1976). *The Biology and Cultivation of Edible Mushrooms.* Academic Press, New York.

Clark, W. C., ed. (1982). *Carbon Dioxide Review: 1982.* Clarendon Press, Oxford.

Committee on Atmospheric Transport and Chemical Transformation in Acid Precipitation (1983). *Acid Deposition.* National Academy Press, Washington, D.C.

Committee on Nuclear and Alternative Energy Systems (1980). *Energy in Transition 1985–2010.* W. H. Freeman, San Francisco.

Committee on the Atmosphere and the Biosphere (1981). *Atmosphere–Biosphere Interactions.* National Academy Press, Washington, D.C.

Cook, E. (1976). *Man, Energy, Society.* W. H. Freeman, San Francisco.

Council on Environmental Quality and the Department of State (1981). *The Global 2000 Report to the President.* USGPO, Washington, D.C.

Danielsen, A. L. (1982). *The Evolution of OPEC.* Harcourt Brace Jovanovich, New York.

Darmstadter, J., J. Dunkerley and J. Alterman (1977). *How Industrial Societies Use Energy.* Resources for the Future, The Johns Hopkins University Press, Baltimore, Maryland.

Dole, J. H. (1970). *Habitable Planets for Man.* Elsevier, New York.

Eckholm, E. P. (1976). *Losing Ground.* W. W. Norton, New York.

Falk, J. (1982). *Global Fission.* Oxford University Press, Melbourne.

FAO (1980). *Carbohydrates in Human Nutrition.* FAO, Rome.

Ferrando, R. (1981). *Traditional and Non-traditional Foods.* FAO, Rome.

Frankel, O. H. and M. E. Soulé (1981). *Conservation and Evolution.* Cambridge University Press, Cambridge.

Goodland, R. J. and H. S. Irwin (1975). *Amazon Jungle: Green Hell to Red Desert?* Elsevier Scientific, New York.

Gordon, R. C. (1981). *An Economic Analysis of World Energy Problems*. The MIT Press, Cambridge, Massachusetts.

Green, M. B. (1978). *Eating Oil: Energy Use in Food Production*. Westview Press, Boulder, Colorado.

Häfele, W., ed. (1981). *Energy in a Finite World: A Global Systems Analysis*. Ballinger, Cambridge, Massachusetts.

Hanson, H., N. E. Borlaug and R. G. Anderson (1982). *Wheat in the Third World*. Westview Press, Boulder, Colorado.

Hodges, M. E. (1979). *Nutrition Metabolic and Clinical Approaches*. Plenum Press, New York.

Holdgate, M. W., M. Kassas and G. F. White, eds. (1982). *The World Environment 1972-1982*. Tycooly International, Dublin.

Hui, Y. H. (1983). *Human Nutrition and Diet Therapy*. Wadsworth Health Sciences Division, Monterey, California.

International Energy Agency (1982). *World Energy Outlook*. IEA, Paris.

Irwin, M. I. (1980). *Nutritional Requirements of Man: A Conspectus of Research*. The Nutrition Foundation, New York.

Jacks, G. V. and R. O. Whyte (1979). *Vanishing Lands: A World Survey of Soil Erosion*. Doubleday, Doran, New York.

Jaidah, A. M. (1983). *An Appraisal of OPEC Oil Policies*. Longman, London.

Jelliffe, D. B. and E. F. P. Jelliffe (1978). *Human Milk in the Modern World*. Oxford University Press, Oxford.

Johnson, G. L. and S. H. Wittwer (1984). *Agricultural Technology Until 2030*. Michigan State University, East Lansing, Michigan.

Joint FAO/WHO *Ad Hoc* Expert Committee (1973). *Energy and Protein Requirements*. WHO, Geneva.

Joint FAO/WHO/UNU Expert Consultation (1985). *Energy and Protein Requirements*. WHO, Geneva.

Kormondy, E. J. and J. F. McCormick, eds. (1981). *Handbook of Contemporary Developments in World Ecology*. Greenwood Press, Westport, Connecticut.

Lanly, J. P. and N. Gillis (1980). *Provisional Results of the FAO/UNEP Tropical Forest Resources Assessment Project*. FAO, Rome.

Levy, R., B. Dennis, E. N. Ernst, B. M. Rifkind (1979). *Nutrition Lipids and Coronary Heart Disease*. Raven Press, New York.

Lieber, R. J. (1983). *The Oil Decade*. Praeger, New York.

Likens, G. E., ed. (1981). *Some Perspectives of the Major Biogeochemical Cycles*. John Wiley & Sons, Chichester.

Lovelock, J. E. (1979). *Gaia*. Oxford University Press, Oxford.

Lovins, A. B. (1977). *Soft Energy Paths*. Friends of the Earth & Ballinger, Cambridge, Massachusetts.

Myers, N. (1979). *The Sinking Ark*. Pergamon Press, Oxford.

Myers, N. (1980). *Report of Survey of Conversion Rates in Tropical Moist Forests*. NAS, Washington, D.C.

Nash, H., ed. (1979). *The Energy Controversy*. Friends of the Earth, San Francisco.

National Academy of Sciences (1977). *Nitrogen Oxides*. NAS, Washington, D.C.

National Academy of Sciences (1978). *Nitrates: An Environmental Assessment*. NAS, Washington, D.C.

Odell, P. R. and K. E. Rosing (1983). *The Future of Oil*. Kogan Page, London.

Odum, Howard T. (1971). *Environment, Power and Society*. Wiley-Interscience, New York.

Odum, H. T. and E. C. Odum (1981). *Energy Basis for Man and Nature.* McGraw-Hill, New York.

Pasqualetti, M. J. and K. D. Pijawka, eds. (1984). *Nuclear Power.* Westview Press, Boulder, Colorado.

Perkins, E. G. and W. J. Visek, eds. (1983). *Dietary Fats and Health.* American Oil Chemist's Society, Champaign, Illinois.

Prance, G. T., ed. (1977). *Extinction is Forever.* New York Botanical Garden, New York.

Prigogine, I. and I. Stengers (1984). *Order out of Chaos.* Bantam Books, New York.

Ramage, J. (1983). *Energy A Guidebook.* Oxford University Press, Oxford.

Rao, K. K. P. N., ed. (1976). *Food Consumption and Planning.* Pergamon Press, Oxford.

Ritchie, G. A. (1979). *New Agricultural Crops.* Westview Press, Boulder, Colorado.

Riva, J. P., Jr. (1983). *World Petroleum Resources and Reserves.* Westview Press, Boulder, Colorado.

Scarpa, I. S., H. C. Kiefer and R. Tatum (1982). *Source Book on Food and Nutrition.* Marquis Academic Media, Chicago.

Schmidt, B. L., R. R. Almaras, J. V. Mannering and R. J. Papendick, eds. (1982). *Determinants of Soil Loss Tolerance.* American Society of Agronomy and Soil Science Society of America, Madison, Wisconsin.

Schneider, S. A. (1983). *The Oil Price Revolution.* The Johns Hopkins University Press, Baltimore, Maryland.

Schonewald-Cox, C. M., S. M. Chambers, B. MacBryde and L. Thomas, eds. (1983). *Genetics and Conservation.* The Benjamin/Cummings Publishing, Menlo Park, California.

Schurr, S. H. and B. C. Netschert (1960). *Energy in the American Economy, 1850–1975.* The Johns Hopkins University Press, Baltimore.

Schurr, S. H., J. Darmstadter, H. Perry, W. Ramsay and M. Russell (1979). *Energy in America's Future.* Resources for the Future, The Johns Hopkins University Press, Baltimore, Maryland.

Scientific American (1970). *The Biosphere.* W. H. Freeman, San Francisco.

Scientific American (1971). *Energy and Power.* W. H. Freeman, San Francisco.

Scientific American (1976). *Food and Agriculture.* W. H. Freeman, San Francisco.

Scientific American (1978). *Human Nutrition.* W. H. Freeman, San Francisco.

Simon, J. L. and H. Kahn, eds. (1984). *The Resourceful Earth.* Basil Blackwell, Oxford.

Slesser, M. (1978). *Energy in the Economy.* Macmillan, London.

Smil, V. (1983). *Biomass Energies.* Plenum Press, New York.

Smil, V., P. Nachman and T. V. Long, II (1983). *Energy Analysis and Agriculture.* Westview Press, Boulder, Colorado.

Smil, V. (1984). *The Bad Earth.* M. E. Sharpe, Armonk, New York.

Smil, V. (1985). *Carbon Nitrogen Sulfur: Human Interference in Grand Biospheric Cycles.* Plenum Press, New York.

Solimano, G. R. and S. A. Lederman, eds. (1983). *Controversial Nutrition Policy Issues.* Thomas, Springfield, Illinois.

Soulé, M. E. and B. A. Wilcox, eds. (1980). *Conservation Biology.* Sinauer, Sunderland, Massachusetts.

Stevenson, F. J., ed. (1982). *Nitrogen in Agricultural Soils.* American Society of Agronomy, Madison, Wisconsin.

Stout, B. A. (1979). *Energy for World Agriculture.* FAO, Rome.

Study of Critical Environmental Problems (SCEP) (1970). *Man's Impact on the Global Environment.* The MIT Press, Cambridge, Massachusetts.

Stumm, W., ed. (1977). *Global Chemical Cycles and Their Alterations by Man*. Dahlem Konferenzen, Berlin.

Subcommittee on Water (1981). *Acidification in the Canadian Aquatic Environment*. National Research Council of Canada, Ottawa.

Swedish Ministry of Agriculture Environment '82 Committee (1982). *Acidification Today and Tomorrow*. Ministry of Agriculture, Stockholm.

Teller, F. (1979). *Energy from Heaven and Earth*. W. H. Freeman, San Francisco.

Trowell, H. C. and D. P. Burkitt (1981). *Western Diseases Their Emergence and Prevention*. Edward Arnold, London.

Woods, R. G., ed. (1981). *Future Dimensions of World Food and Population*. Westview Press, Boulder, Colorado.

INDEX